Lecture Notes in Mathematics

For further volumes:
http://www.springer.com/series/304

Hidetoshi Marubayashi • Fred Van Oystaeyen

Prime Divisors and Noncommutative Valuation Theory

Springer

Hidetoshi Marubayashi
Tokushima Bunri University
Faculty of Science and Engineering
Shido, Sanuki City, Kagawa, Japan

Fred Van Oystaeyen
University of Antwerp
Mathematics and Computer Science
Antwerp, Belgium

ISBN 978-3-642-31151-2 ISBN 978-3-642-31152-9 (eBook)
DOI 10.1007/978-3-642-31152-9
Springer Heidelberg New York Dordrecht London

Lecture Notes in Mathematics ISSN print edition: 0075-8434
ISSN electronic edition: 1617-9692

Library of Congress Control Number: 2012945522

Mathematics Subject Classification (2010): 16W40, 16W70, 16S38, 16H10, 13J20, 16T05

Printed on acid-free paper

Springer is part of Springer Science+Business Media (www.springer.com)

Introduction

The classical theory of valuation rings in fields has natural applications in algebraic geometry as well as in number theory, e.g. as local rings of nonsingular points of curves or, respectively, as prime localizations of rings of integers in number fields. In these applications the rings are Noetherian and the valuations are discrete, that is the value group is the additive group of the integers, $\mathbb{Z}$. The fact that non-discrete valuation rings are not Noetherian may be the reason that general Γ-valuations with totally ordered value group Γ, which need not be discrete, have not been used to the same extent as discrete ones in the aforementioned branches of classical mathematics.

However, function field extensions of arbitrary transcendence degree have been considered; the transcendence degree is recorded as the dimension of some corresponding algebraic variety, but it is also the maximal rank of a valuation ring in the function field. Algebraic geometry easily survives by using the theory of local Noetherian rings, avoiding the non-Noetherian valuation rings dominating them. The algebraic theory in this book deals with generalized Γ-valuations, often related to Γ-filtrations, for totally ordered groups Γ, at places even totally ordered semigroups appear. A source of problems and new ideas is the extension theory for valuations from a subfield to the field. This theory is well developed for number fields in connection with Galois theory, but is perhaps less used for field extensions of finite transcendence degree probably because the (Noetherian) geometric approach is more popular. We want to consider more general algebra extensions including skewfields, matrix rings, etc.

Therefore we develop in Chap. 1 the general theory starting from a noncommutative background, expounding a minimal commutative background when necessary. The primes or prime places introduced provide a very general approach to a valuation theory for associative rings and algebras; this originates in the theory of pseudoplaces as in [70, 71] and the work of Van Geel [68, 69]. The theory is related to the consideration of filtrations and gradations by totally ordered groups; this is the topic of Sect. 1.3 (and Sect. 1.8). Originally pseudoplaces were developed with the aim to apply them to the construction of generic crossed products and generic skewfields, hence it is not really surprising that an interesting class of algebras inviting the development of a generalized valuation theory is the class of central simple algebras, or more generally finitely dimensional algebras over fields. Thus our theory has direct links to the theory of orders and maximal orders; this will be the subject of Chap. 2. The second part of Chap. 1 is devoted to some particular primes: Dubrovin rings. After the basic theory in Sect. 1.4 and the ideal theory in Sect. 1.5 we focus on Dubrovin valuations of finite dimensional (simple) algebras in Sect. 1.6 and on Gauss extensions in Sect. 1.7. To a generalized valuation ring corresponds a reduction of the original algebra and the relation between properties of the reduced algebra and the original one should be clarified. There are properties of homological type or of (noncommutative) geometric type; in Appendix 1.9 we provide some theory related to Auslander regularity of the algebras under consideration.

Chapter 2 deals with orders. It is not our aim to provide a complete theory of orders in (semisimple) Artinian algebras, but we include a short introduction to this important root of the theory relating to the fundamental work of Auslander, Reiner and others (see [57]).

Arithmetical rings with suitable divisorial properties and noncommutative Krull rings are amongst the natural study objects here (Sects. 2.1 and 2.2). Then we consider Ore extensions over Krull orders and noncommutative valuation rings of Ore extensions of simple Artinian rings in Sects. 2.3 and 2.4. To end the chapter we arrive at an arithmetical divisor theory, a noncommutative divisor calculus, leading to a noncommutative Riemann–Roch theorem over a central curve. The latter result fits in the geometry of rings with polynomial identities, cf. [76]; it also fits in this book because it provides a good example of a concrete, almost calculative, application of noncommutative valuation theory, even if here only discrete valuations appear.

In Chap. 3 we leave the biotope of finite dimensional algebras and turn attention to a recently popular class of infinite dimensional algebras containing quantum groups, quantized algebras and deformations. Usually examples of such algebras are given by generators and relations and therefore they are equipped with a standard filtration. The reduction of such an algebra at a central valuation can be "good" or "bad" depending on how the defining relations reduce modulo the maximal ideal of the central valuation ring. Similar phenomena exist in algebraic geometry related to good reduction of (elliptic) curves but now the reducing "equations" are elements of the ideal of relations in the free algebra. The property of good reduction allows the extension of the central valuation to the (skewfield of fractions of the) noncommutative algebra; this is the topic of Sect. 3.1. To establish a case where we can completely calculate all valuations we focus on the Weyl field in Sect. 3.2 and provide some divisor theory related to a very peculiar subring in Sect. 3.3.

Finally we consider finite dimensional Hopf algebras in the last section and extend the valuation filtration of a central valuation to a Hopf filtration on the Hopf algebra. The filtration part of degree zero is a Hopf order and we show how certain Hopf orders, generalizing Larson orders in the case of finite group rings, may be explicitly constructed. The explicit examples over number rings exhibit the importance of ramification properties of the value function corresponding to the Hopf order.

In our opinion noncommutative extensions of valuation theory deserve to be further integrated in noncommutative algebra (and noncommutative geometry too). We do realize that this book does not provide a finished picture of possible noncommutative extensions of valuation theory, but we hope to have opened a few doors for possible new developments pointing at interesting directions for further research.

Contents

1 General Theory of Primes 1
1.1 Prime Places of Associative Rings 1
1.2 Primes in Algebras 8
1.3 Value Functions and Prime Filtration 19
1.4 Dubrovin Valuations 33
1.5 Ideal Theory of Dubrovin Valuation Rings 42
1.6 Dubrovin Valuation Rings of Q with Finite Dimension Over Its Center 58
1.7 Gauss Extensions 70
1.8 Filtrations by Totally Ordered Groups 88
1.9 Reductions of Algebras and Filtrations 96
1.10 Appendix: Global Dimension and Regularity of Reductions 101

2 Maximal Orders and Primes 109
2.1 Maximal Orders 109
2.2 Krull Orders 115
2.3 Ore Extensions Over Krull Orders 132
2.4 Non-commutative Valuation Rings in $K(X;\sigma,\delta)$ 144
2.5 Arithmetical Pseudovaluations and Divisors 153
2.6 The Riemann–Roch Theorem for Central Simple Algebras Over Function Fields of Curves 164

3 Extensions of Valuations to Quantized Algebras 175
3.1 Extension of Central Valuations 175
3.2 Discrete Valuations on the Weyl Skewfield 183
3.3 Some Divisor Theory for Weyl Fields Over Function Fields 193
3.4 Hopf Valuation Filtration 197

Bibliography 213

Index 217

Chapter 1
General Theory of Primes

1.1 Prime Places of Associative Rings

Throughout we fix notation and conventions as follows. By R we denote an associative ring with unit and R' is a subring of R. A **prime in** R is a pair (P, R') such that P is a prime ideal of R' satisfying the condition: for $x, y \in R$, $xR'y \subset P$ entails either $x \in P$ or $y \in P$. A prime (P, R') in R is said to be **completely prime** if P is a completely prime ideal of R' and for $x, y \in R$, $xy \in P$ yields that either $x \in P$ or $y \in P$.

For a given prime (P, R') in R we may consider the idealizer of P in R, that is $R^P = \{x \in R, xP \subset P \text{ and } Px \subset P\}$. A subset S of R is said to be an m**-system** (multiplicative system) whenever $0 \notin S$ and for s_1, s_2 in S there exists an $x \in R$ such that $s_1 x s_2 \in S$. Prime ideals in a ring may be characterized as being ideals having an m-system for their complement in the ring. We extend this definition so that it may be applied with respect to subsets of R. Consider a subset X of R, a subset S of R is said to be an m**-system for** X whenever $0 \notin S$ and for s_1, s_2 in S there is an $x \in X$ such that $s_1 x s_2 \in S$. Obviously, if $X \subset Y$ then any m-system for X is an m-system for Y.

1.1.1 Example

(a) If Q is a prime ideal of R then (Q, R) is a prime in R.
(b) For any prime number p in $\mathbb{Z}$, $((p), \mathbb{Z})$ is a prime in $\mathbb{Z}[T]$, the polynomial ring in one variable over $\mathbb{Z}$. This example shows that in general R' and R need not be very close.

In case $R' \subset R_1 \subset R$, for some subring R_1 of R, then if (P, R') is prime in R it is also a prime in R_1.

Marubayashi and Van Oystaeyen, *Prime Divisors and Noncommutative Valuation Theory*, Lecture Notes in Mathematics 2059, DOI 10.1007/978-3-642-31152-9_1,

1.1.2 Lemma

Let P be an ideal of R', then for (P, R') to be a prime in R it is necessary and sufficient that $R - P$ is an m-system for R'. If (P, R') is a prime in R then so is (P, R^P).

Proof. If $R - P$, the complement of P in R, is an m-system for R' then $R' \cap (R - P) = R' - P$ is also an m-system for R' and thus (P, R') is a prime in R. The converse is trivial. Since R^P is a ring containing P as an ideal, hence as a prime ideal of R^P because (P, R') is a prime in R, the last statement is obvious. □

1.1.3 Convention

If we talk about a **prime** P **of** R then we shall always mean the prime (P, R^P) and we may refer to R^P as the **ring of definition** of the prime P of R. Since R^P is by definition the largest subring of R that may serve as a defining subring for the prime considered, the presentation (P, R^P) may be considered as the canonical presentation for the prime P of R.

It is clear from the foregoing observations that the generality of the definition may put a limitation on the value of results derived in full generality. The theory of primes becomes more rich if we demand a more stringent relation between R' and R. An interesting situation arises when R is (in some well defined sense) a ring of fractions of R'. The following three definitions, increasing in strength, anticipate on the existence of such relation without actually describing a notion of "ring of fractions".

$\mathbb{R}$

1.1.4 Definitions

(a) A prime (P, R') in R is said to be **fractional** if for every $r \in R - R'$ there exist $x, y \in R' - \{0\}$ with x or y in P such that $xry \in R'$.
(b) A fractional prime (P, R') is said to be **localized**, if in (a) we have that x, y are such that $xry \in R' - P$.
(c) A localized prime (P, R') in R is said to be **separated** if for every $r \in R' - \{0\}$ there exist x and y in R such that $xry \in R' - P$.

Observe that in general for $r' \in R'$ there may not exist $x, y \in R' - \{0\}$ such that $xr'y \notin P$, indeed it is possible that P contains a nontrivial ideal of R. Put $P^0 = \{p \in P, RpR \subset P\}$; P^o is the largest ideal of R contained in P. A separated prime is then just a localized prime such that $P^o = 0$.

1.1.5 Proposition

In a simple ring, i.e. an associative ring without nontrivial twosided ideals, localized primes are always separated.

1.1.6 Remark on Terminology

In the publications [69, 71] a localized prime was called a semi-restricted prime, a separable prime was called a restricted prime. This old terminology was completely ad hoc and non descriptive, the terminology we propose to be standardized now is more related to the "meaning" of the properties.

1.1.7 Lemma

If (P, R') is a prime in R that is fractional, respectively localized, respectively separated, then (P, R^P) is a fractional, respectively localized, respectively separated prime in R.

Proof. Since $R' \subset R^P$, $R - R^P \subset R - R'$, $R' - P \subset R^P - P$, the statements are obvious. □

In the lemma the converses need not be true; it is useful that by selecting the canonical representative (P, R^P) for a prime P of R we do not loose the properties defined above if they hold for some presentation (P, R') as a prime in R.

Consider the set $\mathcal{Q}_R = \{(P, R'), R'$ a subring of R, P a prime ideal of R'; elements of $\mathcal{Q}_R$ are called **pairs for** R. On $\mathcal{Q}_R$ we may introduce a partial order $<$, called **domination relation**, by putting: $(P, R') < (P_1, R_1')$ whenever $R' \subset R_1'$ and $P_1 \cap R' = P$. A pair maximal in $\mathcal{Q}_R$ with respect to the domination relation is said to be a **dominating pair** in R. Clearly if (P, R') is dominating then $R' = R^P$. The use of dominating pairs is that they can be used to construct primes in R, see later. For the moment let us look at **localized pairs**, i.e. pairs not necessarily defining a prime of R but satisfying the definition of being localized.

1.1.8 Proposition

A localized pair (P, R') for R is necessarily dominating and it is a prime of R.

Proof. Let (P, R') be a localized pair. Take $z \in R^P - R'$ then there exist $x, y \in R'$ with x or y in P such that $xzy \in R' - P$. Since $z \in R^P$ and either x or y is in P, $xzy \in P$, contradiction. Hence $R^P = R'$. Suppose $(P, R^P) < (Q, R_1)$ such that there is an $x \in R_1 - R^P$. Since (P, R^P) is localized there exist $\lambda, \mu \in R^P - \{0\}$ with λ or μ in P such that $\lambda x\mu \in R^P - P$. Say $\lambda \in P$, the argument for $\mu \in P$ is symmetric. Then $\lambda x\mu \in R^P \cap PR_1 \subset R^P \cap Q = P$, a contradiction, therefore (P, R^P) is dominating. In order to check that (P, R^P) is a prime suppose that $xR^P y \subset P$ with $x \notin R^P$. We find $\lambda, \mu \in R^P - \{0\}$ with $\lambda \in P$ such that $\lambda x\mu \in R^P - P$ but also $x\mu R^P y \subset P$ or $\lambda x\mu R^P y \subset P$. Thus $y \in P$ follows, or (P, R^P) is a prime. □

1.1.9 Proposition

For any prime P, R' in R we have that P^o is a prime ideal of R and $(\overline{P}, \overline{R}')$ is a prime of $\overline{R}$, where $\overline{P} = P/P^o, \overline{R}' = R'/P^o, \overline{R} = R/P^o$. Whenever (P, R') is localized so is $(\overline{P}, \overline{R}')$. Moreover, if (P, R') is localized then $(\overline{P}, \overline{R}')$ is a separated prime of $\overline{R}$.

Proof. If there exist nonzero x and y in R such that $xRy \subset P^o$ then $(RxR)(RyR) \subset P^o$. If neither x nor y is in P^o then neither (RxR) nor (RyR) is in P, so we may select $x_1 \in RxR - P, y_1 \in RyR - P$, such that $x_1 R' y_1 \subset x_1 R y_1 \subset P$, contradicting the prime property of (P, R'). Consequently P^o is a prime ideal of R. In order to check that $(\overline{P}, \overline{R}')$ is a prime of $\overline{R}$ select nonzero $\overline{x}, \overline{y} \in \overline{R}$ such that $\overline{x}\overline{R}'\overline{y} \subset \overline{P}$. For representatives x of $\overline{x}$, y of $\overline{y}$ in R, we have $xR'y \subset P$ because P contains the kernel of $R \to \overline{R}$. Since (P, R') is a prime in R it follows that x or y is in P and thus $\overline{x}$ or $\overline{y}$ is in $\overline{P}$.

Assume that (P, R') is localized and look at $\overline{r} \in \overline{R} - \overline{R}'$ with representative $r \in R - R'$, we may find $\lambda, \mu \in R' - \{0\}$ with λ or μ in P such that $\lambda r\mu \in R' - P$. Since $\lambda r\mu \notin P$ it is clear that $\lambda \notin P^o, \mu \notin P^o$, hence $\overline{\lambda} \neq \overline{0}$ and $\overline{\mu} \neq \overline{0}$; then we observe that $\overline{\lambda}\overline{r}\overline{\mu} \in \overline{R}' - \overline{P}$ because P contains P^o the kernel of $R \to \overline{R}$. That $\overline{P}^o = \overline{0}$ is easily verified hence a localized prime of $\overline{R}$ is necessarily separated. □

1.1.10 Corollary

A separated prime is necessarily defined in a prime ring R.

Proof. We observed that a localized prime (P, R') is separated exactly when $P^o = 0$. The foregoing proposition stated that P^o is a prime ideal of R, hence R has to be a prime ring. □

1.1.11 Example

Let C be any commutative ring. A **valuation** v on C with value Γ where Γ is a totally ordered group is a surjective map $v : C \to \Gamma \cup \{\infty\}$ satisfying

V.1 $v(0) = \infty$ and 0 is the only element of C as such.
V.2 For $x, y \in C, v(xy) = v(x) + v(y)$, in particular $v(1) = 0$.
V.3 For $x, y \in C, v(x + y) \geq \inf.\{v(x), v(y)\}$.

Even if Γ need not be an abelian group we denoted its operation by $+$. Consider $O_C = \{c \in C, v(c) \geq 0\}, m_C = \{c \in C, v(c) > 0\}$, then (m_C, O_C) is a localized prime in C. Indeed if $c \in C - O_C$ then $v(c) < 0$ and we may select a $d \in C$ such that $v(d) = -v(c)$ because v is surjective; then V.2 entails $v(cd) = v(c)+v(d) = 0$ or $cd \in O_C - m_C$ and $d \in m_C$. The epimorphism $\pi_v : O_C \to \overline{C} = O_C/m_C$ is called a **primeplace** of C and it is also referred to as a primeplace of v.

We extend this terminology to arbitrary primes (P, R') in R, i.e. the map $R' \to R'/P$ is called the **primeplace of** (P, R') but we have, at the moment, no valuation function associated to this. We come back to this later.

We now consider R as a subring of A and fix some prime (P, R') in R with corresponding primeplace $\pi : R' \to R'/P = \overline{R}$. A π**-pseudoplace** of A is given by a triple $(A', \Psi, \overline{A})$ where: A' is a subring of A such that $A' \cap R = R', \Psi : A' \to \overline{A}$ is a ring homomorphism such that $\text{Ker}\Psi \cap R = P$ and $\Psi|R' = \pi$, $\overline{R}$ being a subring of $\overline{A}$. A pseudoplace is called a **preplace** if $\ker\Psi$ defines a prime $(\text{Ker}\Psi, A')$ of A. Obviously a π-pseudoplace is a π-preplace if and only if A-$\text{Ker}\Psi$ is an m-system for A'. We shall now prove a general version of the Krull–Chevalley extension theorem for valuations and derive from this that dominating pairs are necessarily primes (actually this is equivalent to the extension theorem).

1.1.12 Notation

For subsets S and T in A we write $S < T >$ for $\{a \in A, a = \sum_j s_j t_j, s_j \in S, t_j \in \overline{T}\}$ where $\overline{T}$ is the multiplicatively closed set in A generated by T, and $1 \in \overline{T}$.

1.1.13 Theorem (the Extension Theorem [69])

With notation as before, fix a dominating pair (P, R^P) in R and consider subsets $M \subset B \subset A$ such that the following properties hold:

(a) $BP \subset P < B >$ and $BR^P \subset R^P < B >$.
(b) $P < B > \cap R \subset P$.
(c) M is an m-system for B.
(d) $R^P - P \subset M$.
(c) $M \cap P < B > = \emptyset$.

Then there exists a π-preplace $(A', \Psi, \overline{A})$ of A such that $\mathrm{Ker}\Psi = Q, BP \subset Q, PB \subset Q$ and $Q \cap R = P$, where π is the primeplace of (P, R^P).

Proof. Consider the family $\mathcal{F} = \{X, X$ is left and right R^P-submodule of A, X is multiplicatively closed and $M \cap X = \emptyset, BX \subset X, XB \subset X, R \cap X = P\}$. Let us first verify $\mathcal{F} \neq \emptyset$. Indeed we claim that

$$L(B) \in \mathcal{F}, L(B) = \{a \in A, aR^P < B > \subset P < B > B\}$$

(i) $L(B)$ is obviously a left and right R-submodule of A.
(ii) $L(B)$ is multiplicatively closed because $P < B > \subset R^P < B >$.
(iii) If $a \in M \cap L(B)$ then $a \in P < B >$ since $1 \in R^P < B >$ but $M \cap P(B) = \emptyset$, hence $M \cap L(B) = \emptyset$.
(iv) For $a \in L(B)$ and $b \in B$ we obtain:

$$baR^P < B > \subset bP < B > \subset P < B > \text{ (the latter from a.)}$$

$abR^P < B > \subset aR^P < B > \subset P < B >$ (also from a).
(v) Look at $L(B) \cap R$. Since $P \subset L(B), P \subset L(B) \cap R$.

We already observed before that $L(B) \subset P < B >$, hence by (b) we arrive at $L(B) \cap R = P$. Zorn's lemma allows us to select a maximal element Q in $\mathcal{F}$; it suffices to establish that $A - Q$ is an m-system for A^Q. Take $x, y \in A - Q$ and suppose that $xA^Q y \subset Q$. Now we define $Q_x = \sum_{i=0}^{\infty} Q_{x,i}$ where $Q_{x,o} = Q$ and for every i, $Q_{x,i} = \{z \in A, z$ is a finite sum of elements of the form $a_1 x a_2 x \dots x a_i$ with $a_j \in A^Q\}$.

It is clear that Q_x is a multiplicatively closed left and right R^P-module containing P as well as x. Define Q_y in exactly the same way. Since by definition $B \subset A^Q$ it follows that $BQ_x \subset Q_x, Q_x B \subset Q_x$ and similarly $BQ_y \subset Q_y, Q_y BB \subset Q_y$. Since Q is maximal in $\mathcal{F}$ the longer Q_x, resp. Q_y, must satisfy either $Q_x \cap M \neq \emptyset$ or $Q_x \cap R \neq P$, resp. $Q_y \cap M \neq \emptyset$ or $Q_y \cap R \neq P$. Put $Q_x \cap R = P_x$. If B satisfies the conditions of the theorem then so does $R^P < B >$ and we may replace B by $R^P < B >$ from the beginning; thus without loss of generality we may assume that $R^P \subset B$. Since now $R^P \subset B$ and B normalizes Q_x as well as Q_y it follows that $R^P \subset R^{P_x}$ and P_x is an ideal in R^{P_x}. In case $P_x \cap R^P \neq P$ we must have $P_x \cap M \neq \emptyset$. In case $P_x \cap R^P = P$ then $P_x = P$ because (P, R^P) was a dominating pair in R; indeed if $(P_x, R^{P_x}) \neq (P, R^P)$ then we may look at the family $\mathcal{S}$ of pairs of (X, R^X) where X is an ideal, not necessarily prime, of R^X such that $R^P \subset R^X, X \cap R^P = P$. Since $(P_x, R^{P_x}) \in \mathcal{S}$ we may apply Zorn's lemma again and select a maximal element of $\mathcal{S}$ say (I, E). If I is not prime that $J_1 J_2 \subset I$ for some ideals J_1, J_2 of E with $J_1 \supsetneqq I, J_2 \supsetneqq I$. Hence $J_1 \cap R^P \supsetneqq P, J_2 \cap R^P \supsetneqq P$ and $I \cap R^P = P$ contradicts the fact that P is a prime ideal of R^P. Consequently I is a prime ideal of E but then $(P, R^P) < (I, E)$ yields that $P = I$ and $R^P = E$ because (P, R^P) is dominating. Since we excluded $P = P_x(P \subset P_x \subset I)$ we are left with the conclusion

$P_x \cap M \neq \emptyset$. In formally the same way we obtain $P_y \cap M \neq \emptyset$. So we may pick $m_x \in Q_x \cap M$ and $m_y \in Q_y \cap M$. Say, $m_x = f_o(x) + f_1(x), + \ldots + f_n(x)$ with $f_i(x) \in Q_{x,i}, m_y = g_o(y)+q_1(y)+\ldots+g_t(y)$ with $g_j(y) \in Q_{y,j}$ we may assume m_x and m_y are chosen such that n and t are minimal. Since $yA^Q x \subset Q$ we have $Q_{x,i}Q_{y,j} \subset Q_{x,i-j}$ for $i \geq j \geq 1$. Suppose that $n \geq t$. By (c) we may select an element $s \in B$ such that $m_x s m_y \in M$. Then: $h = m_x s(m_y - g_o(y)) \in \sum_{k=0}^{n-1} Q_{x,k}$ or

$$m_x s m_y = m_x s g_o(y) + h \textit{ where } m_x s g_o(y) \in Q \qquad (*)$$

(**Note:** we use $M \subset B \subset A^Q, s \in B \subset A^Q$).

Now, (*) contradicts the minimality assumption on n and t unless $m_x = f_o(x) \in Q, m_y = g_o(y) \in Q$ but that is contradiction. □

1.1.14 Corollary

Dominating pairs (P, R^P) of R are primes of R.

Proof. In the theorem take $A = R, B = R^P, M = R^P - P$. We obtain a prime (Q, R^Q) such that $Q \cap R^P = P, R^P \subset R^Q$. Hence $(Q, R^Q) > (P, R^P)$ but this yields $Q = P, R^Q = R^P$, hence (P, R^P) is a prime. □

1.1.15 Example

If R is a field K then (P, R^P) is a valuation pair (M_v, O_v). If L is a field extension of K the theorem reduces to the Krull–Chevalley extension theorem. See also further in Sect. 1.2.

For primes (P, R') and (P_1, R'_1) of R we say that (P_1, R'_1) is a **specialization** of (P, R'), and we denote this by $(P, R') \to (P_1, R'_1)$, if $P \subset P_1 \subset R'_1 \subset R'$. Observe that P is an ideal of R'_1 as well as of R' but P need not be a prime ideal of R'_1 in general. Let us write π for the primeplace $\pi : R' \to R'/P = \overline{R}$ of R and π_1 for the primeplace $\pi_1 : R'_1 \to R'_1/P_1$ of R.

1.1.16 Lemma

With notation as before, if $(P, R') \to (P_1, R'_1)$ then $(P_1/P, R'_1/P)$ is a prime of $\overline{R}$. In that case there is a primeplace $\xi : R'_1/P \to R'_1/P_1$ of $\overline{R}$ such that $\xi \circ \pi|_{R'_1} = \pi_1$.

Proof. Put $\overline{P} = P_1/P, \overline{R}' = R'_1/P, \overline{R} = R'/P$. If $(P, R') \to (P_1, R_1)$ then we have: $P \subset P_1 \subset R'_1 \subset R'$ and $\overline{P}$ is clearly a prime ideal of $\overline{R}'$. Moreover if $\overline{x}, \overline{y} \in \overline{R}$ are such that $\overline{x}\overline{R}'\overline{y} \subset \overline{P}$ then for any choice of representatives x for $\overline{x}$, y

for $\overline{y}$ in R' we have $xR'_1y \subset P_1$ because $P = \text{Ker}(R' \to \overline{R})$ is contained in P_1. Consequently x or y is in P_1 or $\overline{x}$ or $\overline{y}$ is in $\overline{P}$. The second statement follows from this. □

In the situation of the lemma we shall say that π_1 is a specialization of π, denoted by $\pi \to \pi_1$, and this yields the existence of a prime place ξ on the residue ring of π such that as primeplaces: $\pi_1 = \xi \circ \pi$ (meaning that this equality holds where both members are defined). The converse need not be true in this generality! The situation is better for completely prime primes or for primes in algebras over fields.

1.1.17 Remark

In the foregoing we used left and right multiplications but it is possible to obtain a one-sided theory which may be connected to left quotient rings and left Goldie rings. We do not go into this here.

1.2 Primes in Algebras

In this section we consider C-algebras A, i.e. we view C as a subring of the center $Z(A)$ of A. We aim to study primes of A inducing a fixed prime of C, in other words we look at extensions to the possibly noncommutative C-algebra A of primes in C. A case of particular interest is provided by algebras over fields and primes extending primes of fields; thus we include in this section some basic theory about valuations of fields that will be of fundamental importance throughout the other chapters. Let us fix a prime (p, C') in C, and let $\pi : C' \to C'/p$ be the corresponding primeplace. Following [70] we introduce pseudoplaces of A. A π-pseudoplace of A is given by a triple $(A', \Psi, \overline{A})$ where A' is a subring of A such that: $A' \cap C = C'$, $\Psi : A' \to \overline{A}$ is a ringepiomorphism such that $\Psi|C' = \pi$ and $\text{Ker}\Psi \cap C = p$. In particular $\overline{A}$ is a $\overline{C}$-algebra where $\overline{C} = C/p$. In case $(\text{Ker}\Psi, A')$ is a prime in A we call $(A', \Psi, \overline{A})$ a π-preplace of A.

In general every prime (Q, A') in A defines a prime $(Q \cap C, A' \cap C)$ in C hence (Q, A') defines a π_Q-preplace of A where π_Q is the primeplace of C induced by (Q, A') as above.

1.2.1 Proposition

If C' is a subring of the commutative ring C such that $C - C'$ is multiplicatively closed then there is a prime (p, C') in C. On the other hand, for every prime (p, C') of C we have that $C - C^p$ is multiplicatively closed.

Proof. Suppose $C - C'$ is multiplicatively closed, then take $p = \{c' \in C', ac' \in C'$ for some $a \in C - C'\}$. It is not hard to verify that p is a nontrivial ideal of C'; let us just mention that for $x, x' \in p$, that is ax and $a'x'$ are in C' for some $a, a' \in C - C'$, we have $(ax')(a'x) \in C'$, hence either ax' or $a'x$ must be in C', in case $ax' \in C'$ then $a(x - x') \in C'$ hence $x - x' \in p$. Let us check that $C - p$ is multiplicatively closed. If $x, y \in C$ are such that $xy \in p \subset C'$ then for some $a \in C - C'$ we have $axy \in C'$. In case $ax \in C'$ it follows that $x \in p$, but in case $ax \notin C'$ then $y \in p$ follows. In order to prove the second statement in the proposition look at $x, y \in C$ such that $xy \in C^p$. Then we have: $xyp \subset p$ and thus either $x \in p$ or $yp \subset p$. Consequently if $x \notin p$ then $y \in C^p$. □

We put $\mathrm{Dom}(C) = \{C', C'$ a subring of C such that $C - C'$ is multiplicatively closed$\}$. We have already established above that every element of $\mathrm{Dom}(C)$, called proper domain, appears as the domain of a prime in C; moreover $\mathrm{Dom}(C)$ contains all C^p for primes (p, C') of C.

1.2.2 Proposition (cf. [69])

(a) If $C' \in \mathrm{Dom}(C)$ then C' is integrally closed.
(b) The integral closure of any subring D of C is exactly the intersection of all elements of $\mathrm{Dom}(C)$ containing D.

Proof. (a) Look at $C' \in \mathrm{Dom}(C)$, $x \in C - C'$ and suppose we have a relation $x^n + r_1x^{n-1} + \ldots + r_n = 0$ with $r_i \in C'$. We may assume that such a relation has been selected with minimal n. We obtain: $x(x^{n-1} + r_1x^{n-2} + \ldots + r_{n-1}) = -r_n \in C'$, hence $x^{n-1} + r_1x^{n-2} + \ldots + r_{n-1} \in C'$ but that leads to an integral equation over C' of length strictly shorter than n, a contradiction.

(b) From the first part it is clear that the integral closure $\overline{D}$ of D (in C) is contained in $\cap\{C', D \subset C' \subset C, C' \in \mathrm{Dom}(C)\}$. If $x \in C$ is not integral over D, we aim to find a prime (p, C'), C' a proper domain, such that $D \subset C'$ but $x \notin C'$. In case $x^n = 0$ then x is integral over D so $0 \notin \{1, x, x^2, \ldots, x^n, \ldots\} = M$. Let $j_M : C \to M^{-1}C$ be the canonical localization map and look at $E = j_M(D)[x^{-1}] \subset M^{-1}C$. Now $x \notin j_M(D)[x^{-1}]$ because that would lead to an integral equation for x over $j_M(D)$, and thus for x over D because $\mathrm{Ker}\, j_M$ is the ideal of C consisting of elements $x \in C$ such that $x^n c = 0$ for some $n \geq 1$. It follows that x^{-1} is contained in a proper maximal ideal w of E. Now we apply the extension theorem with R replaced by $M^{-1}C$, B by E and M by $E - w$. This yields the existence of a prime of $M^{-1}C$, say $(Q, (M^{-1}C)^Q)$ with $E \subset (M^{-1}C)^Q, w \subset Q$, hence $x \notin (M^{-1}C)^Q$ (because $x^{-1} \in w \subset Q$). By taking inverse images we arrive at a prime $(j_M^{-1}(Q), j_M^{-1}((M^{-1}C)^Q))$ in C having a domain of definition that is in $\mathrm{Dom}(C)$, because $(M^{-1}C)^Q$ is in $\mathrm{Dom}(M^{-1}C)$, and $x \notin j_M^{-1}((M^{-1}C)^Q)$.

1.2.3 Remark

Let $\mathbb{R}$ be the field of real numbers. In $\mathbb{R}[X]$ every subring $\mathbb{R} \subset R' \subset \mathbb{R}[X]$ defines a prime $(0, R')$ of $\mathbb{R}[X]$ but $\mathbb{R}[X]$ is integral over R' and R' is not a proper domain. On the other hand, $\mathbb{R}$ is integrally closed in $\mathbb{R}[X]$ but it cannot be $\mathbb{R}[X]^p$ for some prime (p, R') of $\mathbb{R}[X]$ because $(0, \mathbb{R})$ is the only prime with definition domain $\mathbb{R}$. Again it follows from the foregoing that the property of having a domain of definition that is integrally closed is detected on C^p and in particular C^p is integrally closed for every prime p of C.

Since p is not maximal (perhaps) in C' it may be a natural idea to localize at the prime ideal p of C' and consider the situation $pC'_p \subset C'_p \subset C_p$, where localization is with respect to the multiplicative set $C' - p$. For a C-algebra A one may then consider A_p over C_p where A_p is obtained by central localization of A at $C - p$; however A_p is not necessarily related to a localization at some Q of A even if some prime (Q, A') lies over (p, C') in A. Localization theory (noncommutative) at general primes exists under some extra hypotheses, e.g. [54], but we do not strive for full generality here.

1.2.4 Lemma

If (p, C') is a prime in C then (pC'_p, C'_p) is a prime in C_p, in case (p, C') is fractional, respectively localized, respectively separated, then so is (pC'_p, C'_p).

Proof. Put $t_p(C) = \{c \in C, sc = 0 \text{ for some } s \in C' - p\}$. Since (p, C') is a prime in C it follows that $t_p(C) \subset p$ hence $t_p(C) = t_p(C') \cap C'$. So we may reduce modulo $t_p(C)$, putting $\overline{C} = C/t_p(C)\overline{C'} = C'/t_p(C'), \overline{p} = p/t_p(C)$, and it is clear that $(\overline{p}, \overline{C}')$ is a prime of $\overline{C}$. Since $t_p(C)$ is an ideal of C it is contained in p°. It is easily verified that the reduced $(\overline{p}, \overline{C}')$ is fractional, localized, separated, in case (p, C') is respectively fractional, localized, separated. Thus without loss of generality we may assume $t_p(C) = 0$, hence $C' \subset C'_p, C \subset C_p$. Take $q_1, q_2 \in C_p$ such that $q_1 q_2 \in pC'_p$, then there are s', s'', s''' in $C' - p$ such that: $s'q_1 \in C, s''q_2 \in C$ and $s'''q_1 q_2 \in p$. Hence: $s'''s'q_1 s''q_2 \in p$ with $s'''s'q_1 \in C$ and $s''q_2 \in C$. Consequently, since (p, C') is a prime in C we have $s''q_2 \in p$ or $s'''s'q_1 \in p$, i.e. $q_2 \in pC'_p$ or $q_1 \in pC'p$. This establishes that (pC'_p, C'_p) is a prime in C_p. In case (p, C') is fractional and $q \in C_p - C'_p$ i.e. for some $s \in C' - p$ we have $sq \in C - C'$ and there exist $x, y \in C' - \{0\}$ such that $xsqy \in C'$ with x or y in p and also $xqy \in C'_p$. In case (p, C') is localized and $q \in C_p - C'_p$, i.e. for some $s \in C' - p$, $sq \in C - C'$, hence there exist $\lambda, \mu \in C'$ one at least in p such that $\lambda sq\mu \in C' - p$ or $\lambda q\mu \in C'_p - pC'_p$ with $\lambda, \mu \in C'_p$ at least one in p. Suppose (p, C') is separated, then (pC'_p, C'_p) is known to be localized by the foregoing. Pick $q \in C'_p - \{0\}$ i.e. $sq \in C' - \{0\}$, for some $s \in C' - p$, hence there exist $\lambda, \mu \in C$ such that $\lambda sq\mu \in C' - p$, thus $\lambda q\mu \in C'_p - pC'_p$ with $\lambda, \mu \in C_p$. □

1.2.5 Remark

It is easily verified that $(C^p)_p = (C'_p)^{pC'_p}$, hence by localization we may restrict attention to primes having a domain of definition that is an integrally closed local ring. In case the Noetherian property holds for C' (and C) the foregoing leads us to considering local Krull domains (local integrally closed Noetherian rings, more precisely).

Now focus attention on a field K.

1.2.6 Definition

A proper domain K' in K is called a **valuation ring of** K. If $\lambda \neq 0$ in K then $\lambda\lambda^{-1} = 1$ yields that K' is a proper domain of K exactly when for $\lambda \neq 0$ in K we have $\lambda \in K'$ or $\lambda^{-1} \in K'$.

1.2.7 Corollary

For a field K, the following statements are equivalent for a subring O_v of K:

(a) O_v is a valuation ring of K with maximal ideal m_v.
(b) The pair (m_v, O_v) in Q_K is dominating.
(c) The pair (m_v, O_v) is a localized prime of K.
(d) The pair (m_v, O_v) is a separated prime of K.

Proof. Follows from Proposition 1.1.8 and the fact that in a valuation ring O_v of K the set $\{\lambda \in O_v, \lambda^{-1} \notin O_v\}$ is the unique maximal ideal of O_v. □

1.2.8 Definition

A place of K (corresponding to a valuation ring O_v of K) is a ring homomorphism from O_v to a field k. In the sequel we understand places to be epimorphisms and in that case the kernel of a place is the maximal ideal m_v of O_v; without this restriction we would have places corresponding to all prime ideals of O_v for all valuation rings O_v of K. By localizing at the kernel such places may be related to epimorphic ones.

If K/k is a field extension, a valuation ring O_v of K is said to be a **k-valuation ring** if $k \subset O_v$, i.e. O_v is a k-algebra too. The residue field $k_v = O_v/m_v$ is then also a field extension of k and the place $O_v \to k_v$ is a k-algebra homomorphism. Places corresponding to k-valuation rings of K are called **places of** K/k. If ρ_1, ρ_2 are places of K, respectively of K/k with residue fields k_1, respectively k_2, then

ρ_1 and ρ_2 are said to be **isomorphic places**, respectively **k-isomorphic places**, if there exists an isomorphism, respectively k-isomorphism $i : k_1 \to k_2$ such that $\rho_2 = i \circ \rho_1$. One easily checks that places ρ_1 and ρ_2 are isomorphic if and only if their valuation rings coincide; so up to isomorphism of places we consider the canonical epimorphism $O_v \to k_v = O_v/m_v$ and call it "the" place associated to O_v, denoted by ρ_v. For a place of K/k we define $\dim\rho_v = td(k_v/k)$, where td refers to the transcendence degree of the residual field extension corresponding to ρ_v.

For a commutative ring C the fact of having its ideals totally ordered by inclusion is easily seen to be equivalent to: for $\lambda, \mu \in C$ either $\lambda \in C\mu$ or $\mu \in C\lambda$. If O_v is a valuation domain of K then for $\lambda, \mu \in O_v$ we either have $\lambda^{-1}\mu \subset O_v$ or $\lambda\mu^{-1} \in O_v$ unless λ or μ is zero, in any case either $\lambda \in O_v\mu$ or $\mu \in O_v\lambda$. So we have established a basic but fundamental fact.

1.2.9 Lemma

If O_v is a valuation ring of K then the set of ideals of O_v is totally ordered by inclusion.

1.2.10 Corollary

A finitely generated ideal of O_v is necessarily a principal ideal. A Noetherian valuation domain O_v is necessarily a principal ideal domain.

1.2.11 Lemma

Consider a valuation ring O_v of K.

(a) If $p \neq 0$ is a finitely generated prime ideal of O_v then $p = m_v$ is the unique nonzero prime ideal of O_v, in this case, O_v is Noetherian, $m_v = O_v\pi_v$ also all ideals are of the form $O_v\pi_v^n$. In this case O_v is said to be a **discrete rank one valuation ring of** K (D.V.R. of K).

(b) A nonzero ideal I of O_v that is idempotent cannot be finitely generated and it is a prime ideal.

(c) For a nonzero ideal I of O_v we have that $\cap\{I^n, n \in \mathbb{N}\}$ is a prime ideal of O_v (it may be zero).

Proof. (a) Suppose $p \neq 0$ is a finitely generated prime ideal of O_v. For $\lambda \in m_v - p$ we must have $p \subset O_v\lambda$. Corollary 1.2.9 entails $p = O_v\pi$, thus $\pi = \mu\lambda$ for some $\mu \in O_v$. Since p is a prime ideal $\mu \in p$ and $\mu = \gamma\pi$ for some $\gamma \in O_v$. Thus $\pi = \gamma\pi\lambda$ or $(1-\gamma\lambda)\pi = 0$ and $\gamma\lambda \in m_v$, contradiction. Consequently O_v satisfies the descending chain condition on prime ideals and is Noetherian. If L

is an ideal of O_v such that $L \subset O_v\pi^n$ for all $n \in \mathbb{N}$ then $L = 0$, if not then there is an n such that we have $O_v\pi^n \subset L \subset O_v\pi^{n-1}$, hence $m_v \subset \pi^{-n+1}L \subset O_v$. The latter means $\pi^{-n+1}L = m_v$ or $\pi^{-n+1}L = O_v$, consequently $O_v\pi^n = L$ or $L = O_v\pi^{n-1}$.

(b) If $I \neq 0$ is idempotent and finitely generated, thus principal say $I = O_v\lambda$ then $\lambda = \mu\lambda^2$ for some $\mu \in O_v$ and thus $(1 - \mu\lambda)\lambda = 0$. Since $\mu\lambda \in I$ either $1 \in I$ and I is not a proper ideal or else we have to contradict the assumptions. To establish I is prime consider $\lambda, \mu \notin I$ such that $\lambda\mu \in I$. Then $O_v\lambda \supsetneqq I$ and $O_v\mu \supsetneqq I$, thus $I \supset O_v\lambda\mu \supset I^2 = I$. It thus follows that $I = O_v\lambda\mu$, a contradiction.

(c) If $\lambda, \mu \notin \cap\{I^n, n \in \mathbb{N}\}$ then $O_v\lambda \supset I^e$, $O_v\mu \supset I^d$ for some $e, d \in \mathbb{N}$ (use Lemma 1.2.9) and $O_v\lambda\mu \supset I^{e+d}$. If $\lambda\mu \in \cap\{I^n, n\}$ then $O_v\lambda\mu = I^{e+d} = \cap\{I^n, n \in \mathbb{N}\}$ follows, but then I^{e+d} is principal and idempotent and therefore a contradiction in view of (b). It follows that $\lambda\mu \notin \cap\{I^n, n \in \mathbb{N}\}$ and we have established that the latter is a prime ideal. □

There is a nice correspondence between rings intermediate between O_v and K and prime ideals of O_v.

1.2.12 Proposition

If we have a ring T such that $O_v \subset T \subset K$ then T is a valuation ring of K and it is $(O_v)_p$ for some prime ideal p of O_v. In particular the valuation domain O_v is a DVR if and only if it is a maximal subring of K, its field of fractions.

Proof. If $O_v \subset T \subset K$ then T is certainly a valuation ring of K, with unique maximal ideal m_T say. Since elements of m_T do not have their inverse in T, certainly not in O_v, it follows that $m_T \subset m_v \subset O_v \subset T$ and m_T is a prime ideal of O_v. The multiplicative set $O_v - m_T$ is inverted in T and for every $t \in T - O_v$ we have $t^{-1} \in O_v - m_T$, hence $t = (t^{-1})^{-1} \in (O_v)_{m_T}$, the localization of O_v at m_T. That a DVR is maximal as a subring of K follows from the foregoing. Conversely, assume that R is maximal as a subring of K. If R is not a field it has a nonzero prime ideal p and $R \subset R_p \subset K$ then yields $R = R_p$ because $R_p = K$ is excluded since pR_p is a nontrivial ideal of R_p. If $q \neq 0$ is another prime ideal of R then $R = R_q$ follows and thus $p = q$ because $q \subset p$ follows from $R = R_p$ and $p \subset q$ from $R = R_q$. Therefore R has a unique nonzero prime ideal p. If $x \in K - R$ then $K = R[x]$ follows, hence x^{-1} either is in R or there is a nontrivial relation $x^{-1} = \sum r_i x^i$ leading to an integral equation for x^{-1} over R. However since the integral closure of R cannot be K (otherwise R is a field and K is an algebraic extension) it must be R and then $x^{-1} \in R$. So if R is not a field then it is a valuation ring of K. We may exclude that R is a field since we restricted attention to K being the field of fractions of R. □

1.2.13 Proposition

Let O_v be a valuation ring of K and consider $a_1, \ldots, a_n \in K$ not all zero, then for some $j \in \{1, \ldots, n\}$ $a_j \neq 0$ and $a_i a_j^{-1} \in O_v$ for all $i \in \{1, \ldots, n\}$.

Proof. Select a_j such that $v(a_j)$ is minimal amongst the $v(a_i), i = 1, \ldots, n$. □

Let us look at skewfields now. Because we are dealing with domains most of the time it is natural to look at completely primes as introduced in Sect. 1.1. A subring Λ of a skewfield (= division algebra), D say, is called a **total subring** if for $x \in D$ either x or x^{-1} is in Λ. Sometimes a total subring is called a total valuation ring.

1.2.14 Proposition

A subring Λ of a skewfield D is a total subring if and only if it the domain of a complete prime of D.

Proof. First suppose Λ is a total subring of D and put $P = \{x \in D, x^{-1} \notin \Lambda\}$. For $\lambda \in \Lambda$ and $x \in P$ we obtain from $(\lambda x)^{-1} \in \Lambda$ that $x^{-1}\lambda^{-1}.\lambda = x^{-1} \in \Lambda$, a contradiction, hence $(\lambda x)^{-1} \notin \Lambda$ or $\lambda x \in P$. For $x, y \in P$ we would obtain from $(x+y)^{-1} \in \Lambda$ that $(x+y)^{-1}x \in P, (x+y)^{-1}y \in P$ thus $x^{-1}(x+y) \notin \Lambda$ and $y^{-1}(x+y) \notin \Lambda$. The latter yields $x^{-1}y \notin \Lambda$ and $y^{-1}x \notin \Lambda$ contradicting totality of Λ, hence $x + y \in P$. We established that P is an ideal of Λ (we did the left ideal property but the right hand version is similar). If for some $u, v \in D$ we have $uv \in P$ then $u \notin P$ yields $u^{-1} \in \Lambda$ and thus $u^{-1}(uv) \in P$, so $v \in P$. This establishes that (P, Λ) is a completely prime of D. From the proof it also follows that (P, D^P) is a completely prime of D. In fact $\Lambda = D^P$ follows because if $z \in D^P - \Lambda$ then $z^{-1} \in P$ and $1 = zz^{-1} \in zP \subset P$ since $z \in D^P$, contradiction. Conversely let $D^P \supset P$ define a completely prime of D and assume there exists an $x \in D$ with $x \notin \Lambda, x^{-1} \notin \Lambda, \Lambda = D^P = \{x \in D, xP \subset P \text{ and } Px \subset P\}$. Suppose $xP \not\subset P$, say $xp \notin P$ for some $p \in P$. From $Px^{-1}xP \subset P$ we derive $Px^{-1}xp \subset P$, thus $Px^{-1} \subset P$. If $xP \not\subset \Lambda$ then $x^{-1}xP \subset P$ with $x^{-1} \notin \Lambda$ contradicts the completely primeness of P in D, thus we may conclude that $xP \subset \Lambda$ (in a similar way one can show that $Px \subset \Lambda, x^{-1}P \subset \Lambda$ and $Px^{-1} \subset \Lambda$). Now look at:$(xPx^{-1})(xPx^{-1}) = (xP)(Px^{-1}) \subset P$ because $xP \subset \Lambda$ and $Px^{-1} \subset P$. Consequently, completely primeness of P yields $xPx^{-1} \subset P$, hence $xpx^{-1} \in P$ or $x^{-1} \in P$, a contradiction. □

1.2.15 Proposition

A total subring Λ of a skewfield D satisfies the following properties:

1. Λ is a local ring with maximal ideal P and Λ/P is a skewfield, called the residual skewfield.
2. Every finitely generated left (right) ideal of Λ is principal. For two finitely generated left (right) ideals I and J in Λ we have either $I \subset J$ or $J \subset I$.

Proof. 1. By definition of P it is clear that $\Lambda - P$ contains only units hence Λ is a local ring with maximal ideal P and Λ/P is a skewfield because images of units are invertible elements.

2. Consider $L = \sum_1^n \Lambda x_i$ in Λ. For x_1, x_2 we have either $x_1 x_2^{-1}$ or $x_2 x_1^{-1}$ in Λ, hence either$\Lambda x_1 x_2^{-1} x_2 = \Lambda x_1 \subset \Lambda x_2$, or respectively $\Lambda x_2 x_1^{-1} = \Lambda x_2 \subset \Lambda x_1$. Hence $\Lambda x_1 + \Lambda x_2 = \Lambda x_{12}$ (where x_{12} is either x_1 or x_2); repeating this argument a finite number of times will lead us to $L = \Lambda x_{12\ldots n}$. If I and J are finitely generated then $I = \Lambda x_i, J = \Lambda x_j$ by the foregoing and then $I \subset J$ or $J \subset I$ follows from the same argument used in the proof of (1). □

We may expand somewhat on point (2) in the foregoing proposition and arrive at a characterization of total subrings of D. Recall that for an order Λ in D (i.e. D is the total ring of fractions of Λ) a **left Λ-ideal** L (of D)is just a left Λ-submodule of D such that $Lz \subset \Lambda$ for some $z \in \Lambda$.

1.2.16 Proposition

For an order Λ of the skewfield D the following properties are equivalent.

1. Λ is a total subring of D.
2. The set of left (right) ideals of Λ is linearly ordered by inclusion.
3. The set of left (right) Λ-ideals of D is linearly ordered by inclusion.

Proof. 1. $\Rightarrow$ 2. Look at left ideals I, J of Λ such that $I \not\supset J$. Pick $a \in J - I$ and look at $b \in I$. From (1) either $ba^{-1} \in \Lambda$, then $b \in \Lambda a$ or $b \in J$, or $ab^{-1} \in \Lambda$, then $a \in \Lambda b \subset I$ is a contradiction. Hence $b \in I$ for all $b \in I$, or $I \subset J$.

2. $\Rightarrow$ 3. Look at Λ-ideals I, J of D, then there is a $z \in \Lambda$ such that $Iz, Jz \subset \Lambda$. Thus either $Iz \subset Jz$ or $Jz \subset Iz$ by (2). Consequently either $I \subset J$ or $J \subset I$.
3. $\Rightarrow$ 1. For any $x \in D - \Lambda$ either $\Lambda x \subset \Lambda$ or $\Lambda \subset \Lambda x$ because Λ and Λx are Λ-ideals of D and (3). Indeed since D is the total ring of fraction of Λ we have $xz \in \Lambda$ for some $z \in \Lambda$, thus $\Lambda xz \subset \Lambda$. From $\Lambda \subset \Lambda x$ it follows that $x^{-1} \in \Lambda$. □

1.2.17 Proposition

For a total subring Λ of a skewfield D the following statements are equivalent:

1. Every one-sided ideal of Λ is two-sided.
2. Λ is invariant under all inner automorphisms of D.

Proof. 1. $\Rightarrow$ 2. If one-sided ideals are two-sided then we obtain for $a \in \Lambda$ that $a\Lambda = \Lambda a \Lambda = \Lambda a$, hence $a\Lambda a^{-1} = \Lambda$ for all $a \in \Lambda$ (then also for all $d \in D$ because d or d^{-1} is in Λ). Hence (2) follows.
2. $\Rightarrow$ 1. Look at Λa for $a \in \Lambda$ nonzero. Pick $\mu \in \Lambda$ and look at $a\mu = (a\mu a^{-1})a$ with $a\mu a^{-1} \in \Lambda$; clearly it follows that $a\Lambda \subset \Lambda a$. The converse can be established in a completely similar way. For $L = \sum_{i \in J} \Lambda x_i$, some index set J, $L = \sum_{i \in J} x_i \Lambda$ follows. □

1.2.18 Definition

A total subring Λ of a skewfield D that is invariant under inner automorphisms of D is called a **valuation ring** of D.

Valuation rings of D may be constructed from specific value functions; we refer to next section for full detail on these.

We paid some special attention to fields and skewfields because very often rings considered are also vector spaces over fields or skewfields. With notation as in the beginning of Sect. 1.2 we consider now C-algebras A and a fixed prime (p, C') of C. Look at a π-pseudoplace of A given by a triple $(A', \Psi, \overline{A})$, i.e. $\Psi|C' = \pi$ where $\Psi : A' \to \overline{A}$ is the ring homomorphism defining the π-pseudoplace. Recall that we say that $(A', \Psi, \overline{A})$ is a π-preplace if $(\mathrm{Ker}\Psi, A')$ is a prime of A.

1.2.19 Definition

A π-pseudoplace $(A', \Psi, \overline{A})$ is said to be π-fractional, π-localized or π-separated if in Definition 1.1.4 the elements x, y may be taken from C. Observe that centrality of C in A then allows to replace the two-sided condition in the definition of localized and separated by a onesided condition i.e. the existence of one $c \in C'$ such that $cr \in A'$- $\mathrm{Ker}\Psi$, which replaces the pair $c, d \in C'$ with $crd \in A' - \mathrm{Ker}\Psi$ (in other words one may choose c or d equal to 1).

It is easily seen that every localized pseudoplace of A such that $\overline{A}$ is a prime ring is a preplace of A. Even if the following arguments may be adapted to the case where A is projective as a C-module we shall restrict attention here to the case where it is free, so we assume from hereon that: C is a commutative domain and A is a free C-algebra, $(A', \Psi, \overline{A})$ is a π-pseudoplace with $\pi = (p, C^p)$ being a localized prime of C. For any C-subalgebra B of A we write $\overline{B}$ for the image under Ψ of $A' \cap B$ in $\overline{A}$. Obviously $\overline{B}$ is a $\overline{C}$-algebra and $\overline{C}$ is again a domain. A subalgebra B of A which is free as a C-module is said to be $\boldsymbol{\Psi}$-spanned if $B \cap A'$ contains a C-basis for B and it is said to be $\boldsymbol{\Psi}$-unramified in case that C-basis may be chosen such that its image in $\overline{B}$ is also a $\overline{C}$-basis for $\overline{B}$. We say $(A', \Psi, \overline{A})$ is **unramified** whenever A is Ψ-unramified.

1.2.20 Proposition

With notation and convention as above:

1. If A is Ψ-spanned then $(A', \Psi, \overline{A})$ is π-fractional.
2. An unramified π-pseudoplace of A is π-localized; in case π is separated then an unramified π-pseudoplace is π-separated.

Proof. 1. Let $\{a_i, i \in \mathcal{J}\}$ be a C-basis for A contained in A'. If $\sum_{i=1}^n c_i a_i \in A - A'$ then some c_j is not in C^p. Since we assumed (p, C^p) to be localized there exists a $\mu_j \in p$ such that $\mu_j c_j \in C^p - p$. In $\mu_j(\sum_{i=1}^n c_i a_i)$ there are strictly less coefficients not in C^p than in $\sum_i^n c_i a_i$; after a finite number of repetitions of this argument we arrive at a $\mu \in p$ such that all $\mu c_i \in C^p$, hence $\mu(\sum_{i=1}^n c_i a_i) \in A'$ (note that we also obtained that at least one μc_i is not in p).
2. Suppose $\{a_i, i \in \mathcal{J}\}$ is a C-basis contained in A' such that $\{\Psi(a_i), i \in \mathcal{J}\}$ is a $\overline{C}$-basis for $\overline{A}$. For $x \in A - A'$ we have $x = \sum c_i a_i$ with some $c_i \notin C^p$ for some $i \in \mathcal{J}$. The assumption that (p, C^p) is localized entails that there is a $\mu_i \in p$ such that $\mu_i c_i \in C^p - p$. After repetition of the argument we obtain $\mu \in P$ such that $\mu x \in A'$ but for some j, $\mu c_j \notin p$. From $\mu x \in \mathrm{Ker}\Psi$ we would obtain $\sum' \Psi(\mu e_i)\Psi(a_i) = 0$, thus $\Psi(\mu c_i) = 0$ for all i but that contradicts $\mu c_j \notin p$. A similar argument applies in the separated case. □

1.2.21 Corollary

If $(A', \Psi, \overline{A})$ is an unramified π-pseudoplace such that $\overline{A}$ is a prime ring, then $(\mathrm{Ker}\Psi, A')$ is a prime of A such that $A' = A^{\mathrm{Ker}\Psi}$. In particular, unramified pseudoplaces in algebras A over fields K such that $\overline{A}$ is a prime ring define localized primes of A.

Proof. A localized prime of a field is a valuation ring of that field. □

1.2.22 Proposition

With assumptions as before and assuming C to be a field, if $\dim_C A < \infty$ then there is a π-prime (P, A^P) containing a C-basis of A.

Proof. Select a C-basis for A with structure constants in C^p, say $\{a_1, \dots, a_n\}$, this means that products of elements of $\{a_1, \dots, a_n\}$ are linear combinations of $\{a_1, \dots, a_n\}$ with coefficients in C^P. Put $B = C^p \cup \{a_1, \dots, a_n\}$ and $M = C^p - p$. Now apply Theorem 1.1.3 (the extension theorem), verification of the conditions is straightforward, in particular (b) and (c) follow from the properties of $\{a_1, \dots, a_n\}$. The extension theorem yields the existence of a π-prime (P, A^P) in A with

$B \subset A^P$. Consequently A^P contains a C-basis of A and then $(A^P, \Psi, \overline{A})$ is a π-fractional preplace (see Proposition 1.2.20.1). □

1.2.23 Proposition

In the situation of the foregoing proposition there exists at least one unramified pseudoplace over $\pi = (p, C^p)$.

Proof. Select a C-basis $\{a_1, \ldots, a_n\}$ for A having structural constants in C^p, then the ring $A' = C^p[a_1, \ldots, a_n]$ equals the C^p-module generated by $\{a_1, \ldots, a_n\}$. Then $P = \sum_{i=1}^n pa_i$ is an ideal of A' and we claim that $\Psi : A' \to A'/P = \overline{A}$ yields a pseudoplace of A. In order to establish $A' \cap C = C^p$. Consider $c = \sum_{i=1}^n c_i a_i \in C - C^p$, with $c_i \in C^p$. Since $(p.C^p)$ is localized there is a $\mu \in p$ such that $\mu c \in C^p - p$. Hence $\mu c = \sum_{i=1}^n \mu c_i a_i$ with $\mu c_i \in p$ for $i = 1, \ldots, n$. We calculate: $\mu c a_j = \sum_{k,i=1}^n \mu c_i \alpha_{ij}^k a_k$, where we wrote α_{ij}^k for the structural constants of $\{a_1, \ldots, a_n\}$. It follows that $\mu c = \sum_{i=1}^n \mu c_i \alpha_{ij}^j$, but the right hand members is in p and that contradicts $\mu c \in C^p - p$. Moreover, if in the foregoing we choose $c \in C^p - p$ with $c = \sum_{i=1}^n c_i a_i$ and $c_i \in p$ for $i = 1, \ldots, n$ then we obtain: $ca_j = \sum_{i,k=1}^n c_i \alpha_{ij}^k a_k$, or $c = \sum_{i=1}^n c_i \alpha_{ij}^j \in p$ and that is a contradiction. Finally, if $\sum_{i=1}^n \overline{c}_i \Psi(a_i) = 0$ for some $\overline{c}_i \in \overline{C} = C^p/p$ then select c_i such that $\Psi(c_i) = \overline{c}_i$ for $i = 1, \ldots, n$; hence $\Psi(\sum_{i=1} c_i a_i) = 0$ means that$\sum_{i=1}^n c_i a_i \in P$ and then it follows that each $c_i \in p$ because $\{a_1, \ldots, a_n\}$ is a C-basis, i.e. $\overline{c}_i = 0$ or $\Psi(a_i)$ is a basis for $\overline{A}$ as Ψ is surjective. □

The π-pseudoplaces have been used in constructions of generic crossed products by the second author, cf. [71], and foregoing results stem from the general theory there. More general statements over domains C, as claimed in [69] may be correct for suitable C but in the generality claimed in loc. cit there seems to be a gap in the proof; the point being that a C-basis $\{b_1, \ldots, b_n\}$ multiplied by some $c \in C$ in order to obtain structural constants in C^p need not be a C-basis unless the factor c is a unit in C.

The foregoing may be used to prove a lifting result for *Azumaya* algebras under unramified pseudoplaces.

1.2.24 Remark

In view of Proposition 1.2.14 a localized prime (P, D^P) of a skewfield D is a total subring of D if and only if $\overline{D}$ is a domain (hence skewfield). In case $K = Z(D)$ and π belongs to (m_v, O_v) then a π-localized prime of D having a skewfield $\overline{D}$ for its residual algebra is a total subring with extra properties. For example if (P, D^P) is the corresponding prime then for $d \in D - D^p$ there exists $r, r' \in D^P$ such that

rd and $dr' \in D^P - M$, in fact one may then even take $r = r'$ in O_v, central in D^P. This means that D^P is a Dubrovin valuation ring of D; we shall study this type of primes later in more detail.

1.3 Value Functions and Prime Filtration

1.3.1 Partially and Totally Ordered Groups

A **partially ordered set**, or poset, is a set with a relation $\leq$ that is reflexive, antisymmetric and transitive; a poset $S, \leq$ is a partially ordered group, which is also a group or po-group, if from $x \leq y$ it follows that $sxd \leq syd$ for all $s, d \in S$ where we have denoted the group operation multiplicativity. A partial order on S is said to be **total**, or S is said to be **totally ordered**, if for each $s, d \in S$ either $s \leq d$ or $d \leq s$: In the sequel Γ is a totally ordered group and we write it as an additive group with neutral element 0, even if Γ need not be abelian. The positive part of Γ is $\Gamma_+ = \{\gamma \in \Gamma, \gamma > 0\}$. To $\gamma \in \Gamma$ we associate $|\gamma| \in \Gamma_+$ where $|\gamma| = \gamma$ if $\gamma > 0$, $|0| = 0$ and $|\gamma| = -\gamma$ if $\gamma < 0$. Observe that $\gamma < 0$ yields $-\gamma > 0$. An **isolated subgroup** $K \subset \Gamma$ is one such that $\kappa \in K$ entails that all $\gamma \in \Gamma$ such that $|\gamma| \leq |\kappa|$ are in K.

1.3.1.1 Lemma

The set of isolated subgroups of Γ is totally ordered by inclusion.

Proof. Consider nonzero proper isolated subgroups K_1 and K_2 in Γ. If $K_1 \neq K_2$ then $K_{1,+} \neq K_{2,+}$ where we denote K_+ for any subgroup K of Γ for the set of positive elements of K. Suppose $\gamma_1 \in K_{1,+}$ and $\gamma_1 \notin K_2$ and look at arbitrary $\gamma_2 \in K_{2,+}$. Since Γ is totally ordered and K_2 is isolated $\gamma_2 < \gamma_1$ follows but then $\gamma_2 \in K_1$ since K_1 is isolated too. Consequently $K_{2,+} \subset K_1$ and thus $K_2 \subset K_1$.

The order type of the set of all proper isolated subgroups of Γ is called the **rank of** Γ, denoted $rk\Gamma$. □

1.3.1.2 Proposition

Let K be a invariant (normal) isolated subgroup of Γ then Γ/K is totally ordered such that the canonical $\Gamma \to \Gamma/K$ is an ordered group morphism. Moreover, $rk\Gamma = rkK + rk\Gamma/K$.

Proof. Pick $\overline{\gamma} \neq \overline{\delta}$ in Γ/K and $\gamma, \delta \in \Gamma$ representatives of $\overline{\gamma}, \overline{\delta}$ respectively. Suppose $\gamma > \delta$ and look at $\gamma' = \gamma + \xi$ and $\delta' = \delta + \eta$ with $\xi, \eta \in K$. If $\gamma' \leq \delta'$ then $\xi < \eta$ since $\delta < \gamma$. Then it follows that $\gamma \leq \delta + \eta - \xi$ or

$-\delta+\gamma \le \eta-\xi$ with $\eta-\xi \subset K$. Since K is an isolated subgroup of Γ and $-\delta+\gamma$ being positive, $-\delta+\gamma \in K$ follows. The latter contradicts $\overline{\gamma} \neq \overline{\delta}$ in Γ/K. This allows to define $\overline{\gamma} > \overline{\delta}$ if $\gamma > \delta$ since the foregoing establishes independence of the chosen representative. It is obvious that this makes Γ/K into a totally ordered group such that $\Gamma \xrightarrow{\pi} \Gamma/K$ is order preserving. For the second statement it is sufficient to observe that an isolated subgroup $\mathcal{L}$ of Γ is given by $\mathcal{L} \cap K$, an isolated subgroup of K and $\pi(\mathcal{L})$, an isolated subgroup of Γ/K. The set of isolated subgroups in Γ is the set-theoretic sum of two ordered sets, hence for the order types it follows that $rk\Gamma = rkK + rk\Gamma/K$. □

1.3.1.3 Definition

A totally ordered group Γ is **Archimedean** if for $\gamma, \delta \in \Gamma$ with $\delta > 0$ say, there is an $n \in \mathbb{N}$ such that $n\delta > \gamma$.

1.3.1.4 Proposition

For a totally ordered group Γ, the following statements are equivalent:

(i) $rk\Gamma = 1$.
(ii) Γ is Archimedean.

Proof. (i) → (ii) Suppose $\delta > \gamma > 0$ in Γ are such that there is no $n \in \mathbb{Z}$ such that $n\gamma > \delta$. Look at $S = \{\gamma' \in \Gamma_+, \gamma' < n\gamma \text{ for some } n\}$. Obviously S is closed under addition in Γ and if $\gamma'' < \gamma'$ for some $\gamma' \in S$ then $\gamma'' \in S$. The elements of S and their inverses generate a proper isolated subgroup of Γ (proper because δ is not in it), contradicting $rk\Gamma = 1$ (0 is the only isolated proper subgroup).

(ii) → (i) Suppose $0 \neq K$ is an isolated proper subgroup in Γ, and pick a positive element $\gamma \in K$. Then $\gamma < \lambda$ for each positive element λ not in K (if $\lambda < \gamma$ then $\lambda \in K$ because K is isolated). But by (ii), $\lambda < n\gamma$ for some $n \in \mathbb{N}$ would lead to $\lambda \in K$, contradiction. Hence such K does not exist. □

The following is a classical result, the proof is based on the use of Dedekind cuts, see [61, Theorem 1, p. 6].

1.3.1.5 Theorem

An Archimedean group is order isomorphic to a subgroup of $\mathbb{R}$.

A group of rank 1 is said to be **discrete** if it is order isomorphic to $\mathbb{Z}, +$. An abelian totally ordered group of finite rank is called **discrete** if the factorgroups of successive isolated subgroups are isomorphic to the additive group $\mathbb{Z}, +$. A discrete group is thus necessary order isomorphic to $\mathbb{Z}^r, +$ for some finite r.

It is easy to see that a totally ordered group Γ is necessarily torsion free; indeed if $n\gamma = 0$ for $\gamma \in \Gamma$, $n \in \mathbb{N}$ then either $\gamma \leq 0$ and $n\gamma \leq \gamma \leq 0$ yields $\gamma = 0$, or else $\gamma \geq 0$ and then $n\gamma \geq \gamma \geq 0$ yields $\gamma = 0$. In case Γ is only partially ordered there may exist nontrivial torsion elements but these are then incomparable to 0.

1.3.2 Value Functions of Primes in Simple Artinian Rings

Starting from a prime (P, A^P) in a simple Artinian ring A we associate to $a \in A$ the set $P : a = \{(x, x') \in A \times A, xax' \in P\}$. We define a relation $\sim$ on A by putting $a \sim a'$ whenever $P : a = P : a'$. It is clear that $\sim$ is an equivalence relation on A; we write $\widetilde{a}$ for the equivalence class of $a \in A$. For $a = 0$ we have that $P : 0 = A \times A$ and $a' \sim 0$ then means that $Aa'A \subset P$, thus $a' = 0$ since A is simple. So we may look at $\sim$ on the semigroup $A^* = A - \{0\}$.

We introduce a partial order in $A^*/\sim = \widetilde{A}$ by putting $\widetilde{a} \leq \widetilde{b}$ whenever $P : a \subset P : b$.

1.3.2.1 Lemma

The multiplication of A^* induces a partially ordered semigroup structure on $\widetilde{A}$ with neutral element $\widetilde{1}$.

Proof. The relation $\sim$ is compatible with multiplication, that is: if $a \sim a', b \sim b'$ then $ab \sim a'b'$. Indeed if $(x, y) \in A \times A$ is such that $xaby \in P$ then $(x, by) \in P : a = P : a'$, thus $xa'by \in P$ or $(x, a') \in P : b = P : b'$, thus $xa'by \in P$ or $P : ab \subset P : a'b'$. The converse inclusion follows by interchanging the role of (a, b) and (a', b'). The relation $\widetilde{a} \leq \widetilde{b}$ whenever $P : a \subseteq P : b$ is obviously reflexive, antisymmetric and transitive. Moreover for $\widetilde{a} = \widetilde{b}$ and $\widetilde{c} \in \widetilde{A}$ we have $\widetilde{ac} \leq \widetilde{bc}$ as well as $\widetilde{ca} \leq \widetilde{cb}$, indeed if $(x, y) \in P : ca$, then $xcay \in P$ or $(xc, y) \in P : a \subseteq P : b$ yields $xcby \in P$ or $(x, y) \in P : cb$ and thus $P : ca \subset P : cb$ or $\widetilde{ca} \leq \widetilde{cb}$ (the right symmetric statement follows in a symmetric way). Now we define $\widetilde{a}.\widetilde{b}$ in $\widetilde{A}$ by putting $\widetilde{a}.\widetilde{b} = \widetilde{ab}$ and this is well defined; this makes $\widetilde{A}$ into a partially ordered semigroup with neutral element $\widetilde{1}$. □

Define a value function $\varphi : A^* \to \widetilde{A}, a \mapsto \widetilde{a}$.

1.3.2.2 Corollary

The value function $\varphi : A^* \to \widetilde{A}$ satisfies:

(i) φ is multiplicative, i.e. $\varphi(ab) = \varphi(a)\varphi(b)$ for $a, b \in A^*$.
(ii) If $\widetilde{a} \geq \widetilde{c}$ and $\widetilde{b} \geq \widetilde{c}$ then $(a + b)^\sim \geq \widetilde{c}$. In particular when $\widetilde{a} \geq \widetilde{b}$ then $(a + b)^\sim \geq \widetilde{b}$ or $\varphi(a + b) \geq \varphi(b)$.

Proof. (i) By definition.
(ii) Take $(x, y) \in P : c$, thus $(x, y) \in P : a$ and $(x, y) \in P : b$. This yields $xay \in P$ and $xby \in P$, thus $x(a + b)y \in P$ or $(x, y) \in P : (a + b)$. Consequently $P : c \subset P : (a + b)$. □

Observe in (ii) that $\varphi(a + b) \geq \min\{\varphi(a), \varphi(b)\}$ whenever that minimum exists, so φ does extend the notion of a valuation function for a totally ordered value group to a partially ordered semigroup. Note that for $a \in U(A)$ we have $\varphi(a^{-1}) = \varphi(a)^{-1}$ because from $aa^{-1} = a^{-1}a = 1$ we obtain $\varphi(a)\varphi(a^{-1}) = \varphi(a^{-1})\varphi(a) = \widetilde{1}$. So for $a \in U(A)$, if $\widetilde{a} \geq \widetilde{1}$ then $\widetilde{a}^{-1} = (a^{-1})^{\sim} \leq \widetilde{1}$.

If we consider an arbitrary ring A in the foregoing, say R. Then $a' \sim 0$ means $Ra'R \subset P$ or $a' \in P^o$. Therefore we replace R by R/P^o and look at the prime $(P/P^o, R^P/P^o)$ in R/P^o. On $\overline{R}$ we define $\widetilde{R} = \overline{R}^* | \sim$. We may check Lemma 1.3.2.1 for $\overline{R}$ and see that $\widetilde{R}$ is a partially ordered semigroup and obtain a value function $\overline{\varphi} : \overline{R}* \to \widetilde{R}$ satisfying the properties of Corollary 1.3.2.2. For separated primes (P, R^P) in R we have $P^o = 0$ and $\varphi : R^* \to \widetilde{R}$ is then well defined.

In case A is a skewfield, D say, then $\widetilde{D}$ is a partially ordered group. In general when we consider a prime (P, R^P) in a ring R, then we call it an **arithmetical prime of** R in case it is separated and $\widetilde{R}$ is a partially ordered group.

Value functions associated to primes in skewfields were introduced by J. Van Geel in [69]; they may be considered as the natural modification of valuation functions.

Now consider a localized prime (P, D^P) in D and look at $O_P = \{x \in D, x = 0$ or $\varphi(x) \geq 0\} \cup \{0\}$ where φ denotes the value function on D^* associated to the prime (P, D^P) and we have written the value group $\widetilde{D}$ additively with neutral element 0 (even if it need not be abelian).

1.3.2.3 Proposition

With notation as above; O_P is a subring of D invariant under inner automorphisms of D, in fact $O_P = \bigcap_{z \in D^*} zD^P z^{-1}$.

Proof. If suffices to prove the second statement after verifying that O_P is a subring of D, but the latter is an immediate consequence of Corollary 1.3.2.2. Now look at x such that $\varphi(x) \geq 0$, thus $(P : 1) \subset (P : x)$ and consider zxz^{-1} for some $z \in D^*$. We calculate: $(a, b) \in (P : 1)$ (i.e. $ab \in P$) implies $(a, b) \in P : x$, $axb \in P$ and $(az, z^{-1}b)$ satisfies $azz^{-1}b = ab \in P$ thus $(az, z^{-1}b)$ is in $P : 1$ hence in $P : x$, consequently $azxz^{-1}b \in P$ and $P : zxz^{-1}$ contains (a, b). This establishes $P : 1 \subset P : zxz^{-1}$, and therefore O_P is invariant under inner automorphisms of D. Consider $x \in \bigcap_{z \in D^*} zD^P z^{-1}$ and let $a, b \in D^*$ be such that $ab \in P$. From $x \in bD^P b^{-1}$, say $x = byb^{-1}$ with $y \in D^P$ we obtain: $abyb^{-1}b = aby \in axb = abyb^{-1}b \in P$. Therefore we have $P : 1 \subset P : x$ or $\bigcap_{z \in D^*} zD^P z^{-1} \subset O_P$. The converse inclusion follows from the fact that $O_P \subset D^P$. Indeed $x \in O_P - D^P$ would lead to the existence of $r, s \in D^P$ such that $rxs \in D^P - P$ with $rs \in P$, by the fact that

(P, D^P) is a localized prime of D. However $rs \in P$ yields $(r, s) \in P : 1 \subset P : x$ and $rxs \notin P$ then leads to a contradiction, so there is no x in $O_P - D^P$. □

1.3.2.4 Lemma

If U stands for the group of units of O_P, then we have the following exact sequence of group morphisms:

$$1 \to U \to D^* \underset{\varphi}{\longrightarrow} \widetilde{D} \to 0$$

Proof. If $x \in D^*$ is such that $\varphi(x) = 0$ then from $x^{-1}x = 1$ it follows that $0 = \varphi(x^{-1}) + 0$, hence $x^{-1} \in \mathrm{Ker}\varphi$. Conversely if $x \in U$ then $x, x^{-1} \in O_P$ and $\varphi(x) + \varphi(x^{-1}) = 0$ with $\varphi(x), \varphi(x^{-1}) \geq 0$ hence if $\varphi(x) > 0$ then in the partial order of $\widetilde{D} : 0 = \varphi(x) + \varphi(x^{-1}) \geq \varphi(x) + 0 > 0$ leads to $\varphi(x) = \varphi(x^{-1}) = 0$. □

A localized prime (P, D^P) of D is called **strict** if from $\varphi(a), \varphi(b) > \widetilde{c}$ for $\widetilde{c} \in \widetilde{D}$ it follows that $\varphi(a + b) > \widetilde{c}$. In particular when $\widetilde{D}$ is totally ordered then (P, D^P) is strict as is easily seen.

1.3.2.5 Corollary

Let (P, D^P) be a strict localized prime in D. The ring O_P is a local ring with maximal ideal $p = \{x \in D, \varphi(x) > 0 \text{ or } x = 0\}$.

Proof. Obviously $p = O_P - U$. If $x, y \in p$ then $\varphi(x + y) > 0$ hence $x + y \in p$. For $x \in p$ and $z \in O_P$ we have $\varphi(zx) = \varphi(z) + \varphi(x)$ with $\varphi(x) > 0, \varphi(z) \geq 0$, thus $\varphi(z) + \varphi(x) \geq 0 + \varphi(x) > 0$ entails $zx \in p$, similarly it is established that $xz \in P$. Therefore p is an ideal of O_P and the statement follow. □

We do not know in general whether (p, O_P) is a prime of D but we can summarize its properties as follows.

1.3.2.6 Proposition

Let (P, D^P) be a strict localized prime in D. The ring O_P is local and O_P/p is a skewfield. Every one sided ideal of O_P is two-sided and $O_P = zO_Pz^{-1}$, $p = xpz^{-1}$ for every $z \in D^*$. The set $O_P^* = O_P - \{0\}$ is a left and right Ore set of O_P and so O_P has a skewfield of fractions $D(P)$ obtained by inverting (left or right) the Ore set O_P^*. If (p, O_P) is a prime of $D(P)$ with domain O_P, then O_P is a valuation ring of $D(P)$ with value group Γ in $\widetilde{D}$ obtained as the subgroup $\{v_1 - v_2, -v_2 + v_1$, with $v_1 \geq 0, v_2 \geq 0\}$ being the subgroup generated by elements comparable to 0 in the partial order of $\widetilde{D}$.

Proof. From the fact that O_P is local (Corollary 1.3.2.5.) it follows that O_P/p is a skewfield since U maps to invertible elements. From Proposition 1.3.2.3 it follows that $zO_Pz^{-1} \subset O_P$ hence in fact $zO_Pz^{-1} = O_P$, because $z^{-1}O_Pz \subset O_P$. Since

$O_P \subset D^P$ we have $P \cap O_P \neq O_P$ hence as an ideal $P \cap O_P$ is contained in the unique maximal ideal p. For any $z \in D^*, zpz^{-1}$ is an ideal of $O_P = zO_Pz^{-1}$ and so $zpz^{-1} \subset p$. Hence $zpz^{-1} = p$ follows. If $a \in O_P$ then $O_Pa = (aO_Pa^{-1})a = aO_P$ hence all onesided ideals of O_P are twosided. Since O_P is a domain we only have to verify one condition for $O_P - \{0\} = S$ to be a left (similarly: right) Ore set of O_P: given $\lambda \in O_P, s \in S$ there exist $\lambda' \in O_P, s' \in S$ such that $s'\lambda = \lambda' s$. Take $s' = s$ and pick $\lambda' \in O_P$ such that $\lambda' s = s\lambda$, this is possible because $O_P s = sO_P \ni s\lambda$. The skewfield of fractions $D(P)$ is the left and right quotient ring of O_P with respect to the left and right Ore set S (see [63, 72, 76] and many papers about localization theory). If we assume that (p, O_P) is a prime in $D(P)$ with domain O_P then if $x \in D(P)$ such that $x^{-1} \notin O_P$ we have $pxO_Px^{-1} \subset p$ because $xO_P = O_Px$ (write $x = s^{-1}a$ for $s \in S, a \in O_P$ then $s^{-1}aO_P = s^{-1}O_Pa$ and from $sO_P = O_Ps$ it follows that $O_Ps^{-1} = s^{-1}O_P$, hence $s^{-1}aO_P = s^{-1}O_Pa = O_Ps^{-1}a$). Since (p, O_P) is a prime by assumption $px \subset p$ follows from $x^{-1} \notin p$. In a completely symmetric way it follows that $xp \subset p$, hence $x \in P_p$. This proves that O_P is a total subring of $D(P)$ and as it is invariant under all inner automorphisms of $D(P)$ it is a valuation ring by definition. The values of $s^{-1}a$ with $s \in S, a \in O_P$ are $-v(s) + v(a)$ with $v(s) \geq 0, v(a) \geq 0$, so the value group for $D(P)$ is the subgroup of $\widetilde{D}$ generated by the elements as claimed in the proposition. It follows that this is a totally ordered subgroup of $\widetilde{D}$. □

1.3.2.7 Remark

If in the foregoing proposition we only assume that (p, O_P) is a prime of $D(P)$ then it does not follow that O_P is the domain of definition of (p, O_P). Let us say that p is **full** if $v(\delta) > -v(\rho)$, for some $\delta \in D(P)$ and all $\rho \in p$, implies $v(\delta) \geq 0$. In case p is full then the domain of (p, O_P) is exactly O_P. Indeed if Δ is the domain of (p, O_P), i.e. $\Delta = \{\delta \in D(P), \delta p \subset p \text{ and } p\delta \subset p\}$, then $v(\delta\rho) = v(\delta) + v(\rho) > 0$ for every $\delta \in \Delta$ and $\rho \in p$. From $v(\delta) > -v(\rho)$ for all $\rho \in p$ it then follows by the assumption that $v(\delta) \geq 0$ or $\delta \in O_P$.

1.3.2.8 Lemma

Consider $a, b \in D^*$ then $a \in O_Pb$ if and only if $\varphi(a) \geq \varphi(b)$.

Proof. If $a = \lambda b$ with $\lambda \in O_P$ then $\varphi(a) = \varphi(\lambda) + \varphi(b) \geq \varphi(b)$. Conversely, if $\varphi(a) \geq \varphi(b)$ then $\varphi(ab^{-1}) \geq 0$, hence $ab^{-1} \in O_P$ and $a \in O_Pb$. □

For any partially ordered group G we define **positive filters** of G to be the subsets of positive elements such that if g is in it then every $h \in G, h \geq g$, is also in it. A positive filter is said to be **pointed** if it has a least element. If F_1 and F_2 are positive filters in G then $F_1 + F_2$ is the positive filter consisting of all $f_1 + f_2$ with $f_1 \in F_1, f_2 \in F_2$. Note that if $f \geq f_1 + f_2$ then $f = f - f_1 + f_1$ with $f - f_1 \geq f_2$ hence $f - f_1 \in F_2$ and thus $f \in F_1 + F_2$.

1.3.2.9 Proposition

With notation as before assume $\widetilde{D}$ is totally ordered.

Ideals of O_P correspond bijective with the positive filters in $\widetilde{D}$. Principal ideals correspond to pointed filters. The multiplicative system of ideals of O_P is isomorphic to the additive system of positive filters in $\widetilde{D}$. Since $\widetilde{D}$ is totally ordered, the set of ideals of O_P is totally ordered.

Proof. To I, an ideal of O_P, we associate $\varphi(I) \subset \widetilde{D}$. As a consequence of Lemma 1.3.2.8, $\varphi(I)$ is a filter of positive elements. Conversely to a positive filter F in $\widetilde{D}$ we associate $\mathcal{J}(F) = \{a \in O_P, \varphi(a) \in F\}$; the properties of φ entail that $\mathcal{J}(F)$ is an ideal of O_P. Since φ is surjective, $\varphi(\mathcal{J}(F)) = F$; we also have $\mathcal{J}(\varphi(I)) = I$. If $I = O_P a$ then $\varphi(I)$ is the positive filter consisting of $\widetilde{x} \in \widetilde{D}$ such that $\widetilde{x} \geq \varphi(a)$, hence it has least element $\varphi(a)$. Conversely if $F = \{\widetilde{x} \in \widetilde{D}_+, \widetilde{x} \geq \widetilde{x}_0\}$ for some fixed $\widetilde{x}_o \in \widetilde{D}_+$ then taking $x_o \in O_P$ such that $\varphi(x_0) = \widetilde{x}_o$ yields that every $x \in \mathcal{J}(F)$ has $\varphi(x) \geq \varphi(x_o)$ hence $x \in O_P x_o$ and the ideal $\mathcal{J}(F)$ is generated by x_o. For ideals I, J in O_P we have $\varphi(IJ) = \varphi(I) + \varphi(J)$. Indeed if $z = \sum_{i=1}^{n} a_i b_i \in IJ$ then $\varphi(z) \geq \min\{\varphi(a_i b_i), i = 1, \ldots, n\}$, say: $\varphi(z) \geq \varphi(a_{i_o}) + \varphi(b_{i_o})$ for some $i_o \in 1, \ldots, n$. This yields $\varphi(z) \in \varphi(I) + \varphi(J)$. Conversely if $h = \varphi(a) + \varphi(b) \in \varphi(I) + \varphi(J)$ with $a \in I, b \in J$, then we have $h = \varphi(ab) \in \varphi(IJ)$. Since $\widetilde{D}$ is totally ordered the set of positive filters is totally ordered too and so is the set of ideals of O_P in view of the bijective correspondence established before. □

1.3.2.10 Remark

For a finitely generated ideal $O_P a_1 + \ldots + O_P a_n = I$ in O_P it is true, that I is principal because we can select one of the $\varphi(a_i), i = 1, \ldots, n$ as a minimal one! If $\widetilde{D}$ is not totally ordered, then this property fails.

Next we look at a totally ordered group Γ. In the sequel we link value functions onto Γ to valuation rings.

1.3.2.11 Definition

For a skewfield D, let $U(D) = D - \{0\}$ be its multiplicative group. A surjective map $v : U(D) \to \Gamma$, where Γ is a totally ordered group, is called a Γ-valuation of D if v satisfies the following conditions:

(i) For $a, b \in U(D)$, $v(ab) = v(a) + v(b)$ (where we denote the operation in Γ by $+$ even if it need not be an abelian group).
(ii) For $a, b \in U(D)$, $v(a + b) \geq \min\{v(a), v(b)\}$.

In case Γ is abelian then v is called an **abelian valuation**. The set $\Lambda = \{d \in D, v(d) \geq 0\}$ is a subring of D with $P = \{d \in D, v(d) > 0\}$ being an ideal of Λ (note that 0 denotes the neutral element of Γ). The elements $u \in U(D)$ with $v(u) = 0$ form an invariant subgroup of $U(D)$ contained in Λ and in fact it is exactly $\Lambda - P$. Therefore Λ is a local ring with unique maximal ideal P. Since v is by definition surjective it defines an exact sequence of groups $0 \to \Lambda - P \to U(D) \xrightarrow{v} \Gamma \to 0$, actually defining Γ in terms of $U(D)$.

1.3.2.12 Remark

If $a, b \in U(D)$ are such that $v(a) \neq v(b)$ then we actually have $v(a+b) = \min\{v(a), v(b)\}$. Indeed if we assume $v(a) > v(b)$ say then from $v(a+b) > \min\{v(a), v(b)\}$ would lead to $v(b) = v(a+b-a) \geq \min\{v(a+b), v(a)\} > v(b)$, a contradiction. Therefore $v(a+b) = \min\{v(a), v(b)\}$ follows. Observe that for $d \in D, v(d^{-1}) = -v(d)$ in particular Λ is invariant under all inner automorphisms of D.

Let us write Λ_v for the subring of D corresponding to a valuation function $v : U(D) \twoheadrightarrow \Gamma$.

1.3.2.13 Proposition

With notation as before: Λ_v is a valuation ring of D.

Proof. By definition Λ_v is obviously a total subring of D. It is also obvious that Λ_v is invariant under inner automorphisms of D. □

We can add the converse to the foregoing proposition and arrive at the following theorem (a version of Theorem 3 of [61]).

1.3.2.14 Theorem

The valuation rings of a skewfield D are exactly the Λ_v corresponding to a valuation function v.

Proof. In view of the foregoing proposition we only have to show that a valuation ring Λ in D is of the form Λ_v for some Γ-valuation function $v : U(D) \twoheadrightarrow \Gamma$. Write U for $\Lambda - P$ and consider $v : U(D) \twoheadrightarrow U(D)/U$, the canonical epimorphism. Since U is invariant under inner automorphisms of D it follows that $\lambda U = U\lambda$ for every $\lambda \in \Lambda$ and we have $v(a) = aU = Ua$ in $U(D)/U$. For $a, b \in U(D)$ either $a^{-1}b$ or $b^{-1}a$ is in Λ, say $a^{-1}b \in \Lambda$. Then $ba^{-1} \in \Lambda$ too because $ba^{-1} = a(a^{-1}b)a^{-1}$. Define $\alpha > \beta$ in $\Gamma = U(D)/U$ whenever $\alpha = v(a), \beta = v(b)$ such that $ab^{-1} \in P$ (then also $b^{-1}a \in P$). It is easy to verify that this makes Γ into a totally ordered group and also verifies in a straightforward way that v is a valuation function $U(D) \twoheadrightarrow \Gamma$. That Λ is exactly Λ_v is obvious. □

Valuation rings of D can also be characterized via primes as in the following proposition.

1.3.2.15 Proposition

The primes (P, D^P) of D with $\widetilde{D}$ a totally ordered group such that D^P is invariant under inner automorphisms of D are exactly the valuation rings of D.

The proof is a modification of an argument in the proof of Proposition 1.3.2.6, taking into account that (P, D^P) is strict because $\widetilde{D}$ is totally ordered.

Proof. If (P, D^P) defines a valuation ring $\Lambda_v = D^P$ then D^P is by definition invariant under inner automorphisms of D. Conversely, suppose D^P is invariant and assume $x \in D$ such that $x^{-1} \notin D^P$. From $xD^Px^{-1} \subset D^P$ it follows that $PxD^Px^{-1} \subset P$ and because (P, D^P) is a prime of D it follows that $Px \subset P$. Similarly, from $x^{-1}D^PxP$ with $x^{-1} \notin D^P$ it follows that $xP \subset P$. But $xP \subset P$ and $Px \subset P$ means $x \in D^P$. Thus D^P is a total subring of D. Consequently, since D^P is assumed to be invariant under inner automorphisms it follows that D^P is a valuation ring of D. □

Abelian valuations have better structural properties than general ones. An easy characterization of abelian valuations is that Λ_v contains $U(D)'$, the commutator subgroup of $U(D)$. As an example of the good behaviour of abelian valuations let us mention Krasner's theorem.

1.3.2.16 Proposition

Let $D \subset E$ be skewfields and v an abelian valuation on D given by $\Lambda_v \subset D$, $P_v \subset \Lambda_v$. Then v extends to an abelian valuation w on E if and only if $P_vU(E)'$ is a proper subset of $\Lambda_vU(E)'$.

Proof. If c is in $U(E')$ and $d \in D$ then $dc = (dcd^{-1})d$ and the conjugate of a commutator is again a commutator. It follows that $\Lambda_vU(E)' = U(E)'\Lambda_v$ and $P_vU(E)' = U(E)'P_v$. Now we apply the extension Theorem 1.1.3 with $B = U(E)'\Lambda_v$ and $M = \Lambda_v - P_v$ (conditions (a),…,(e) are verified by definition here) since $P_vU(E)'$ is a proper subset of $\Lambda_vU(E)'$. This leads to the existence of a prime of E, (P_1, E^{P_1}) say, such that $U(E)'\Lambda_v \subset E^{P_1}$ and $P_1 \cap D = P_v$. For any $x \in E$ we have that $xE^{P_1}x^{-1} \subset E^{P_1}$ because for every $\lambda \in E$, $x\lambda x^{-1}\lambda^{-1}$ is in $U(E)' \subset E^{P_1}$. Thus E^{P_1} is invariant under inner automorphisms of E therefore a valuation ring in view of Proposition 1.3.2.15. □

A deeper understanding of the connections between the ideal theory of a valuation ring in a skewfield and the group theory of the totally ordered value group may result by restricting attention to prime ideals.

Let Λ_v be a valuation ring of the skewfield D with corresponding valuation function $v : D^* \to \Gamma$. If Δ is an isolated subgroup of Γ we denote the sets of positive elements of Γ and Δ by Γ_+ and Δ_+ respectively. Then $\Gamma_+ - \Delta_+$ is a filter of Γ; indeed if $\gamma \in \Gamma_+ - \Delta_+$ and $\delta \geq \gamma$, say $\delta \neq \gamma$ hence $\delta > \gamma$, then $\delta \in \Gamma_+$ and if $\delta \notin \Gamma_+ - \Delta_+$ we must have $\delta \in \Delta_+$ and so $\gamma \in \Delta_+$ because Δ is an isolated subgroup, a contradiction, hence $\delta \in \Gamma_+ - \Delta_+$ as desired.

1.3.2.17 Proposition

The nonzero prime ideals of Λ_v correspond bijectively to the isolated subgroup of Γ and prime ideals invariant under inner automorphisms of D correspond to invariant subgroups of Γ.

Proof. First observe that prime ideals of Λ_v are in fact completely prime. Indeed, from Proposition 1.3.2.6 we retain that one-sided ideals of Λ_v are two-sided hence from $ab \in P$ for some prime ideal P of Λ_v it follows that $\Lambda_v ab = a\Lambda_v b \subset P$ or a or $b \in \Lambda_v$. Put $F = \Gamma_+ - \Delta_+$ then $J(F) = \{a \in \Lambda_v, v(a) \in F\}$ is an ideal of Λ_v (Proposition 1.3.2.9). If $x, y \notin J(F)$ we have $v(x), v(y) \in \Delta_+$ and then $v(xy) \in \Delta_+$ hence $xy \notin J(F)$, consequently $J(F)$ is a prime ideal. Conversely, look at the positive filter for some prime ideal P of Λ_v, say $\mathcal{F}(P)$. It suffices to show that $\Gamma_+ - \mathcal{F}(P)$ is the positive part Δ_+ of an isolated subgroup Δ of Γ. Suppose we have $0 < \gamma < \delta$ with $\delta \in \Gamma_+ - \mathcal{F}(P)$. Then $\gamma \in \Gamma_+ - \mathcal{F}(P)$ too (because otherwise $\gamma \in \mathcal{F}(P)$ yields $\delta \in \mathcal{F}(P)$). Now put $\Delta_+ = \Gamma_+ - \mathcal{F}(P), \Delta_- = \{-\gamma, \gamma \in \Delta_+\} \cup \{0\}$. Observe that for δ, δ' in Δ_- also $\delta + \delta' \in \Delta_+$ because $\delta = v(a), \delta' = v(a')$ for $a, a' \in \Lambda_v - P$, hence $v(aa') = \delta + \delta'$ and $aa' \notin P$. It is easily checked that the compositum of Δ_+ and Δ_-, given by finite sums of elements $\gamma + \delta, -\gamma + \delta, -\delta - \gamma, -\delta + \gamma$, for $\gamma, \delta \in \Delta_+$, is a group Δ such that Δ_+ is exactly its positive part. Argumentation as just above establishes that Δ is isolated. It is clear that invariant prime ideals correspond to isolated subgroups invariant under conjugation in Γ. □

1.3.2.18 Corollary

If Γ is abelian then every prime ideal of Λ_v is invariant under inner automorphisms of D.

1.3.2.19 Proposition

Every nonzero prime ideal p of Λ_v defines a quotient ring of functions $\Lambda_{v,p} \subset D$ containing Λ_v such that p is again a prime ideal in $\Lambda_{v,p}$ such that $\Lambda_{v,p}/p$ is a skewfield, hence p is the Jacobson radical of $\Lambda_{v,p}$.

Proof. Since left ideals of Λ_v are two-sided the multiplicative set $S = \Lambda_v - p$ is an Ore set (left and right) so we have a localized domain $S^{-1}\Lambda_v$ of left (right) fractions.

Now write $\Lambda_{v,p} = S^{-1}\Lambda_v$, $\Lambda_v \subset \Lambda_{v,p}$ and $p\Lambda_{v,p} \neq \Lambda_{v,p}$ because $p \cap S = \emptyset$. By general localization results $p\Lambda_{v,p}$ is a maximal ideal of $\Lambda_{v,p}$, say $M = p\Lambda_{v,p}$. If $x \in M - \Lambda_v$, then there are $\lambda, \mu \in \Lambda_v$ such that $\lambda x \mu \in \Lambda_v - p$ but $\lambda x \mu \in M$ and $\lambda x \mu \in \Lambda_v$ must imply $\lambda x \mu \in p$ (otherwise M would contain a unit of Λ_v), a contradiction. Consequently $M = p$.

For $y \in \Lambda_{v,p} - p$, if $y \in \Lambda_v$ then $y \in S$ and then y is a unit in $\Lambda_{v,p}$. On the other hand if $y \notin \Lambda_v$ then $y^{-1} \in \Lambda_v$ and thus again y is a unit in $\Lambda_{v,p}$. Therefore p is the unique maximal ideal of $\Lambda_{v,p}$ (hence completely prime) and $\Lambda_{v,p}/p$ is a skewfield. □

1.3.2.20 Corollary

In the situation of foregoing theorem, $\Lambda_{v,p}$ is a total subring of D, in case p is an invariant prime of Λ_v then $\Lambda_{v,p}$ is a valuation ring of D.

Proof. If $x \notin \Lambda_{v,p}$ then $x^{-1} \in \Lambda_v \subset \Lambda_{v,p}$. If p is invariant then the localization of Λ_v at p is also invariant (in fact $S = \Lambda_v - p$ is also invariant), in combination with the first statement $\Lambda_{v,p}$ is then a valuation ring. □

Finally let us return to the notion of specialization as introduced before Lemma 1.1.16 for arbitrary primes.

The situation $p \subset \Lambda_v \subset \Lambda_{v,p}$ with (p, Λ_v) and $(p, \Lambda_{v,p})$ being complete primes of D yields that (p, Λ_v) is a specialization of $(p, \Lambda_{v,p})$ in the sense of Lemma 1.1.16, denoted by $(p, \Lambda_{v,p}) \mapsto (p, \Lambda_v)$. It is clear that Λ_v/p is a total subring of $\Lambda_{v,p}/p = D(p)$, the residual skewfield for $\Lambda_{v,p}$. In case p is invariant in Λ_v then Λ_v/p is a valuation ring of $D(p)$. The prime place $\Lambda_v \to F(v) = \Lambda_v/P_v$, where P_v is the maximal ideal of Λ_v, is called the **place** corresponding to the valuation v, the prime place $\Lambda_{v,p} \to \Lambda_{v,p}/p = D(p)$ is called the **place** corresponding to p even if this is not a place of a valuation in general (unless p is invariant). Now $(P_v/p, \Lambda_v/p)$ is the induced complete prime in $D(p)$ and the primeplace $\rho_p : \Lambda_{v,p} \to D(p)$ may be composed with $\rho_v : \Lambda_v/p \to \Lambda_v/P_v = F(v)$ to a morphism defined on $\rho_p^{-1}(\Lambda_v/p) \subset \Lambda_{v,p}$. Of course $\rho_p^{-1}(\Lambda_v/p)$ contains the valuation ring Λ_v and it is therefore again a total subring. We now obtain Theorems 3 and 6 as in [61, p. 16, 17].

1.3.2.21 Theorem

1. With notation as above assume that p is invariant and let $\Delta \subset \Gamma$ be the invariant isolated subgroup of Γ corresponding to p. Then $\Lambda_{v,p}$ defines a Γ/Δ-valued valuation of D and Λ_v/p defines a Δ-valuation for $D(p)$.
2. Let v be a Γ_1-valuation of the skewfield D with residual skewfield $D(v)$ and w is a Γ_2-valuation of $D(v)$ with residual skewfield $D(w)$. If the inverse image of $\Lambda_w \subset D(v)$ in Λ_v under $\rho_v : \Lambda_v \to D(v)$ is invariant then it defines a valuation,

let us denote it by $w.v$, with value group Γ having an isolated subgroup Γ_2' with $\Gamma_2' \cong \Gamma_2$ and such that:

$$0 \to \Gamma_2' \to \Gamma \to \Gamma_1 \to 0, \text{ i.e. } \Gamma/\Gamma_2' \cong \Gamma_1$$

Proof. 1. The valuation $v_{\Gamma/\Delta}$ is defined by $v_{\Gamma/\Delta}(a) = v(a)\text{mod}\Delta \in \Gamma/\Delta$. Observe that Γ/Δ is totally ordered (Proposition 1.3.1.2). This is obviously a valuation. If $v_{\Gamma/\Delta}(a) \geq 0$ then either $v(a) \geq 0$ or $v(a) = -\delta$ for some $\delta \in \Delta_+$. In the second case $a = b^{-1}$ for some b with $v(b) \in \Delta_+$, then $b \in \Lambda_v - p$ and thus $b^{-1} \in \Lambda_{v,p}$; hence in both cases $a \in \Lambda_{v,p}$. Conversely if $ad^{-1} \in \Lambda_{v,p}$ with $v(a) \geq 0, v(d) \in \Delta_+$ then $v_{\Gamma/\Delta}(ad^{-1}) = v_{\Gamma/\Delta}(a) = v(a)\text{mod}\Delta$, therefore $\Lambda_{v,p}$ is contained in the valuation ring of $v_{\Gamma/\Delta}$ and thus equal to it. For $\overline{x} \in D(p)$ we define $v_\Delta(\overline{x}) = v(x)$ where x is any representative for $\overline{x}$ (the definition of v_Δ does not depend on the choice of representative). Clearly $v_\Delta(\overline{x}) \in \Delta$ and all elements of Δ are values of elements in $D(p)$ because only elements of $\Lambda_{v,p} - p$ have values in Δ where $(\Lambda_{v,p} - p)\text{mod}p = D(p) - \{0\}$. Standard verification leads to the observation that v_Δ is a valuation and its valuation ring in $D(p)$ is Λ_v/p.
2. If $\rho_v^{-1}(\Lambda_w)$ is invariant then it is a valuation ring because it is a total subring (see remarks before the theorem) and the corresponding valuation is $w.v$. If Γ is the value group of $w.v$. then the isolated subgroup of Γ corresponding to $\rho_v^{-1}(0) = P_v$ is order isomorphic to Γ_2 and $\Gamma/\Gamma_2' = \Gamma_1$,where Γ_2' is the copy of Γ_2 identified in Γ. □

Let φ be the value function for an arithmetical prime (P, R^P) for any ring R. We may consider for any $\gamma \in \Gamma = \widetilde{R}$ in the value group (only assumed to be a partially ordered group for now) $F_\gamma R = \{r \in R, \varphi(r) \geq \gamma\}$. We write the operation in Γ additively and write 0 for its neutral element.

1.3.2.22 Lemma

With notations as above:

1. For $\gamma \in \Gamma$, $F_\gamma R$ is an additive subgroup of R.
2. For $\gamma, \delta \in \Gamma$, $F_\gamma R F_\delta R \subset F_{\gamma+\delta} R$.
3. The unit of R, say 1, is in $F_0 R$.
4. $\cup_{\gamma\in\Gamma} F_\gamma R = R$.
5. For the set of elements $r \in R$ such that $\varphi(r) \geq \gamma$ for all $\gamma \in \Gamma$ we write $F_\infty R$. Then $F_\infty R$ is an ideal of R.
6. If $\gamma > \delta$ in Γ then we have $F_\gamma R \subset F_\delta R$.

Proof. 1. Take $x, y \in F_\gamma R$ for a certain $\gamma \in \Gamma$, then $\varphi(x), \varphi(y) \geq \gamma$, thus $\varphi(x + y) \geq \gamma$ (see Corollary 1.3.2.2.(ii) and the remarks following it), hence $x + y \in F_\gamma R$.
2. Take $x \in F_\gamma R$, $y \in F_\delta R$ then $\varphi(xy) = \varphi(x) + \varphi(y) \geq \gamma + \delta$, then $xy \in F_{\gamma+\delta} R$, so if $\sum_{i=1}^n x_i y_i \in F_\gamma R F_\delta R$ then we obtain $\varphi(\sum_{i=1}^n x_i y_i) \geq \gamma + \delta$ because each $\varphi(x_i y_i) \geq \gamma + \delta$ and thus $\sum_{i=1}^n x_i y_i \in F_{\gamma+\delta} R$.

3. Obvious from $\varphi(1) = 0$.
4. Clear because φ is surjective.
5. It is clear that $F_\infty R$ is an additive subgroup of R. Take $r \in R, x \in F_\infty R$, since $r \in F_\gamma R$ for some $\gamma \in \Gamma$, we have $rx \in F_{\gamma+\delta} R$ for all $\delta \in \Gamma$ and since $\gamma + \Gamma = \Gamma$, because Γ is a group as (P, R^P) is arithmetical, it follows that $rx \in F_\infty R$. In a similar way $xr \in F_\infty R$ and consequently $F_\infty R$ is an ideal of R.
6. Obvious. □

1.3.2.23 Definition

For a partially ordered group Γ and a ring A, a **Γ** -filtration of A is obtained from an ascending chain of additive subgroups $f_\gamma A, \gamma \in \Gamma$, i.e. if $\delta \leq \gamma$ then $f_\delta A \subset f_\gamma A$, such that $f_\gamma A f_\tau A \subset f_{\gamma+\tau} A$ for $\gamma, \tau \in \Gamma$ and $1 \in f_0 A$. We say that the Γ-filtration fA is **exhaustive** if $\cup_{\gamma\in\Gamma} f_\gamma A = A$.

1.3.2.24 Corollary

The value function of an arithmetical prime (P, R^P) of R defines an exhaustive Γ-filtration $f_\gamma R = F_\gamma R$, with respect to the opposite partial order of Γ in the indexes.

1.3.2.25 Remark

For an arithmetical prime (P, R^P) of R we have a subring O_P of R given as $O_P = F_0 R$ (with notation as above). Just like in Proposition 1.3.2.3. it follows that O_P is invariant under inner automorphisms of R but it need not be equal to $\cup_{z\in U(R)} z R^P z^{-1}$, where $U(R)$ is the group of units of R!

The restriction of $\varphi : R^* = R - \{0\} \to \widetilde{R}$ to $U(R)$, say $\varphi|U(R) : U(R) \to \widetilde{R}$ has image a partially ordered subgroup Γ of $\widetilde{R}$ (this works for any separated prime of R, it need not be arithmetical) and we obtain an exact sequence of groups: $1 \to U(O_P) \to U(R) \xrightarrow{\varphi} \Gamma \longrightarrow 0$ (this may be checked along the lines of Lemma 1.3.2.4. This extends the notion of a valuation function to value functions of arbitrary separated primes, but it is only defined on $U(R)$ which is different from R^* in general if R is not a skewfield.

The relation between $F_0 R$ and R^P is not too clear in general, even in the case of a skewfield $R = D$ we obtain $F_0 D = O_P$ and from Proposition 1.3.2.3 it is clear that the relation between O_P and D^P may be traceable sometimes but it need not be as strict as one might hope for in an optimistic mood.

On the other hand, the theory of filtered rings and algebras is well developed, see [40] for example; therefore it is plausible to start from the idea that suitable filtrations on R forcing convenient relations between R and $F_0 R$ are natural replacements for value functions or valuations. In the remainder of this section we merely introduce some basic facts postponing full detail on filtration methods to Sect. 1.7.

For a ring A with Γ-filtration fA we may view the filtration on a basis of neighbourhoods of o that allows us to view A as a topological ring. From this point of view $f_{-\infty}A = \cap_{\gamma\in\Gamma} f_\gamma A \neq 0$ would obstruct the Hausdorf property, hence we are interested in a condition that would correspond to the separation conditions (like Hausdorf, $T_2, \ldots$). In case φ derives from a value function there is a special property of the filtration stemming just from the fact that φ is (well-) defined, that is: for every $x \in A$ there is a unique $\gamma \in \Gamma$ (in fact $\varphi(x) = \gamma$) such that $x \in F_\gamma A$ but for every $\gamma' > \gamma$ we have $x \notin F_{\gamma'} A$. Indeed, $x \in F_{\gamma'} A$ would mean $\varphi(x) \geq \gamma' > \gamma$, contradiction. Uniqueness of γ follows from: $x \in F_\delta A$ means $\varphi(x) \geq \delta$ hence either $\gamma = \varphi(x) = \delta$ and $F_\gamma A = F_\delta A$ or $\varphi(x) > \delta$ and $x \in F_\gamma A \subset F_\delta A$ and $F_\delta A$ does not have the minimality property.

1.3.2.26 Definition

A Γ-filtration of A as above is said to be **separated** if for every $x \in A$ there exists a (unique) $\gamma \in \Gamma$ such that $x \in f_\gamma A$ but for every $\gamma' < \gamma$ we have $x \notin f_{\gamma'} A$. Obviously this is a version of the property of FA as above but taking into account that then $f_\gamma A = F_\gamma A$, passing to the opposite partial order on Γ.

When Γ is a totally ordered subgroup of the additive group $\mathbb{R}_+^N$ then a more intrinsic characterization of separatedness is available.

1.3.2.27 Proposition

Let $\Gamma \subset \mathbb{R}_+^N$ be totally ordered and fA a Γ-filtration of a ring A. The following statements are equivalent to one another:

(i) The filtration fA is separated.
(ii) If for $\gamma \in \Gamma$ we have that $\gamma = \inf\{\tau \in \Gamma, \tau \in \mathcal{J}\}$, for some family $\mathcal{J}$, then $f_\gamma A = \cap_{\tau\in\mathcal{J}} f_\tau A$, in particular we demand: $\cap_{\gamma\in\Gamma} f_\gamma A = 0$.

Proof. (i) $\Rightarrow$ (ii) Suppose $\gamma = \inf\{\tau \in \mathcal{J} \subset \Gamma\}$ and put $B = \cap_{\tau\in\mathcal{J}} f_\tau A$. Suppose $x \in B - f_\gamma A$. Separatedness of fA entails that there is an element $\delta \in \Gamma$ such that $x \in f_\delta A - f_{\delta'} A$, for all $\delta' < \delta$. Since $x \notin f_\gamma A$ we cannot have $\delta \leq \gamma$ thus $\delta > \gamma$ because Γ is totally ordered. Since $\gamma = \inf\{\tau \in \mathcal{J}\}$ there is some $\tau \in \mathcal{J}$ such that $\delta > \tau \geq \gamma$. From $\tau \in \mathcal{J}$ it follows that $x \in f_\tau A$ but from $\delta > \tau$ it follows that $x \notin f_\tau A$, a contradiction. Consequently $B = f_\gamma A$. The second statement is easy (and very similar).

(ii) $\Rightarrow$ (i) Consider any $x \in A$ and put $J_x = \{\gamma \in \Gamma, x \in f_\gamma A\}$. Put $\delta_x = \inf\{J_x\}$, then $f_{\delta_x} A = \cap_{\gamma\in J_x} f_\gamma$ contains x, i.e.$\delta_x \in J_x$ is the least element of J_x. □

For a separated filtration fA it is possible to define the **principal symbol map**, $\sigma : A \to \Gamma, x \mapsto \delta_x$, where δ_x is such that $x \in f_{\delta_x} A$ but $x \notin f_{\delta'} A$ when $\delta' < \delta_x$.

In case fA derives from a value function φ of an arithmetical prime (P, A^P) of A then σ and φ coincide. Indeed if $a \neq 0$ in A then $\varphi(a) = P : a$ in $\Gamma = \widetilde{A}$ and clearly $a \in f_{\varphi(a)} A$. If $a \in f_\gamma A$ with $\gamma < \varphi(a)$ in the opposite ordering of Γ, i.e. $a \in F_\gamma A \subset F_{\varphi(a)} A$ then $\varphi(a) \geq \gamma$ in the original ordering hence $\varphi(a) \leq \gamma$ in the opposite ordering, or $\varphi(a) = \gamma$ follows. Thus $\sigma(a) = \varphi(a)$. If $x \in A^*$ is in all $F_\gamma A$ then $\varphi(x) \geq \gamma$ for every $\gamma \in \Gamma$ hence Γ has a maximal element $\varphi(x)$; but $\varphi(x) + \varphi(x) > \varphi(x)$ in the partial ordered group Γ if $\varphi(x) > 0$ in Γ. On the other hand if $\varphi(x) < 0$ then $-\varphi(x) > 0$ and $-\varphi(x) = \varphi(y)$ for some $y \in A^*$ by definition of φ (surjective!). Thus unless $\varphi(x) = 0$ for all $x \in A^*$ there is no maximal element in Γ. However for $p \in P$ we have that $\varphi(p) = P : p$ contains all (x, y) with $x, y \in A^P$, then $\varphi(P) = 0$ leads to $xy \in P$ for all $x, y \in A^P$ and thus $1 \in P$, a contradiction. In conclusion, the filtrations fA stemming from an arithmetical prime (in particular separated) is a separated filtration! Note that in general, the principal symbol map σ is not necessarily multiplicative: more about filtrations in Sect. 1.7.

1.4 Dubrovin Valuations

In this section Q is a simple Artinian ring, hence a matrix ring over a skewfield. A pair (M, R) defines a **Dubrovin** valuation ring R of Q if the following properties hold:

DV.1 The ring R/M is a simple Artinian ring.
DV.2 For any $q \in Q - R$ there exists $r, r' \in R$ such that we obtain $rq, qr' \in R - M$.

1.4.1 Lemma

If R is a Dubrovin valuation ring of Q then (M, R) is a localized prime of Q.

Proof. Look at $a, b \in Q$ such that $aRb \subset M$ and suppose that $b \notin M$. If both a and b are not in R take r, s such that $ra, bs \in R - M$; then $raRbs \subset rMs \subset M$ yields a contradiction because M is prime. hence if $a \notin R$ then $b \in R$, select r such that $ra \in R - M$, again $raRb \subset M$ entails $b \in M$. In case $a \in R$ and $b \notin R$ then select s such that $bs \in R - M$ and if follows that $aRbs \subset M$ yields $a \in M$; the case where both $a, b \in R$ is clear. So we have established that (R, M) is a prime. In order to check the localized property look at $q \in Q - R$ and we have to find $x, y \in R - \{0\}$ such that $xqy \in R - M$, but this is clear because we may take $(x, y) = (r, 1)$ or $(1, r')$ with r and r' as in DV.2. □

1.4.2 Lemma

If R is a Dubrovin valuation ring then M is the Jacobson radical of R i.e. $M = J(R)$, in particular M is the unique maximal ideal of R.

Proof. It suffices to prove that for all $m \in M, 1 + m \in U(R)$. If $(1 + m)x = 0$ for some $x \in Q$ then either $x \in R$ and $x = -mx$ or $x \notin R$ but then there is a $b \in R$ such that $xb \in R - M$ and $xb = -mxb$ is then a contradiction. Hence we only have to deal with the first case $x \in R$ and $x = -mx \in R$. Since Q is simple $l_Q(1 + m)r_Q(1 + m) = 0$ or else $1 \in l_Q(1 + m)r_Q(1 + m)$, where for any $r \in R$ $l_R(a)(r_Q(a))$ is a left (right) annihilator of a in Q. In the latter case $1 = \sum_i x_i y_i$ with $x_i(1 + m) = 0, (1 + m)y_i = 0$, thus: $0 = \sum x_i(1 + m)y_i = 1+\sum_i x_i m y_i$. The foregoing entails that both $l_Q(1+m)$ and $r_Q(1+m)$ are contained in R, thus $\sum_i x_i m y_i \in M$ cannot be equal to -1. Thus we only have to consider the case where $l_Q(1 + m)r_Q(1 + m) = 0$. Since Q is Artinian hence Noetherian, the ascending chain: $l_Q(1 + m) \subset l_Q((1 + m)^2) \subset \ldots \subset l_Q((1 + m)^n) \subset \ldots$ must terminate, say at n. Since $(1+m)^n = 1+m'$ with $m' \in M$ we may assume from the start that $l_Q(1 + m) = l_Q((1 + m)^2) = \ldots$ because from $l_Q(1 + m') = 0$ it follows that $l_Q(1 + m) = 0$. In Q we write $Q(1 + m) = Qe$ for some idempotent element e of Q. Then: $r_Q(1 + m) = r_Q(Q(1 + m)) = (1 - e)Q$ and also $l_Q(r_Q(1 + m)) = Qe = Q(1 + m)$. From $l_Q(1 + m)r_Q(1 + m) = 0$ it follows that $l_Q(1 + m) \subset l_Q(r_Q(1+m))$, hence any $x \in l_Q(1+m)$ necessarily has the form $x = q(1+m)$ for a suitable $q \in Q$ in view of the above deduction. Then from $q(1 + m)(1 + m) = 0$, or $q \in l_Q((1 + m)^2) = l_Q(1 + m)$ it follows that $x = 0$, thus we have established $l_Q(1 + m) = 0$. Since a (left) regular element in Q is invertible in Q it follows that $(1+m)^{-1} \in Q$. Suppose $(1+m)^{-1} \notin R$, then also $(1+m)^{-1} - 1 \notin R$ so there exists $y \in R$ such that $((1+m)^{-1} - 1)y \in R - M$. We calculate: $(1+m)((1+m)^{-1} - 1)y = -my$ and also, $((1+m)^{-1} - 1)y = -my - m((1+m)^{-1} - 1)y \in M$, a contradiction and $(1 + m)^{-1} \in R$ leads to the conclusion. □

Denote $\overline{R} = R/M, \pi : R \to \overline{R}$ being the canonical ring epimorphism. For any simple Artinian ring A any finitely generated A-module is completely reducible i.e. a finite direct sum of simple A-modules. For a finitely generated A-module M we have $M = S_1 \oplus \ldots \oplus S_d$ where $S_i, i = 1, \ldots, d$, are simple A-modules; the number $d = d_A(M)$ depends only on M and not on the chosen decomposition of M into simple components. In case $d_A(M) \leq d_A(A)$ then M is a cyclic A-module; on the other hand if $d_A(M) > d_A(A)$ then $M \cong A \oplus N$ with $d_A(N) = d_A(M) - d_A(A)$. These basic facts about finitely generated modules over simple Artinian rings may be applied to the study of Dubrovin valuation rings.

1.4.3 Lemma

Let R be a Dubrovin valuation ring of Q and N a finitely generated R-submodule of Q, then N is a cyclic R-module.

Proof. First we establish $d_{\overline{R}}(N/J(R)N) \leq d_{\overline{R}}(\overline{R})$. Suppose the opposite, then $\overline{N} = N/J(R)N$ decomposes as $\overline{R} \oplus \overline{T}$ and so we may select $a, b \in N$ such that $Ra + J(R)N/J(R)N \cong \overline{R}$ and $(Ra + Rb + J(R)N)/J(R)N =$

$(Ra + J(R)N)/J(R)N \oplus (Rb + J(R)N)/J(R)N$ with $b \notin J(R)N$. Since $Ra \cap J(R)N \supset J(R)a$ and we have that $Ra/(Ra \cap J(R)N) \cong (Ra + J(R)N)/J(R)N \cong \overline{R}$ it follows that $Ra \cap J(R)N = J(R)a$, hence $Ra/J(R)a \cong \overline{R}$ and thus $l_R(a) \subset J(R)$. For $q \in Q$, $Rq \cap R \subset J(R)$ entails $q \in R$ and $q \in J(R)$ (note that if $q \notin R, rq \in R - J(R)$ for some $r \in R$ with $rq \in Rq \cap R$). Therefore $l_R(a) \subset J(R)$ entails $l_Q(a) \subset J(R)$. Then $l_Q(a) = 0$ because otherwise this left ideal of Q contains an idempotent e but then $e(1-e) = 0$, with $1-e$ being a unit in R since $e \in J(R)$, yields $e = 0$. The latter means that a is unit in Q. Now $Ra \cap J(R)N = J(R)a$ yields $R \cap J(R)Na^{-1} = J(R)$ i.e. $R \cap Rx \subset J(R)$ for every $x \in J(R)Na^{-1}$, leading to $x \in J(R)$ hence $J(R)Na^{-1} \subset J(R)$ or $J(R)N \subset J(R)a$. Thus $J(R)N = J(R)a$. By the choice of a, b we have $Ra \cap Rb \subset J(R)N = J(R)a$, therefore $R \cap Rba^{-1} \subset J(R)$ with $ba^{-1} \in Q$. Again the latter means $ba^{-1} \in J(R)$ and $b \in J(R)a = J(R)N$, a contradiction. Hence $d_{\overline{R}}(N/J(R)N) \leq d_{\overline{R}}(\overline{R})$ or $N/J(R)N$ is a cyclic $\overline{R}$-module as $\overline{R}$ is simple Artinian. Since N is finitely generated we may apply the Nakayama lemma and conclude that N is cyclic too. □

1.4.4 Proposition

Let R be a Dubrovin valuation ring of Q. Then:

1. R is an order of Q.
2. Finitely generated left (right) ideals of R are principal.
3. R is a prime Goldie ring.

Proof. 1. Let us check that R is a right order in Q, the left symmetric statement is easily checked in a similar way (using the right module version of Lemma 1.4.3).

For $q \in Q$, $R + Rq = Rs$ for some $s \in Q$. In particular $1 = rs$ for some $r \in R$ and $q = ts$ for some $t \in R$. It follows that $q = tr^{-1}$ with $t \in R$ and r regular in R. Conversely if r regular in R and $qr = 0$ for some $q \in Q - R$, then there is an $r' \in R$ such that $r'q \in R - J(R)$ and $r'qr = 0$; the latter contradicts $l_R(r) = 0$. Hence $l_R(r) = 0$ entails $l_Q(r) = 0$ or r is invertible in Q since Q is simple Artinian.

2. Follows from Lemma 1.4.3.
3. It is a well-known result that prime Goldie rings are exactly orders in simple Artinian rings, cf. [47]. □

1.4.5 Definition

A subring R of S is called an **n-chain ring of S** on the left if for any $n+1$ elements $s_0, \ldots, s_n$ of S there is an i such that $s_i \in \sum_{j \neq i} Rs_j$. A right n-chain ring of S is defined similarly and R is said to be an **n-chain ring of S** if it is an n-chain ring of S on the left and on the right.

If $R \subset T \subset S$ and R is an n-chain ring of S, then it is obvious that R is an n-chain ring of T and T is an n-chain ring of S. Also if an ideal I of S is contained in R and R is a (left) n-chain ring of S, then R/I is a (left) n-chain ring of S/I.

1.4.6 Lemma

Let R be semisimple, i.e. $J(R) = 0$. Then R is (right) Artinian if and only if R is a (right) n-chain ring for some n.

Proof. If R is a right n-chain ring in some n, then for every $a_0, \dots, a_n \in R$ some a_i must be in $\sum_{j \neq i} Ra_j$. The left Artinian property of R will follow if we establish that a finite intersection of maximal left ideals of R is zero. Consider the descending chain:

$$R \supset M_1 \supset M_1 \cap M_2 \supset \dots \supset M_1 \cap \dots \cap M_s \supset \dots$$

where each M_i is a maximal left ideal of R. We have

$$R/M_1 \cap \dots \cap M_s = (R/M_1) \oplus \dots \oplus (R/M_s)$$

with each R/M_j a simple left R-module. Select $a_1, \dots, a_s \in R$, preimages of some nonzero elements $\overline{a}_1, \dots, \overline{a}_s$ in $R/M_1, \dots, R/M_s$, respectively. Clearly, no a_i can be in the left ideal generated by the others. The remark starting the proof then learns that $M_1 \cap \dots \cap M_n = 0$. Conversely look at a semisimple Artinian ring R and put $n = d(R)$. For any $a_0, \dots, a_n \in R$, different elements of R, the chain of left ideals: $Ra_0 \subset Ra_0 + Ra_1 \subset \dots \subset Ra_0 + \dots + Ra_n$ has length at most n, thus there is an i such that $Ra_0 + \dots + Ra_i = Ra_0 + \dots + Ra_{i+1}$, hence $a_{i+1} \in \sum_{j=0}^{i} Ra_j$. □

An order R in Q is called a **left Bezout order** if any finitely generated left ideal is principal. A **right Bezout order** in Q is defined similarly and R is called **Bezout** if it is **left and right Bezout**. R is called **left semi-hereditary** if any finitely generated left ideal is projective. A right semi-hereditary order in Q is defined similarly and R is said to be **semi-hereditary** if it is **left and right semi-hereditary**. □

1.4.7 Theorem

For a subring R of a simple Artinian ring Q, the following are equivalent:

1. R is a Dubrovin valuation ring of Q.
2. R is a local Bezout order in Q.
3. R is a local semi-hereditary order in Q.
4. R is a local n-chain ring in Q for some n with $n = d(\overline{R})$, where $\overline{R} = R/J(R)$.

1.4.10 Lemma

Let R be a Dubrovin valuation ring of Q then for regular elements $x, y \in R$ the left (right) ideal $Rx \cap Ry$ (resp. $xR \cap yR$) is principal.

Proof. Let $x, y \in R$ be regular then there are $r, s \in R$ with r regular such that $xs = yr$ ($Q = Q(R)$!). We have $xR + yR = zR$ for some $z \in R$, then z is regular and $x = zx_0$, $y = zy_0$ for some $x_0, y_0 \in R$. Moreover $z = xu + yv$ for some $u, v \in R$, thus $1 = x_0u + y_0v$. Pick $w \in xR \cap yR$, then $w = xw_1 = yw_2$ for some $w_1, w_2 \in R$. Thus we obtain

$$w = zx_0w_1 = z(x_0u + y_0v)x_0w_1 = zx_0ux_0w_1 + zy_0vx_0w_1$$

$$= xux_0w_1 + yvx_0w_1 = xuy_0w_2 + yvx_0w_1.$$

The last equality follows from $x_0w_1 = y_0w_2$ since $xw_1 = yw_2$ with $x = zx_0$, $y = zy_0$ and z is regular. It follows that $xR \cap yR \subset xuy_0R + yvx_0R$. However $xuy_0 = (z - yv)y_0 = y - yvy_0 = y(1 - vy_0)$ is in $xR \cap yR$; similarly $yvx_0 \in xR \cap yR$ Thus we obtain that $xR \cap yR = xuy_0R + yvx_0R$ is finitely generated and then it is principal. □

1.4.11 Proposition

Let R be a Dubrovin valuation ring of Q. Let P be a Goldie prime ideal of R then $\mathcal{C}(P)$ is an Ore set of R and R_P is a Dubrovin valuation ring with $J(R_P) = PR_P$.

We postpone the proof till after the following reduction of Dubrovin valuations to Dubrovin valuations in skewfields.

1.4.12 Lemma

Let R be a Dubrovin valuation ring of $M_n(\Delta)$, Δ a skewfield. Then there exists a subring S of Δ and $q \in U(M_n(\Delta))$ such that $qRq^{-1} = M_n(S)$.

Proof. Let $\{e_{ij}\}$ be a complete set of matrix units for $M_n(\Delta)$. There is a regular $a \in R$ such that $e_{ij}a \in R$ for $i, j = 1, \cdots, n$. Consider $\{a^{-1}e_{ij}a\} = X$. Clearly $aX \subset R$ and X is also a complete set of matrix units for $M_n(\Delta)$. Since aXR is finitely generated there exists a regular $b \in R$ such that $aXR = bR$, i.e. $XR = a^{-1}bR$. Since $XR \supset R$, the foregoing entails $a^{-1}bR \supset R$ and $b^{-1}aR \subset R$ or $b^{-1}a \in R$. Put $A = \{r \in R, rX \subset R\} = \{r \in R, ra^{-1}b \in R\} = Rb^{-1}a$ From $AX = A$ it follows that $Rb^{-1}aX = Rb^{-1}a$, thus $Rb^{-1}aXa^{-1}b = R$. Since

$(b^{-1}e_{ij}b)$ is a complete set of matrix units of R there exists a subring S of Δ such that $bRb^{-1} = M_n(S)$. □

1.4.13 Lemma

Let R be a Dubrovin valuation ring of $Q = M_n(\Delta)$, Δ a skewfield.

1. $M_m(R)$ is a Dubrovin valuation ring of $M_m(Q)$ for any $m \in \mathbb{N}$.
2. Let e be an idempotent of R, then eRe is a Dubrovin valuation ring of eQe.
3. There exists a Dubrovin valuation ring S of Δ such that $R = q^{-1}M_n(S)q$ for some $q \in Q$.

Proof. (1) and (2) are easy enough.

3. There is a subring S of Δ such that $qRq^{-1} = M_n(S)$ in view of Lemma 1.4.12. Clearly qRq^{-1} is a Dubrovin valuation ring of Q, say R_1. Then $S \cong e_{11}R_1e_{11}$ is a Dubrovin valuation ring of Δ in view of (2). □

1.4.14 Corollary

The property of being a Dubrovin valuation ring is Morita invariant.

1.4.15 Lemma

Let R be a Dubrovin valuation ring of Q and P a Goldie prime ideal of R. Then for regular elements $r, s \in R$ with $s \in \mathcal{C}(P)$, there exists $r', s' \in R$, with $s' \in \mathcal{C}(P)$, such that $rs' = sr'$.

Proof. We have: $rR + sR = dR$ for some $d \in R$, hence $r = du, s = dv$, where u, v are regular and $uR + vR = R$. Write $\overline{x} \in R/P$ for the canonical image $x \in R$. From $\overline{s} = \overline{d}\overline{v}$ it follows that $\overline{v}$ is regular in R/P, hence $v \in \mathcal{C}(P)$. By Lemma 1.4.10: $uR \cap vR = uxR$ for some $x \in R$, thus for some $y \in R$, $ux = vy$. Therefore $rx = dux = dvy = sy$. It is enough to prove that $x \in \mathcal{C}(P)$ then put $s' = x, r' = y$. From $uR + vR = R$ we obtain:

$$R/vR = (uR + vR)/vR \cong uR/(uR \cap vR) = uR/uxR.$$

Since u is regular in R it follows that $R/vR \cong R/xR$. Let $\pi : R/vR \to R/xR$ be the latter isomorphism. There exist $a, b \in R$ such that $\pi(a + vR) = 1 + xR$ and $\pi(1 + vR) = b + xR$. Since R/P is a Goldie ring, $\overline{a}\overline{e} = \overline{v}\overline{f}$ for some $e, f \in R$

with $\overline{e}$ regular in R/P. Put $p = ae - vf \in P$, then $\pi(p + vR) = \pi(1 + vR)p = (b + xR)p = bp + xR$. Also we obtain:

$$\pi(p + vR) = \pi(ae + vR) = \pi(a + vR)e = e + xR.$$

Consequently $bp = e - xc$ for some $c \in R$, hence $x \in \mathcal{C}(P)$. □

Proof of Proposition 1.4.11

We write $Q = M_n(\Delta)$ and we may assume that $R = M_n(S)$ for some Dubrovin valuation ring S of Δ, $P = M_n(p)$ for some prime ideal p of S. In view of the foregoing lemma $\mathcal{C}_S(p)$ is an Ore set of S. Then S_p is an overring of S in Δ hence a Dubrovin valuation ring of Δ and $J(S_p) = p$. In view of Lemma 1.4.13 we have that $M_n(S_p)$ is a Dubrovin valuation ring of Q containing R and $J(M_n(S_p)) = M_n(p) = P$. As a consequence of Theorem 1.4.9(3) it follows that $\mathcal{C}(P)$ is an Ore set and $M_n(S_p) = R_P$. □

1.4.16 Corollary

There is a one-to-one correspondence between the set of all Goldie prime ideals of R and the set of all overrings of R, which is given by $P \to R_P$ and $S \to J(S)$, where P is a Goldie prime ideal of R and S is an overring of R. Moreover if P and P' are Goldie prime with $P \supsetneqq P'$, then $R_{P'} \supsetneqq R_P$.

1.4.17 Proposition

Let S be a Dubrovin valuation ring of Q and let $\widetilde{R}$ be a Dubrovin valuation ring of $\overline{S} = S/J(S)$. Then $R = \{r \in S, r + J(S) \in \widetilde{R}\}$ is a Dubrovin valuation ring of Q.

Proof. Let $M = \{m \in R, m + J(S) \in J(\widetilde{R})\}$. M is an ideal of R such that R/M is simple Artinian. Let $q \in Q - R$. First suppose $q \in S$. There are r and $r' \in R$ such that $qr + J(S)$ and $r'q + J(S) \in \widetilde{R} - J(\widetilde{R})$. Hence qr and $r'q \in R - M$. Suppose $q \notin S$. Then $I = J(S)q \cap S \not\subset J(S)$ the image $\overline{I}$ of I in $\overline{S}$ contains a non-zero idempotent. If $I \subset M$, then $\widetilde{I} \subset J(\widetilde{R})$, a contradiction. Thus $I \not\subset M$ and there is $s \in J(S) \subset R$ such that $sq \in S - M$. Hence $xsq \in R - M$ for some $x \in R$. Similarly $qr \in R - M$ for some $r \in R$. Hence R is a Dubrovin valuation ring of Q. □

1.5 Ideal Theory of Dubrovin Valuation Rings

In this section we study ideal theory of Dubrovin valuation rings R. We first show that the set of R-ideals is linearly ordered by inclusions. We define the prime segments to study prime ideals and primary ideals. As an application we show that any ideal is a product of a primary ideal and a stabilizing element under a mild condition.

1.5.1 Lemma

Let R be a Dubrovin valuation ring of Q with $R \neq Q$ and X a left R-submodule of Q. If X contains a regular element, then any element of X is sum of two regular elements. In particular this holds for $X = R$.

Proof. By assumption we have $R \neq Q$ thus $J(R) \neq 0$ because $R/J(R)$ simple Artinian. Hence $J(R)$ contains a regular element (an ideal of a prime Goldie ring), z say. Let $x \in X$ be a regular element then for any $y \in X$ there is a $c \in U(Q)$ such that: $R(y-x) + Rzx = Rc$, being contained in X.

Now (see claim) $Rc = R(y - x + rzx)$ for some $r \in R$, and $y - x - rzx$ is regular as $c \in U(Q)$. Then it follows that $y = (y - x + rzx) + (1 - rz)x$ is a sum of two regular elements. □

Proof of Claim

If $X + Ra = Rb$ for some $a, b \in Q$ then there is an $x \in X$ such that $R(x + a) = Rb$. There are two cases to consider:

1. There is a regular element in Rb.
2. There is no regular element in Rb.

1. If Rb contains a regular element then b is regular and $Xb^{-1} + Rab^{-1} = R$. Then there exists $x \in X$ such that $xb^{-1} + ab^{-1} = u$ unit in R (a general fact for (semi-)local rings [41, Lemma 6.1]). Hence $R(x + a) = Rub = Rb$.
2. Then $Qb \oplus Qc = Q$ for some $c \in Q$ hence $Rb \oplus Rc$ contains a regular element and $Rb \oplus Rc = Rd$ for some regular element $d \in Q$. We then arrive at: $X + Ra + Rc = Rb + Rc = Rd = R(x + rc + a)$ for some $x \in X$ and $r \in R$ by the part (1). Therefore: $b = y(x + rc + a) = y(x + a) + yrc = y(x + a)$, some $y \in R$, because $Rb \cap Rc = 0$. This leads to $Rb = R(x + a)$ and finishes the proof of the claim. □

1.5.2 Lemma

Let R be a Dubrovin valuation ring of Q and let $X_2 \subset X_1$ be left R-submodules of Q such that

1. X_1 contains a regular element in Q.
2. $S = O_r(X_2) = \{q \in Q, X_2 q \subset X_2\}$ is right Bezout.

Then either $X_1 = X_2$ or $J(R)x_1 \supset X_2$ for some regular $x_1 \in X_1$.

Proof. Let $x_1 \in X_1$ be regular. Then $R \cap X_2 x_1^{-1} + J(R) = Re + J(R)$ for some $e \in R$. If $e \in J(R)$, then $R \cap X_2 x_1^{-1} \subset J(R)$, which implies $X_2 x_1^{-1} \subset J(R)$ and $X_2 \subset J(R)x_1$ follows. In the case where $e \notin J(R)$ for any regular $x_1 \in X_1$, choose $x_1 \in X_1$ such a way that $d(Re + J(R)/J(R))$ is minimal. Let $x_2 \in X_1$ be any regular. There is a regular $x \in X_1$ such that $x_1 S + x_2 S = xS$ and hence $(S : xS)_l = Sx^{-1} \subset Sx_1^{-1} \cap Sx_2^{-1}$. Consequently $R \cap X_2 x^{-1} = R \cap X_2 S x^{-1} \subset R \cap X_2 x_1^{-1} \cap X_2 x_2^{-1}$. The choice of x_1 yields $(R \cap X_2 x^{-1}) + J(R) = R \cap X_2 x_1^{-1} + J(R) = Re + J(R)$. Thus $e \in R \cap X_2 x^{-1} + J(R) \subset R \cap X_2 x_2^{-1} + J(R)$ for any regular $x_2 \in X_1$. Let $r \in R$ be any regular. Then $rx_1 \in X_1$ is also regular and we obtain:

$$\begin{aligned} er &\in (R \cap X_2 x_1^{-1} r^{-1} + J(R))r \subset R \cap X_2 x_1^{-1} + J(R) \\ &\subset R \cap X_2 x_2^{-1} + J(R) \subset X_2 x_2^{-1} + J(R) \end{aligned}$$

Hence from Lemma 1.5.1, we have $eR \subset X_2 x_2^{-1} + J(R)$. Since $e \notin J(R)$, $R = ReR \subset X_2 x_2^{-1} + J(R)$, hence $x_2 \in Rx_2 \subset X_2 + J(R)x_2$, i.e. $x_2 = x' + zx_2$ for some $x' \in X_2, z \in J(R)$. Consequently $x_2 = (1-z)^{-1}x'$. Since x_2 is any regular element of X_1, $X_1 = X_2$ follows. □

Let R be an order in Q. A left R-submodule I of Q is said to be a **(fractional) left R-ideal** if $I \cap U(Q) \neq \emptyset$ and $Ia \subset R$ for some $a \in U(Q)$. In a similar fashion (fractional) right R-ideals and (fractional two-sided) R-ideals are defined. If I is both a left R-ideal and a right S-ideal for some order S in Q then I is called an (R, S)-ideal.

1.5.3 Proposition

Let R be a Dubrovin valuation ring of Q and S be a Bezout order in Q.

1. The set of left R and right S-ideals of Q is linearly ordered by inclusion. In particular, the set of R-ideals is linearly ordered by inclusion.
2. Let I be an (R, R)-bimodule in Q with $I \neq Q$. Then I is an R-ideal.

Proof. 1. Let X_1 and X_2 be left R and right S-ideals of Q and $X = X_1 \cap X_2$. If $X = X_1$, then $X_1 \subset X_2$. So we may assume that $X \subsetneqq X_1$. Then, by Lemma 1.5.2, $J(R)x_1 \supset X$ for some regular $x_1 \in X_1$. Thus $X_2 \cap Rx_1 \subset X_2 \cap X_1 = X \subset J(R)x_1$ and $X_2 x_1^{-1} \subset J(R)$, which implies $X_2 x_1^{-1} \subset R$. Hence $X_2 \subset Rx_1 \subset X_1$ follows.

2. I contains a regular element in Q since $I \cap R$ is a non-zero ideal of R. Applying Lemma 1.5.2 to $I = X_2$ and $Q = X_1$, we have $J(R)a \supset I$ for some regular $a \in Q$. Hence $R \supset J(R) \supset Ia^{-1}$, showing I is a left R-ideal. Similarly I is a right R-ideal. □

For any R-ideal A, define $O_l(A) = \{q \in Q, qA \subset A\}$ called a **left order** of A, $O_r(A) = \{q \in Q, Aq \subset A\}$, a **right order** of A and $A^{-1} = \{q \in Q, AqA \subset A\}$, the **inverse** of A.

1.5.4 Lemma

Let R be a Dubrovin valuation ring of Q.

1. $O_l(J(R)) = R = O_r(J(R))$ and either $J(R) = aR = Ra$ or $J(R) = J(R)^2$.
2. If $J(R) = J(R)^2$ then $J(R)^{-1} = R$.

Proof. 1. Suppose $R \subsetneqq O_l(J(R))$. Then $R \subset J(R)c$ for some $c \in O_l(J(R))$ and $c^{-1} \in J(R)$. Thus $1 = cc^{-1} \in O_l(J(R))J(R) = J(R)$, a contradiction. Hence $O_l(J(R)) = R$ and similarly $O_r(J(R)) = R$. Suppose $J(R) \supsetneqq J(R)^2$. Then there is an $a \in J(R)$ with $J(R)^2 \subseteq J(R)a$, which implies $J(R)a^{-1} \subset O_r(J(R)) = R$ and $J(R) \subset Ra$. Hence $J(R) = Ra$. $R = O_r(J(R)) = a^{-1}Ra$ entails $J(R) = Ra = aR$.

2. Since $J(R)J(R)^{-1}J(R) \subset J(R)$, we have $J(R)J(R)^{-1} \subset O_l(J(R)) = R$ and $J(R)J(R)^{-1} \subset J(R)$, because $J(R) = J(R)^2$. Thus $J(R)^{-1} \subset R$ and hence $J(R)^{-1} = R$. □

1.5.5 Corollary

Let P be a Goldie prime ideal of R. Then $O_l(P) = R_P = O_r(P)$.

Proof. Follows from Corollary 1.4.16 and Lemma 1.5.4. □

1.5.6 Lemma

Let R be a Dubrovin valuation ring of Q and A be an R-ideal of Q with $S = O_l(A)$. Then the following are equivalent:

1. A is principal as a left S-ideal.
2. $AA^{-1} = S$.
3. $A \supsetneqq J(S)A$.

Proof. 1. $\Rightarrow$ 2. and 2. $\Rightarrow$ 3. are obvious.
3. $\Rightarrow$ 1. There is a regular $a \in A$ with $J(S)a \supset J(S)A$ and $Aa^{-1} \subset O_r(J(S)) = S$. Hence $A \subset Sa$ and $A = Sa$ follows.

1.5.7 Lemma

Let R be a Dubrovin valuation ring of Q and A be an R-ideal of Q. Then $O_l(A) = O_r(A^{-1})$ and $O_r(A) = O_l(A^{-1})$.

Proof. Put $S = O_l(A)$ and $T = O_r(A^{-1})$. We may assume that $R \subset S \subset T$ by Proposition 1.5.3. If $S = AA^{-1}$, then $A = Sa$ for some regular $a \in A$ by Lemma 1.5.6 and $A^{-1} = a^{-1}S$. Hence $O_r(A^{-1}) = S$ follows. If $S \supsetneqq AA^{-1}$ then $AA^{-1} \subset J(T)$, otherwise $AA^{-1} \supsetneqq J(T)$ implies $T = (AA^{-1})_{J(T)} = AA^{-1}T = AA^{-1} \subsetneqq S$, a contradiction. Suppose $T \supsetneqq S$. Then $TA \supsetneqq A$ and $TA \supset Sc \supset J(S)c \supset A$ for some regular $c \in TA$. Multiplying T on the relations from the left, it follows that $TA = Tc$. $Sc \supset A$ entails $c^{-1} \in A^{-1}$. Thus $T = TAc^{-1} \subset TAA^{-1} \subset J(T)$, a contradiction. Hence $S = T$ and similarly, $O_r(A) = O_l(A^{-1})$.

□

Let A be an R-ideal with $S = O_l(A)$ and $T = O_r(A)$. Note that $A^{-1} = (T : A)_l = \{q \in Q, qA \subset T\}$ and $A^{-1} = (S : A)_r = \{q \in Q, Aq \subset S\}$. Define ${}^*A = (S : (S : A)_r)_l$ and $A^* = (T : (T : A)_l)_r$.

1.5.8 Proposition

Let R be a Dubrovin valuation ring of Q and let A be an R-ideal of Q. Then

1. ${}^*A = \cap\{Sc, Sc \supset A \text{ for some } c \in U(Q)\}$.
2. ${}^*A = (A^{-1})^{-1} = A^*$.
3. ${}^*({}^*A) = {}^*A$ and ${}^*(A^{-1}) = A^{-1}$.
4. If A is not a principal left S-ideal then $AA^{-1} = J(S)$ and $J(S) = J(S)^2$.
5. If $A \subsetneqq {}^*A$, then ${}^*A = Sc$ and $A = J(S)c$ for some regular $c \in {}^*A$.

Proof. 1. It is clear that ${}^*A \subset \cap\{Sc, Sc \supset A \text{ for some } c \in U(Q)\}$. Since $(S : A)_r = \sum xS$, where x runs over all regular in $(S : A)_r$, it follows that ${}^*A = (S : \sum xS)_l = \cap Sx^{-1}$ and $Sx^{-1} \supset A$. Hence the statement follows.
2. $(A^{-1})^{-1} = \{q \in Q, A^{-1}qA^{-1} \subset A^{-1}\} = \{q \in Q, A^{-1}q \subset T\} = (T : A^{-1})_r = (T : (T : A)_l)_r = A^*$ by Lemma 1.5.7. Similarly $(A^{-1})^{-1} = {}^*A$.
3. These are obvious.

4. It follows from Lemma 1.5.6 that $S \supsetneqq AA^{-1}$ and $AA^{-1} \subset J(S)$. If $AA^{-1} \subsetneqq J(S)$, then $AA^{-1} \subset J(S)c$ for some regular $c \in J(S)$ and $AA^{-1}c^{-1}A \subset J(S)A \subset A$. Thus $A^{-1}c^{-1} \subset A^{-1}$ and $c^{-1} \in O_r(A^{-1}) = O_l(A) = S$, a contradiction. Hence $AA^{-1} = J(S)$. If $J(S) \supsetneqq J(S)^2$, then $J(S) = Sx$ for some $x \in J(S)$. So $AA^{-1}x^{-1} = S$ and $A^{-1}x^{-1} \subset (S : A)_r = A^{-1}$, which entails $x^{-1} \in O_r(A^{-1}) = S$, a contradiction. Hence $J(S) = J(S)^2$.
5. Note that A is not a principal left S-ideal. ${}^*A \supsetneqq A$ implies $J(S)c \supset A$ for some regular $c \in {}^*A$ and $J(S)c \subset J(S)\,{}^*A = (AA^{-1}){}^*A = A(A^{-1}(A^{-1})^{-1}) \subset AT = A$. Hence $A = J(S)c$. Furthermore, $A^{-1} = c^{-1}J(S)^{-1} = c^{-1}S$ by Lemma 1.5.4 and ${}^*A = (A^{-1})^{-1} = Sc$. □

An R-ideal A is called **divisorial** if $A = {}^*A$ (note: ${}^*A = A = A^*$ if A is divisorial).

1.5.9 Lemma

Let R be a Dubrovin valuation ring of Q and A, B be R-ideals with $O_r(A) = S = O_l(B)$. Then $O_l(AB) = O_l(A)$ and $O_r(AB) = O_r(B)$.

Proof. It is clear that $O_l(AB) \supset O_l(A)$ and $A \supset ABB^{-1}$ since $S \supset BB^{-1}$. If $A = ABB^{-1}$, then $O_l(A) = O_l(ABB^{-1}) \supset O_l(AB)$. If $A \supsetneqq ABB^{-1}$, then $BB^{-1} \neq S$ and B is not a principal left S-ideal. Thus $BB^{-1} = J(S)$ by Proposition 1.5.8 and $A \supsetneqq AJ(S)$. Hence A is a principal right S-ideal by the right version of Lemma 1.5.6, say, $A = aS$ for some regular $a \in A$. It follows that $O_l(AB) = O_l(aB) = aO_l(B)a^{-1} = aSa^{-1} = O_l(A)$. Similarly $O_r(AB) = O_r(B)$. □

For divisorial R-ideals A, B with $O_r(A) = O_l(B)$, define a product "$\circ$" with $A \circ B = {}^*(AB)$.

1.5.10 Lemma

Let R be a Dubrovin valuation ring of Q and A, B, C be divisorial R-ideals such that $O_r(A) = O_l(B)$ and $O_r(B) = O_l(C)$. Then

1. $A \circ (B \circ C) = (A \circ B) \circ C$.
2. $A \circ A^{-1} = O_l(A)$ and $A^{-1} \circ A = O_r(A)$.

Proof. 1. $A \circ (B \circ C) = A \circ (BC)^* = (A(BC)^*)^* = (ABC)^* = {}^*(ABC) = {}^*({}^*(AB)C) = {}^*((A \circ B)C) = (A \circ B) \circ C$.
2. Since $S = O_l(A) \supset AA^{-1}$, it follows that $S \supset A \circ A^{-1}$. To prove the converse inclusion, suppose that $AA^{-1} \subset Sc$ for some $c \in U(Q)$. Then $AA^{-1}c^{-1} \subset S$ and $A^{-1}c^{-1} \subset (S : A)_r = A^{-1}$. Thus $c^{-1} \in O_r(A^{-1}) = S$ implies $S \subset cS$. Hence $A \circ A^{-1} \subset S$ and $A \circ A^{-1} = S$ follows. Similarly $A^{-1} \circ A = O_r(A)$. □

1.5.11 Corollary

Let S be an overring of R. Then $D_S(R) = \{A : \text{divisorial } R\text{-ideal}, O_r(A) = S = O_l(A)\}$ is a group.

1.5.12 Remark

See [21,41] for more detail results on divisorial R-ideals.

We will classify prime ideals of Dubrovin valuation rings in terms of prime segments.

1.5.13 Lemma

Let R be a Dubrovin valuation ring of Q and I be an idempotent ideal of R. Then:

1. I is not principal as a left $O_l(I)$-ideal and a right $O_r(I)$-ideal.
2. $O_l(I) = I^{-1} = O_r(I)$ and $I = J(S)$, where $S = O_l(I)$. In particular, I is a Goldie prime.

Proof. 1. Put $S = O_l(I)$ and $T = O_r(I)$. If $I \supsetneqq J(S)$, then $S = SI = I$ since $S = R_{J(S)}$, which is a contradiction. Thus $I \subset J(S)$. Suppose I is principal as a left S-ideal, say, $I = Sa$ for some $a \in I$. $I = I^2 = SaSa$ entails $S = SaS$. Write $1 = \sum s_i a t_i$ for $s_i, t_i \in S$. There are $c, d \in \mathcal{C}(J(S))$ with $cs_i, t_i d \in R$ and $cd = c(\sum s_i a t_i)d \in I$, a contradiction. Hence I is not principal as a left S-ideal. Similarly I is not principal as a right T-ideal.

2. Note that $(S : I)_r = I^{-1} = (T : I)_l$ and $I^{-1} \supset S, T$. By Proposition 1.5.3, either $S \supset T$ or $S \subset T$. Suppose $S \subset T$ and let $x \in I^{-1}$. Then $Ix \subset S$ and $Ix = I^2 x \subset IS \subset IT = I$, which entails $x \in T$ and $I^{-1} = T$ follows. Since $I^{-1}I = J(T) \subset R$ by Lemma 1.5.6 and (1), it follows that $I^{-1}I \subset I$ and $I^{-1} \subset O_l(I) = S$. Hence $S = I^{-1} = T$ and $I = J(S)$. □

1.5.14 Lemma

Let R be a Dubrovin valuation ring of Q and let $\{P_i, i \in J\}$ be a set of Goldie primes of R.

1. $P = \cap P_i$ is a Goldie prime and $R_P = \cup R_{P_i}$.
2. $P' = \cup P_i$ is a Goldie prime and $R_{P'} = \cap R_{P_i}$.

Proof. 1. Put $S = \cup R_{P_i}$, an overring of R such that $P = \cap P_i \supset J(S)$. If $P \supsetneqq J(S)$ then $S = PS$ and write $1 = \sum p_i s_i$ for some $p_i \in P$ and $s_i \in S$.

There is an i_0 such that $s_i \in R_{P_{i_0}}$ for all i and $1 = \sum p_i s_i \in P_{i_0} R_{P_{i_0}} = P_{i_0}$, a contradiction. Hence $P = J(S)$ and $R_P = O_l(P) = O_l(J(S)) = S$ by Corollary 1.5.5.

2. If there is an i_0 such that $P_{i_0} \supset P_i$ for all i, then $P' = P_{i_0}$ is a Goldic prime and $R_{P'} = \cap R_{P_i}$. For any i, there is a j with $P_j \supsetneqq P_i$. Then $P' \supset P'^2 \supset P_i$ for all i which entails $P' = P'^2$. Hence P' is a Goldie prime and $R_{P'} \subset \cap R_{P_i} = T$. Let $x \in T$ and $p \in P'$. Then there is a P_i with $p \in P_i$ and $px \in P_i R_{P_i} = P_i \subset P'$. Thus $P'x \subset P'$ and $x \in O_r(P') = R_{P'}$. Hence $R_{P'} = \cap R_{P_i}$. □

Let A be a proper ideal of R that is not a Goldie prime. It follows from Lemma 1.5.14 that there is a pair of primes $P \supsetneqq P_1$ such that $P \supsetneqq A \supsetneqq P_1$ and there are no further Goldie primes between P and P_1. Such a pair of primes is called a **prime segment**. The maximum length of sequences of Goldie primes is called the **rank** of R.

1.5.15 Lemma

Let R be a Dubrovin valuation ring of Q and A be an ideal of R. Then $P = \cap_{n=1}^{\infty} A^n$ is a Goldie prime.

Proof. Suppose that P is not a Goldie prime. By Lemma 1.5.13, $A \supsetneqq A^2$. There is a prime segment $P_1 \supsetneqq P_2$ such that $P_1 \supsetneqq A \supsetneqq P_2$. Put $B = P_1 A P_1$. Then $A^3 \subset B$ and $P = \cap A^n = \cap B^n$. In addition, P and B are R_{P_1}-ideals. After localizing at P_1 we obtain $R_{P_1} \supsetneqq P_1 \supset B \supsetneqq \cap B^n = P \supsetneqq P_2$ and P is not a Goldie prime in R_{P_1}. We therefore can consider R_{P_1}/P_2 and assume that R has rank one with $R \supset J(R) = P_1 \supset B \supsetneqq P = \cap B^n \supsetneqq (0)$. We consider the following set W of ideals in R:

$$W = \{L, P_1 \supsetneqq L \supsetneqq P\},$$

which is non-empty, because $W \ni B^n$ for $n \geq 2$. In the first case we assume that W contains an ideal L with $L \neq {}^*L$. Since R is of rank one, it follows that $O_l(L) = R$ and so ${}^*L = Ra$, $L = J(R)a = P_1 a$ for some regular $a \in {}^*L$ by Proposition 1.5.8. Since $R = O_r({}^*L) = a^{-1}Ra$, we have $L^* = Ra = aR$ and $a^n R = Ra^n$ for $n \geq 1$. Hence the set $\mathcal{C} = \{a^n, n \in \mathbb{N}\}$ is an Ore set in R and the localization $R_{\mathcal{C}}$ of R at $\mathcal{C}$ is a proper overring of R and hence $Q = R_{\mathcal{C}}$. For any regular element $c \in P, c^{-1} = a^{-n}r$ for some $r \in R$ and $n \geq 1$. Hence $P \ni a^n = rc$ implies $({}^*L)^n \subset P$. If $P \subsetneqq {}^*L$, then $B^m \subset {}^*L$ for some m and $(B^m)^n \subset ({}^*L)^n \subset P$. Thus $P = B^{mn}$ and $P^2 = (B^{mn})^2 = P$, a contradiction. Hence $P \supset {}^*L \supsetneqq L$, which is also a contradiction. Therefore P is a Goldie prime. In the second case we have $L = {}^*L$ for every ideal L in W. By Corollary 1.5.11, the set of all divisorial R-ideals forms a group and hence $L^{-1} \supsetneqq R$ for every $L \in W$. Hence $Q = \cup\{L^{-1}, L \in W\}$ as before. Let c be a regular element in P. There is an $L \in W$ with $c^{-1} \in L^{-1}$ and $c^{-1}L \subset L^{-1}L \subset R$, i.e. $L \subset cR \subset P$, a contradiction. Hence P is a Goldie prime. □

Let C be a prime ideal of R that is not Goldie. There is a prime segment $P \supsetneqq P_1$ in R with $P \supsetneqq C \supsetneqq P_1$. Such a prime segment is called **exceptional**. It follows from Lemmas 1.5.13 and 1.5.14 that $P = P^2$ and $P_1 = \cap C^n$. On the other hand, we say that a prime segment $P \supsetneqq P_1$ of R is **Archimedean** if for every $a \in P - P_1$ there is an ideal $I \subsetneqq P$ with $a \in I$ and $\cap I^n = P_1$. It follows from Lemma 1.5.15 that this exactly the case when either $P \neq P^2$ or $P = P^2$ and $P = \cup\{I : \text{ ideal}, P \supsetneqq I \supsetneqq P_1\}$. The following theorem shows that there are exactly three types of prime segments of R.

1.5.16 Theorem

For a prime segment $P \supsetneqq P_1$ of a Dubrovin valuation ring R of Q exactly one of the following possibilities occurs:

1. The prime segment $P \supsetneqq P_1$ is Archimedean.
2. The prime segment $P \supsetneqq P_1$ is simple, i.e. there are no further ideals between P and P_1.
3. The prime segment $P \supsetneqq P_1$ is exceptional.

Proof. We consider $L(P) = \cup\{I : \text{ideals}, P \supsetneqq I\}$. If $L(P) = P_1$, then the prime segment $P \supsetneqq P_1$ is simple, characterizing the possibility (2). The prime segment is exceptional if and only if $P \supsetneqq L(P) \supsetneqq P_1$ and $P = P^2$. If these conditions are satisfied, and A, B are ideals of R properly containing $L(P)$, then $A \supsetneqq P$ and $B \supsetneqq P$ and $AB \supset P^2 = P$, which implies that $L(P)$ is prime but not Goldie. The converse was proved before starting the theorem. Finally suppose $P = L(P)$. This is exactly the case when the prime segment is Archimedean. □

1.5.17 Remark

1. If $P = \cup\{P_i, P \supsetneqq P_i \text{ Goldie primes}\}$, then we can not construct the prime segment $P \supsetneqq P_1$. Such a prime ideal is called **upper prime**.
2. See [6] for examples of the prime segments. If Q is a finite dimension over its center, then any prime segment is Archimedean which follows from Proposition 1.6.3, because any prime segments of commutative rings are Archimedean.

If we define $K(P) = \{a \in P, PaP \subsetneqq P\}$, then it is an ideal of R_P as well as R and the following result holds:

1.5.18 Corollary

The prime segment $P \supsetneqq P_1$ is Archimedean if and only if $K(P) = P$, it is simple if and only if $K(P) = P_1$, and it is exceptional if and only if $P \supsetneqq K(P) \supsetneqq P_1$. In the last case $K(P) = C$ is a non-Goldie prime.

1.5.19 Lemma

Let R be a Dubrovin valuation ring of Q and let C be a non-Goldie prime with the prime segment $P \supsetneqq C \supsetneqq P_1$. Then $O_l(C^n) = R_P = O_r(C^n)$ for any $n \in \mathbb{N}$ and $C = {}^*C$.

Proof. C is an ideal of R_P since $C = \{a \in P, PaP \subsetneqq P\}$ and hence $S = O_l(C^n) \supset R_P$. If $S \supsetneqq R_P$, then $P \supsetneqq J(S)$ and $C \supsetneqq J(S)$, because $J(S)$ is Goldie prime. Thus $C^n \supsetneqq J(S)$ and $C^n \cap \mathcal{C}(J(S)) \neq \emptyset$, which shows $C^n = SC^n = R_{J(S)}C^n = R_{J(S)}$, a contradiction. Hence $S = R_P$ and similarly $O_r(C^n) = R_P$. If ${}^*C \supsetneqq C$, then $C = J(R_P)\ {}^*C = P\ {}^*C$ by Proposition 1.5.8, which is a contradiction, because C is prime. Hence ${}^*C = C$ follows. □

1.5.20 Lemma

Let R be a Dubrovin valuation ring of Q. Suppose the prime segment $J(R) \supsetneqq P$ is Archimedean.

1. If $J(R) = RxR$ for some $x \in J(R)$, then $J(R)$ is a principal left and right ideal.
2. If R is of rank one and $J(R) = J(R)^2$, then for any $a \in J(R)$, $I = RaR$ is a principal left and right ideal with $J(R) \supsetneqq I$.

Proof. 1. We consider $\mathcal{F} = \{A, A$ an ideal and $x \notin A\}$ that is non-empty and inductive set. It contains a maximal element in $\mathcal{F}$, say, B. Since there are no ideals between $J(R)$ and B properly, B is prime if $J(R) = J(R)^2$. In this case, it follows $B = P$, which contradicts the Archimedean property. Hence $J(R) \supsetneqq J(R)^2$ and $J(R)$ is principal as left and right ideal by Lemma 1.5.6.

2. It follows form (1) and the assumption that $J(R) \supsetneqq I$. Suppose I is not principal as a left ideal. Then $I = J(R)I$. Write $a = \sum r_i a x_i$ for some $r_i \in J(R)$ and $x_i \in R$ and $Rr = \sum Rr_i$ for some $r \in J(R)$. It follows that $I = RrRI$ and $RrR \subsetneqq J(R)$. There is a regular $s \in J(R)$ such that $RrR \subset sJ(R)$. Thus $I = RrRI \subset sJ(R)I \subset sI$ and $s^{-1} \in O_l(I) = R$. This implies that s is a unit in R, a contradiction. Hence I is a principal left ideal and similarly it is a principal right ideal. □

Suppose R is of rank one. Since Q is the only proper overring of R, we have $O_l(I) = R = O_r(I)$ for every non-zero R-ideal I of Q. The set $D(R)$ of divisorial R-ideals is a group by Corollary 1.5.11. Let I be an R-ideal of Q such that $I = Ra$ for some $a \in Q$ Then $R = O_r(I) = a^{-1}Ra$ and hence $aR = Ra$. We denote by $H(R)$ the set of principal R-ideals. An R-ideal I of Q is either divisorial or ${}^*I = Ra = aR, I = J(R)a$ and $J(R)$ is not a principal left R-ideal. The lattice of R-ideals is therefore known completely if $D(R)$ and $H(R)$ are known.

1.5.21 Lemma

Let R be a Dubrovin valuation ring of Q with rank one. $D(R)$ is order isomorphic to a subgroup of $(\mathbb{R}, +)$.

Proof. Let $I \subsetneqq R$ be a divisorial ideal. Then $\cap I^n = (0)$ by Lemma 1.5.15 if $I \subsetneqq J(R)$ or $I = J(R) \neq J(R)^2$ since R has rank one. If $J(R) = J(R)^2$, then it is not divisorial by Lemma 1.5.4. If $K = \cap\, {}^*(I^n) \neq 0$, then it is clearly divisorial and $K \supset I^k$ for some $k \in \mathbb{N}$. Thus ${}^*K = K \supset {}^*(I^k)$ and ${}^*(I^k) \supsetneqq {}^*(I^{k+1})$ since $D(R)$ is a group, which is a contradiction. Hence $K = 0$ and therefore Hölder's Theorem (cf. [23]) shows that $D(R)$ is order isomorphic to a subgroup of $(\mathbb{R}, +)$. □

1.5.22 Proposition

Let R be a Dubrovin valuation ring of Q with rank one. Then exactly one of the following possibilities occurs:

1. The prime segment $J(R) \supsetneqq (0)$ is Archimedean and either:
   (i) $J(R) \supsetneqq J(R)^2$ and then $D(R) \cong \langle J(R) \rangle \cong H(R)$ is an infinite cyclic group or
   (ii) $J(R) = J(R)^2$ and then $D(R) \cong (\mathbb{R}, +)$ and $H(R)$ is a dense subgroup of $D(R)$
2. The prime segment $J(R) \supsetneqq (0)$ is simple and then $D(R) = H(R) = \langle R \rangle$ is the trivial group.
3. The prime segment $J(R) \supsetneqq (0)$ is exceptional and C is a non-Goldie prime with $J(R) \supsetneqq C \supsetneqq (0)$. Then $D(R) = \langle C \rangle$ is the infinite cyclic group generated by $C = {}^*C$ and $H(R) = \langle {}^*(C^k) \rangle$ for some integer $k \geq 0$.

Proof. 1.(i) Since $J(R) \supsetneqq J(R)^2$, it follows that $J(R) = aR = Ra$ for some $a \in J(R)$ and hence $J(R)^n = a^n R = Ra^n$ for any $n \in \mathbb{N}$. Let I be any non-zero ideal. There is an $n \in \mathbb{N}$ with $J(R)^{n-1} \supsetneqq I \supset J(R)^n$ and $R \supsetneqq IJ(R)^{-(n-1)} \supset J(R)$. Hence $I = J(R)^n$ which shows the statement.

1.(ii) Note that $D(R)$ is order isomorphic to a subgroup of $(\mathbb{R}, +)$ by Lemma 1.5.21. For any non-zero element $a \in J(R)$, there is an ideal I with $I \ni a$, i.e. $I \supset RaR$ which is principal by Lemma 1.5.20. For every $RaR \subsetneqq J(R)$ there is a $b \in J(R) - RaR$ and $RaR \subsetneqq RbR \subsetneqq J(R)$. Hence $H(R)$ and $D(R)$ are isomorphic to a dense subgroups of $(\mathbb{R}, +)$. Let I_i be divisorial ideals of R. Then $L = \cap I_i$ is also divisorial if $L \neq 0$ and hence $D(R)$ is complete and $D(R) \cong (\mathbb{R}, +)$ follows.

2. This is clear.
3. By Lemma 1.5.19, C is a maximal divisorial ideal. For any $I \in D(R)$ with $R \supsetneqq I$, there is a minimal $n \in \mathbb{N}$ such that $I \supset C^n$ and hence $C^{n-1} \supsetneqq I \supset$

$^*(C^n)$. It follows that $R = {}^*(C^{n-1}) \circ {}^*(C^{-(n-1)}) \supsetneqq I \circ {}^*(C^{-(n-1)})$ and, by the maximality of C, $C = I \circ {}^*(C^{-(n-1)})$ which entails $I = {}^*(C^n)$. Hence $D(R) = \langle C \rangle$ and $H(R)$ is equal to $\langle {}^*(C^k) \rangle$ for some $k \geq 0$. □

We will describe primary ideals by using the prime segments which are applied to obtain the structure of ideals of Dubrovin valuation rings under a natural condition.

1.5.23 Lemma

Let R be a Dubrovin valuation ring of Q with rank one. The prime segment $J(R) \supsetneqq (0)$ is not simple if and only if R is bounded, i.e. any essential one-sided ideal contains a non-zero ideal.

Proof. Suppose the prime segment $J(R) \supsetneqq (0)$ is simple. It is clear that R is not bounded. Suppose the prime segment $J(R) \supsetneqq (0)$ is not simple. If it is exceptional with non-Goldie prime C and put $S = \cup_{n=1}^{\infty}({}^*(C^n))^{-1}$ which is a proper overring of R and hence $S = Q$. Let c be a regular element in R. Then $c^{-1} \in ({}^*(C^n))^{-1}$ for some $n \in \mathbb{N}$ and $C^n c^{-1} \subset C^n\, {}^*(C^n)^{-1} \subset R$. Hence $C^n \subset Rc$ and $C^m \subset cR$ for some $m \in \mathbb{N}$. Hence R is bounded. Suppose the prime segment is Archimedean. If $J(R) \supsetneqq J(R)^2$, then $J(R) = bR = Rb$ for some $b \in J(R)$. If $J(R) = J(R)^2$, then for any non-zero $a \in J(R)$, $RaR = bR = Rb$ for some $b \in RaR$. In any cases, $\mathcal{C} = \{b^n, n \in \mathbb{N}\}$ is an Ore set with $Q = R_{\mathcal{C}}$. Hence as in the case of the exceptional prime segment, we have that R is bounded. □

For any ideal A, we denote by $P(A)$ the prime radical of A. If R is a Dubrovin valuation ring, then $P(A)$ is a prime ideal of R. An ideal A is called **left** $P(A)$-**primary** if $aRb \subset A$, where $a, b \in R$, implies either $a \in P(A)$ or $b \in A$.

1.5.24 Lemma

Let R be a Dubrovin valuation ring of Q and let A be a non-zero ideal of R with $P = P(A)$ Goldie prime. Then $S = O_l(A) \subset R_P$. In particular if A is left P-primary, then $S \supset R_{P_1}$ for any Goldie prime P_1 with $P_1 \supsetneqq P$.

Proof. Either $A \supsetneqq J(S)$ or $A \subset J(S)$. If the former holds then $A \cap \mathcal{C}(J(S)) \neq \emptyset$ and $A = SA = R_{J(S)}A = R_{J(S)}$, a contradiction. Thus $A \subset J(S)$ which entails $P \subset J(S)$. Hence $R_P \supset R_{J(S)} = S$. Suppose A is left P-primary and P_1 is a Goldie prime with $P_1 \supsetneqq P$. Then $P = R_P P = R_{P_1} P \supset R_{P_1} A$. Let $x = c^{-1}a \in R_{P_1}A$, where $c \in \mathcal{C}(P_1)$ and $a \in A$. Then $P_1 x = P_1 c^{-1} a = P_1 a \subset A$ implies $x \in A$ and hence $R_{P_1}A = A$, i.e. $R_{P_1} \subset O_l(A) = S$. □

It is not necessarily true that $O_l(A) = R_P$ for any ideal A with $P = P(A)$; even if A is left P-primary (cf. [9]).

In order to classify left primary ideals of a Dubrovin valuation rings, for a fixed Goldie prime P, we consider the following four cases:

$\mathcal{P}_1$ There is a Goldie prime P_1 such that $P_1 \supsetneqq P$ is a prime segment.

(a) $P_1 \supsetneqq P$ is Archimedean.
(b) $P_1 \supsetneqq P$ is exceptional such that $P_1 \supsetneqq C \supsetneqq P$ with C non-Goldie prime.
(c) $P_1 \supsetneqq P$ is simple.

$\mathcal{P}_2$ P is lower limit, i.e. for any Goldie prime P_1 with $P_1 \supsetneqq P$, there is a Goldie prime P_2 such that $P_1 \supsetneqq P_2 \supsetneqq P$. This is equivalent to $P = \cap\{P_i, P_i$ is a Goldie prime with $P_i \supsetneqq P\}$.

1.5.25 Lemma

Let R be a Dubrovin valuation ring of Q and A be a non-zero ideal of R with $P = P(A)$ Goldie prime. Suppose one of $\mathcal{P}_1$ (a), $\mathcal{P}_1$ (b) or $\mathcal{P}_2$ holds. Then A is left P-primary if and only if $O_l(A) = R_P$.

Proof. Suppose $O_l(A) = R_P$ and $aRb \subset A$ with $a \in R - P$ and $b \in R$. Then $RaR \supsetneqq P$, $R_P RaR = R_P$ and hence $b \in R_P b = R_P aRb \subset A$. Hence A is left P-primary. Conversely suppose A is left P-primary. First assume that $\mathcal{P}_1$ (a) or $\mathcal{P}_1$ (b) holds. Then $R_{P_1} A = A$ by Lemma 1.5.24. To prove that $O_l(A) = R_P$, let $x = c^{-1}a \in R_P A$, where $c \in \mathcal{C}(P)$ and $a \in A$. Then $cR \supsetneqq P$ by right version of Proposition 1.5.3, because cR and P are both left cRc^{-1} and right R-modules (note $cR_Pc^{-1} = R_P$). Put $\overline{R}_{P_1} = R_{P_1}/P \supsetneqq \overline{P}_1 = J(\overline{R}_{P_1}) \supsetneqq (0)$. Then $\overline{R}_{P_1}$ is of rank one and the prime segment is either Archimedean or exceptional. Since $\overline{c}$ is regular, there is a non-zero ideal $\overline{B}$ such that $\overline{B} \subset \overline{R}_{P_1}\overline{c}$ by Lemma 1.5.23, $P \subsetneqq B + P \subset R_{P_1}c + P$ and $(B + P)x \subset (R_{P_1}c + P)x \subset R_{P_1}a + Pa \subset A$. Hence $x \in A$, i.e. $R_P A = A$ and therefore $O_l(A) = R_P$ by Lemma 1.5.24. Next assume that $\mathcal{P}_2$ holds. Then $R_P = \cup R_{P_i}$ by Lemma 1.5.14. Let $x = c^{-1}a \in R_P A$, where $c \in \mathcal{C}(P)$ and $a \in A$. There is an i with $c^{-1} \in R_{P_i}$ and $P_i x = P_i a \subset A$. Hence $x \in A$ and $R_P A = A$ follows. Hence $O_l(A) = R_P$. □

1.5.26 Lemma

Let R be a Dubrovin valuation ring of Q and A be a non-zero ideal of R with $P = P(A)$ Goldie prime. Suppose $\mathcal{P}_1$ (c) holds. Then A is left P-primary if and only if either $O_l(A) = R_P$ or $O_l(A) = R_{P_1}$ and $A = {}^*A$.

Proof. Assume that A is left P-primary. Then $R_{P_1} \subset S = O_l(A) \subset R_P$ and hence either $R_{P_1} = S$ or $R_P = S$. In the case $R_{P_1} = S$, if $A \neq {}^*A$, then $A = J(S)\ {}^*A = P_1\ {}^*A$ by Proposition 1.5.8, which is a contradiction. Hence

$A = {}^*A$ follows. In order to prove the if part, first assume that $O_l(A) = R_P$. Then an argument similar to the one in Lemma 1.5.25 shows that A is left P-primary. Assume that $O_l(A) = R_{P_1}$ and $A = {}^*A$. Suppose $aRb \subset A$, where $a \in R - P$ and $b \in R$. Since $RaR \supsetneqq P$, we have either $RaR = P_1$ or $RaR \supsetneqq P_1$, i.e. $R_{P_1}aR = R_{P_1}$. Furthermore we may assume that b is regular element in R, because RbR is generated by regular elements in RbR. If $RaR = P_1$, then $A \supset RaRb = P_1 b$ and $A = {}^*A \supset {}^*(P_1 bR) = {}^*({}^*P_1 RbR) \supset R_{P_1}b \ni b$, because ${}^*P_1 = R_{P_1}$ by Lemma 1.5.4 and Theorem 1.5.16 (note that $J(R_{P_1}) = P_1$). If $R_{P_1}aR = R_{P_1}$, then $b \in R_{P_1}aRb \subset A$. Hence A is left P-primary. □

1.5.27 Lemma

Let R be a Dubrovin valuation ring of Q and C be a non-Goldie prime such that $P \supsetneqq C \supsetneqq P_0$ is the exceptional prime segment. Let A be an ideal of R with $C = P(A)$. Then A is left C-primary if and only if $O_l(A) = R_P$ and $A = {}^*A$.

Proof. First we prove that $S = O_l(A) \subset R_P$. If $A \supsetneqq J(S)$, then $R_{J(S)} = R_{J(S)}A = SA = A$, a contradiction. Hence $J(S) \supset A$ and $J(S) \supset P$ since $J(S)$ is a Goldie prime. This implies $R_P \supset S = R_{J(S)}$. Suppose that A is left C-primary. To prove that $S = R_P$, on the contrary, assume that $R_P \supsetneqq S$. Then $R_P A \supsetneqq A$ and for any $x = c^{-1}a \in R_P A - A$, where $c \in \mathcal{C}(P)$ and $a \in A$, we have $Px = Pa \subset A$, a contradiction. Hence $O_l(A) = R_P$. If ${}^*A \supsetneqq A$, then ${}^*A = R_P d$ and $A = Pd$ for some $d \in {}^*A - A$ by Proposition 1.5.8, which is again a contradiction. Hence ${}^*A = A$ follows. Conversely suppose $O_l(A) = R_P$ and ${}^*A = A$. If $aRb \subset A$ with $a \in R{-}C$ and $b \in R$, then $RaR \supset P$ and as before we may assume that b is regular. It follows that $A \supset PbR$ and $A = {}^*A = {}^*PbR = {}^*({}^*PRbR) \supset R_P b \ni b$. Hence A is left C-primary. □

We denote by $l - st_P(R) = \{a \in R, R_P a$ is an ideal of $R\}$. The following is a complete description of left primary ideals of Dubrovin valuation rings (cf. [9] for examples).

1.5.28 Theorem

Let R be a Dubrovin valuation ring of a simple Artinian ring Q.

1. Let P be a Goldie prime of R satisfying one of the $\mathcal{P}_1 a)$,$\mathcal{P}_1 b)$, or $\mathcal{P}_2$.
   (i) If $P \supsetneqq P^2$, then $\{P^n, n \in \mathbb{N}\}$ is the set of left P-primary ideals of R and $P^n = p^n R_P = R_P p^n$ for some $p \in P$.
   (ii) If $P = P^2$, then $\{P, Pa, A, a \in l - st_P(R)$ with $P(R_P a) = P$ and A is an ideal with $P = P(A)$, $O_l(A) = R_P$ and $A = {}^*A\}$ is the set of left P-primary.

2. Let P be a Goldie prime of R satisfying $\mathcal{P}_1c)$.
   (i) If $P \supsetneqq P^2$, then $\{P^n, R_{P_1}a$ and $A, a \in l - st_{P_1}(R)$ with $P = P(R_{P_1}a)$ and A is an ideal with $P = P(A)$, $O_l(A) = R_{P_1}$ and $A = {}^*A\}$ is the set of left P-primary ideals. In particular, $P^n = p^n R_P = R_P p^n$ for some $p \in P$ with $P = pR_P = R_P p$.
   (ii) If $P = P^2$, then $\{P, Pa, R_{P_1}b$, and $A, R_{P_1}a$ with $a \in l - st_P(R)$ with $P = P(R_P a), b \in l - st_{P_1}(R)$ with $P = P(R_{P_1}b)$ and A is an ideal with $P = P(A), A = {}^*A$ and either $O_l(A) = R_P$ or $O_l(A) = R_{P_1}\}$ is the set of left P-primary.
3. Let C be a non-Goldie prime of R. Then $\{{}^*(C^n), n \in \mathbb{N}\}$ is the set of left C-primary ideals of R.

Proof. 1.(i) Since $P = pR_P = R_P p$ for some $p \in P$ by Lemma 1.5.4, $P^n = p^n R_P = R_P p^n$ for all $n \in \mathbb{N}$ and P^n are all (left) P-primary by Lemma 1.5.25. Let A be left P-primary. Then $P \supset A \supset P_0 = \cap P^n$. We may assume that $P^n \supset A \supsetneqq P^{n+1}$ for some $n \in \mathbb{N}$ and $R_P \supset p^{-n}A \supsetneqq pR_P = P$. If $p^{-n}A \supsetneqq R$, then $A \supsetneqq p^n R$, a contradiction. Thus $p^{-n}A \subset R$ and $C(P) \cap p^{-n}A \neq \phi$ which entails $p^{-n}A = R_P$, i.e. $A = p^n R_P = P^n$.

1.(ii) It is easy to see from Lemma 1.5.25 that P, Pa and A stated in (ii) are left P-primary. Let B be left P-primary. Then $R_P = O_l(B)$. If $B = {}^*B$, then there is nothing to do. If $B \subsetneqq {}^*B$, then $B = Pb$ and ${}^*B = R_P b$ for some $b \in^* B - B$ by Proposition 1.5.8 (note $b \in R_P$). If $b \in U(R_P)$ then $B = P$. If $b \notin U(R_P)$ then we prove $P \supset R_P b$, i.e. $b \in l - \mathrm{st}_P(R)$. Suppose $P \subsetneqq R_P b$. Then $Pb^{-1} \subset P$ since $P = P^2$ and so $b^{-1} \in O_r(P) = R_P$, a contradiction. Hence $B = Pb$ with $b \in l - \mathrm{st}_P(R)$ and $P = P(R_P b)$ follows.

2.(i) It follows from Lemma 1.5.26 and the proof of 1.(i) that P^n, $R_{P_1}a$ and A are all left P-primary, where $a \in l - st_{P_1}(R)$ with $P = P(R_{P_1}a)$, and A is an ideal with $P = P(A)$, $O_l(A) = R_{P_1}$ and $A = {}^*A$. Let B be left P-primary. If $O_l(B) = R_P$, then $B = P^n$ for some $n \in \mathbb{N}$ as in 1.(i). Suppose that $O_l(B) = R_{P_1}$. If B is a finitely generated left R_{P_1}-ideal, i.e. $B = R_{P_1}a$ with $a \in l - st_{P_1}(R)$. If B is not a finitely generated left R_{P_1}-ideal, there are two cases: either $B = {}^*B$ or $B \subsetneqq {}^*B$ with $B = P_1 b$ and ${}^*B = R_{P_1}b$ for some $b \in {}^*B - B$. In the latter case, B is not left P-primary.

2.(ii) P, Pa, $R_{P_1}b$ and A in 2.(ii) are all left P-primary. Let B be left P-primary. Suppose $O_l(B) = R_P$. It is shown as in 1.(ii) that B is one of P, Pa or A, where $a \in l - st_P(R)$ with $P = P(R_P a)$ and $A = {}^*A$ with $O_l(A) = R_P$. Suppose $O_l(B) = R_{P_1}$. Then B is one of $B = {}^*B$ or $B = R_{P_1}b$ with $b \in l - st_{P_1}(R)$ and $P = P(R_{P_1}b)$ as in 2.(i).

3. Since $C = {}^*C \supset {}^*(C^n) \supset C^n$ for all $n \in \mathbb{N}$, it follows that $C = P({}^*(C^n))$ and $O_l({}^*(C^n)) = R_P$ by Lemmas 1.5.7 and 1.5.19. Hence ${}^*(C^n)$ is left C-primary by Lemma 1.5.27. Let B be left C-primary. Then $B = {}^*B$ and $O_l(B) = R_P$. Since $P_0 = \cap C^n$, there is an $n \in \mathbb{N}$ such that $C^n \supset B \supset C^{n+1}$ and $R_P \supset C^{-n}C^n \supset C^{-n}B \supset C^{-n}C^{n+1}$.

Taking $*$-operation, $R_P \supset {}^*(C^{-n}B) \supset {}^*C = C$. If ${}^*(C^{-n}B) \supset P$, then ${}^*(C^{-n}B) \supset {}^*P = R_P$ and hence $R_P = {}^*(C^{-n}B)$. Hence $B = {}^*({}^*(C^nC^{-n}))B) = {}^*(C^n(C^{-n}B)) = {}^*(C^nR_P) = {}^*(C^n)$. If $P \supsetneq {}^*(C^{-n}B)$, then $C \supset^* (C^{-n}B), {}^*(C^{n+1}) \supset B$ and hence $B = {}^*(C^{n+1})$ follows. □

1.5.29 Lemma

Let R be a Dubrovin valuation ring of Q and I be an R-ideal of Q. If I is not a finitely generated as a left S-ideal, where $S = O_l(I)$, then $I^{-1} = (S : I)_r = (R : I)_r$ and $I(R : I)_r = J(S)$.

Proof. It follows from Lemma 1.5.6 and Proposition 1.5.8 that $I = J(S)I$, $J(S) = J(S)^2$ and $II^{-1} = J(S)$. Hence $(R : I)_r = (R : J(S)I)_r = ((R : J(S))_r : I)_r = (O_r(J(S)) : I)_r = (S : I)_r = I^{-1}$. □

1.5.30 Lemma

Let R be a Dubrovin valuation ring of Q and $I = RqR$ for some $q \in Q$ with $S = O_l(I)$ and $T = O_r(I)$. Suppose the prime segment $J(S) \supsetneq P$ is Archimedean, then $I = Sa = aT$ for some $a \in I$.

Proof. To prove that $I = Sa$, it suffices to prove that $I \supsetneq J(S)I$. Suppose, on the contrary, $I = J(S)I$. Then $q = x_1qr_1 + \cdots + x_nqr_n$ and $xS = x_1S + \cdots + x_nS$, where $x_i, x \in J(S)$, and $r_i \in R$. It is clear that $I = SxSI$. If $J(S) = SxS$, then $J(S) = sS = Ss$ for some $s \in J(S)$ by Lemma 1.5.20. Thus $I = sI$ and $s^{-1} \in O_l(I) = S$, a contradiction. Hence $J(S) \supsetneq SxS$ which implies $tJ(S) \supset SxS$ for some $t \in J(S)$. Then $I = SxSI \subset tJ(S)I = tI$ and $t^{-1} \in O_l(I) = S$, a contradiction. Hence $I = Sa$ for some $a \in I$ and $I = aa^{-1}Sa = aO_r(I) = aT$. □

1.5.31 Proposition

Let R be a Dubrovin valuation ring of Q and let I be an R-ideal with $O_l(I) = S = O_r(I)$ and I is not finitely generated as a left R-ideal. Suppose $J(S)$ is Archimedean, i.e. the prime segment $J(S) \supset P$ is Archimedean. Then either $I = aS = Sa$ or $I = Aa$ for some $J(S)$-primary A and some a with $aS = Sa$.

Proof. First suppose I is finitely generated as a left S-ideal, say $I = Sa$. Then $S = O_r(I) = a^{-1}Sa$ and hence $aS = Sa = I$. Suppose I is not finitely generated as a left S-ideal. By Lemma 1.5.29, $I(R : I)_r = J(S)$. Consider $\mathcal{F} = \{x, x \in (R : I)_r\}$ with $J(S) = P(IxS)$, which is non-empty, because $J(S)$ is Archimedean. We claim that $B = IxS$ is $J(S)$-primary for any $x \in \mathcal{F}$. It suffices to prove that $O_l(B) = S = O_r(B)$ by Lemma 1.5.25 and its right version. It is clear that $O_l(B) \supset S$ and assume that $O_l(B) \supsetneqq S$. Then $O_l(B) \supset R_P$, because there are no Goldie primes between $J(S)$ and P and then $B = R_P B = R_P$, a contradiction. Hence $O_l(B) = S$ and similarly $O_r(B) = S$. Since $(R : I)_r = (S : I)_r$ and $I(R : I)_r = J(S)$, we have $(R : I)_r = \bigcup\{SxS, x \in (R : I)_r\} \supset \bigcup\{SxS, x \in \mathcal{F}\}$. If $(R : I)_r \supsetneqq \bigcup\{SxS, x \in \mathcal{F}\}$, then there is a $y \in (R : I)_r$ such that $y \notin SxS$ for any $x \in \mathcal{F}$, i.e. $SyS \supsetneqq SxS$. However $y \notin \mathcal{F}$ implies $P \supset IyS \supset IxS$ for any $x \in \mathcal{F}$ and $P \supset P(IxS) = J(S)$, a contradiction. Hence $\bigcup\{SxS, x \in \mathcal{F}\} = (R : I)_r$ holds. We shall prove that $S = O_l(SxS)$ for some $x \in \mathcal{F}$. Suppose $O_l(SxS) \supsetneqq S$ for all $x \in \mathcal{F}$. Then $O_l(SxS) \supset R_P$ and hence $R_P(R : I)_r = (R : I)_r$ follows. Since $I^{-1} = (R : I)_r$, we have $R_P I^{-1} = I^{-1}$ and $R_P \subset O_l(I^{-1}) = O_r(I) = S$, a contradiction. Thus there is some $x \in \mathcal{F}$ such that $O_l(SxS) = S$. Put $A = ISxS$, a $J(S)$-primary. It is clear that $O_r(SxS) = S$ since $O_r(A) = S$. By Lemma 1.5.30, $SxS = cS = Sc$ for some $c \in SxS$. Hence we have $I = Ac^{-1}$ with $c^{-1}S = Sc^{-1}$. □

Note If Q is a finite dimension over its center K, then $O_r(I) = O_l(I)$ for any R-ideal I of Q and $J(S)$ is Archimedean (cf. Corollary 1.6.6). However these properties do not necessarily hold (cf. [6, 33]).

We end this section with the following which will be used in Sect. 1.7.

1.5.32 Proposition

Let S be a proper overring of a Dubrovin valuation ring R.

1. $(R : S)_l = J(S) = (R : S)_r$.
2. $(R : J(S))_l = S = (R : J(S))_r$.

Proof. 1. It is clear that $(R : S)_l \supset J(S)$. If $(R : S)_l \supsetneqq J(S)$, then $(R : S)_l = (R : S)_l S = S$, because $(R : S)_l$ is an ideal of R and $S = R_{J(S)}$ by Theorem 1.4.9, which is a contradiction. Hence $(R : S)_l = J(S)$ and similarly $(R : S)_r = J(S)$.

2. It is clear that $(R : J(S))_l \supset S$. To prove the converse inclusion, first assume that $J(S) = sS = Ss$ for some $s \in J(S)$. Then we have $(R : J(S))_l = (R : S)_l s^{-1} = S$ by (1). Next, assume that $J(S)$ is not finitely generated as a one-sided S-ideal. Then, by Lemma 1.5.4, $J(S) = J(S)^2$ and so $(R : J(S))_l \subset (S : J(S))_l = O_l(J(S)) = S$ by Lemma 1.5.4. Hence $(R : J(S))_l = S$ and similarly $(R : J(S))_r = S$.

1.6 Dubrovin Valuation Rings of Q with Finite Dimension Over Its Center

Throughout this section, Q will be a simple Artinian ring with finite dimension over its centre K and R will be a Dubrovin valuation ring of Q with $V = Z(R)$.

1.6.1 Lemma

Let M and N be V-submodules of an K-module X with $N \subset M$ and $J(V)M \subset N$. Then $[\overline{M} : \overline{N}] \leq [X : K]$, where $\overline{M} = M/N$ and $\overline{V} = V/J(V)$.

Proof. It suffices to prove that $m_1, \cdots, m_n \in M$ is linearly independent over K if $\overline{m_1}, \cdots, \overline{m_n}$ is linearly independent over $\overline{V}$. Assume that $k_1 m_1 + \cdots + k_n m_n = 0$ with some $k_i \neq 0$, where $k_i \in K$. We may assume that $k_1 \neq 0$ and $k_1 V \supset k_i V$ for all i, $1 \leq i \leq n$. Then $k_i = k_1 v_i$ for some $v_i \in V$ and $\overline{m_1} + \overline{v_2 m_2} + \cdots + \overline{v_n m_n} = \overline{0}$, a contradiction. □

1.6.2 Lemma

Let $q \in Q$ such that either $qR \subset Rq$ or $qR \supset Rq$. Then

1. $Rq^n = Rk$ for some $n \in \mathbb{N}$ and $k \in K$.
2. $Rq = qR$ and $Sq = qS$ for any overring S of R.

Proof. 1. Suppose $n = [Q : K]$. There is a non-trivial linear dependence $k_n q^n + \cdots + k_1 q + k_0 = 0$, where $k_i \in K$. Let $Rq \supset qR$. Then $Rk_n q^n, \cdots, Rk_1 q, Rk_0$ are all R-ideals so that there is a maximum one, say, $Rk_i q^i$. $Rk_i q^i = \sum_{j \neq i} Rk_j q^j \subset Rk_l q^l$ for some $l \neq i$, i.e. $Rk_i q^i = Rk_l q^l$. Hence $Rq^e = Rk$ for some $k \in K$ and some $e \leq n$.

2. Suppose $Rq \supsetneqq qR$. It follows that $q^{-n} Rq^n \supsetneqq R$ for any $n \in \mathbb{N}$. But $q^{-e} Rq^e = k^{-l} Rk = R$, a contradiction. Hence $Rq = qR$. Since qSq^{-1} and S are both R-ideals, we have either $qSq^{-1} \subset S$ or $qSq^{-1} \supset S$. Hence $Sq = qS$ as before. □

1.6.3 Proposition

Let R be a Dubrovin valuation ring of Q with $V = Z(R)$. The following hold:

1. V is a valuation ring of K and $J(V) = J(R) \cap V$.

2. Any prime ideal P of R is Goldie prime such that $R_P = R_p$ and $P = J(R_p)$, where $p = P \cap V$ is a prime ideal of V.
3. The map: $P \longrightarrow p$ is a bijection between the set of prime ideals of R and the set of prime ideals of V, where P is a prime ideal of R.
4. The map: $S \longrightarrow J(S)$ is a bijection and anti-inclusion preserving between the set of overrings of R and the set of prime ideals of R, where S is any overring of R.
5. For any prime ideal P of R, there is an $l \in \mathbb{N}$ such that $P^l = pR$, where $p = P \cap V$.
6. Let S be an overring of R with $V = Z(S)$. Then $S = R$.

Proof. 1. For any $x \in K$, we have $Rx \cap V = Vx$. This shows the set of principal ideals is linearly ordered by inclusion. Hence V is a valuation ring of K. Since $J(V)R$ is a proper ideal of R, we have $J(V)R \subset J(R)$. Hence $J(V) \supset J(R) \cap V \supset J(V)R \cap V \supset J(V)$ and $J(V) = J(R) \cap V$ follows.

2. It is clear that $p = P \cap V$ is a prime ideal of V and $P_p \cap R = P$. Thus $P_p \subset J(P_p) \subsetneqq R$ by Theorem 1.4.9 and $P_p = P$ follows. Since $[R_p/P : V_p/p] \leq [Q : K]$, R_p/P is an Artinian ring and there is an $n \in \mathbb{N}$ such that $J(R_p)^n \subset P$, which entails $J(R_p) = P$. Hence P is Goldie prime and $R_p = R_P$.
3. For any prime ideal p of V, $P = J(R_p)$ is a prime ideal of R with $P \cap V = p$. Hence the statement follows form (2).
4. This follows from Corollary 1.4.16 and (2).
5. By (2) $P = J(R_p)$ and $pR = pV_pR = pR_p$, an ideal of R_p. Thus $P^l \subset pR_p$ for some $l \in \mathbb{N}$ by Lemma 1.6.1. If $P = P^2$, then $P^l = P = pR_p$. Otherwise $P = aR_p = R_pa$ for some $a \in P$ by Lemma 1.5.4. Let l be the smallest natural number such that $P^l \subset pR$, i.e. $a^lR_p \subset pR_p \subsetneqq a^{l-1}R_p$. Then $aR_p \subset a^{-l+1}pR_p \subsetneqq R_p$ and $P^l = pR_p$ follows.
6. Suppose that $S \supsetneqq R$. Then $p = J(S) \cap V$ is a prime ideal of V with $p \neq J(V)$. Hence $S = R_p$ and $Z(S) = V_p \supsetneqq V$, a contradiction. Hence $S = R$. □

1.6.4 Lemma

Let R be a Dubrovin valuation ring of Q and I be a finitely generated R-ideal of Q, i.e. $I = Rx_1R + \cdots + Rx_nR$ for some $x_i \in I$. Then $I = aS = Sa$ for some $a \in I$ and $O_l(I) = O_r(I)$, where $S = O_l(I)$.

Proof. Since R is Bezout, we may assume that $I = RxR$ for some regular $x \in I$.

(i) In the case $I = J(R)$. It follows from Lemma 1.5.4 that $O_l(I) = R = O_r(I)$. As in Lemma 1.5.20, let B be an ideal maximal with $x \notin B$ and assume that $J(R) = J(R)^2$. Then B is a prime ideal, the prime segment $J(R) \supset B$ is simple and $\overline{R} = R/B$ is a Dubrovin valuation ring of $Q(R/B)$ with rank one such that the prime segment $J(\overline{R}) \supsetneqq (\overline{0})$ is simple. However $\overline{R}$ is a prime

P.I.-ring (cf. [47, (13.6.5)]), i.e. $\overline{R}$ is bounded, which contradicts Lemma 1.5.23. Thus $J(R) \supsetneqq J(R)^2$ and hence $J(R) = aR = Ra$ for some $a \in J(R)$.

(ii) In the case $I \neq J(R)$, to prove that $I = Sa$, assume that $I = J(S)I$. We can write $q = \sum_{i=1}^n x_i q r_i$, where $x_i \in J(S)$ and $x_i \in R$. Let $xS = \sum_{i=1}^n x_i S$, where $x \in J(S)$ and $SxSI = I$. Suppose $J(S) = SxS$. Then $J(S) = yS$ for some regular $y \in J(S)$ and $yI = I$, i.e. $y^{-1} \in O_l(I) = S$, a contradiction. Hence $SxS \subsetneqq J(S)$ and $SxS \subset sJ(S)$ for some regular $s \in J(S)$ by Lemma 1.5.2. This implies $I = SxSI \subset sJ(S)I \subset sI$ and $s^{-1} \in O_l(I) = S$, a contradiction. Hence $I = Sa$ for some regular $a \in I$ and $O_r(I) = a^{-1}Sa$. Since S and $O_r(I)$ are overrings of R, we have either $S \subset a^{-1}Sa$ or $S \supset a^{-1}Sa$. Hence $Sa = aS$ by Lemma 1.6.2 and $S = O_r(I)$. □

1.6.5 Proposition

Let R be a Dubrovin valuation ring of Q with finite dimension over its center K. Then the set of R-ideals forms a commutative semi-group.

Proof. It suffices to prove that $RyRzR = RzRyR$ for any non-zero elements $y, z \in Q$. By Lemma 1.6.4, $RyR = sS = Ss$ and $RzR = tT = Tt$ for some regular $s, t \in Q$, where $S = O_l(RyR)$ and $T = O_l(RzR)$. We may suppose that $T \subset S$. Then $tS = St$ by Lemma 1.6.2. Hence $RyRzR = sSTt = sSt = stS$ and similarly $RzRyR = Sts$. There is an $n \in \mathbb{N}$ and $u \in K$ such that $s^n S = uS$ and $(t^{-1}stS)^n = t^{-1}s^n tS = t^{-1}s^n St = t^{-1}uSt = s^n S = (sS)^n$. Since S-ideals are linearly ordered, $t^{-1}stS = sS$, i.e. $stS = tsS$. Hence $RyRzR = RzRyR$. □

For any R-ideal I of Q, we have $I = IO_r(I) = O_r(I)I$ and $I = O_l(I)I = IO_l(I)$. Thus we have

1.6.6 Corollary

For any R-ideal I of Q, $O_l(I) = O_r(I)$.

1.6.7 Lemma

1. Let R be a subring of Q with $Z(R) = V$. If R is an Azumaya algebra, then it is a Dubrovin valuation ring of Q.
2. Suppose that V has no minimal prime ideals.

(i) For any K-basis $a_1, \cdots, a_n$ of Q, there is an Azumaya algebra S in $Q (S \neq Q)$ such that $S = a_1 W \oplus \cdots \oplus a_n W$ and $W = Z(S)$.
(ii) If R is a Dubrovin valuation ring of Q, then there is an Azumaya algebra S_1 of Q such that $S_1 = a_1 W_1 \oplus \cdots \oplus a_n W_1$ such that $S_1 \supset R$ and $W_1 \supset V$.

Proof. 1. It follows from the properties of Azumaya algebras that $J(R) = J(V)R$ and $R/J(R)$ is a simple Artinian ring with finite dimension over the center $V/J(V)$.

Thus for any $q \in Q - R$, there is a $v \in V$ such that $qv \in R - J(R)$. Hence R is a Dubrovin valuation ring.

2.(i) Write $a_i a_j = \sum a_t \alpha_{ijt}$, where $\alpha_{ijt} \in K$. It follows that $N = (a_i a_j) \in M_n(Q)$ has its inverse N^{-1}, say, $N^{-1} = (c_{ij})$, where $c_{ij} \in Q$ (cf. [41, Proposition A.10]). Write $c_{ij} = \sum a_t \beta_{ijt}$ and $1 = \sum a_i \gamma_i$, where $\beta_{ijt}, \gamma_i \in K$. Since V has no minimal prime ideals, there is an overring $W (\neq K)$ of V with $\alpha_{ijt}, \beta_{ijt}, \gamma_i \in W$ for all i, j, t. Then $S = a_1 W \oplus \cdots \oplus a_n W$ is a subring of Q and $M = a_1 J(W) \oplus \cdots \oplus a_n J(W)$ is an ideal of S. It is clear that $Z(S) \supset W$. To prove the converse inclusion, let $\alpha \in Z(S)$ and $f \in \mathrm{Hom}_K(Q, Q)$ with $f(a_1) = 1$ and $f(a_i) = 0$ for $i = 2, \cdots, n$. Write $a_1 \alpha = a_1 w_1 + \cdots + a_n w_n$, where $a_i \in W$. We have $\alpha = f(a_1 \alpha) = w_1$ and $Z(S) = W$ follows. Hence S is an Azumaya algebra over W because $N^{-1} \in M_n(S)$ (cf. [41, Proposition A.10]).

(ii) Let $b_1, \cdots, b_n$ be a K-basis of Q being contained in R and write $b_i = \sum a_j \delta_j$, where $\delta_j \in K$. As in (i), put $b_i b_j = \sum b_t \gamma_{ijt}$, $(d_{ij}) \in M_n(Q)$ is the inverse of $(b_i b_j)$, $d_{ij} = \sum b_t \upsilon_{ijt}$ and $1 = \sum b_i \varepsilon_i$. Then there is an overring $W_1 (\neq K)$ of V such that W_1 contains all $\delta_j, \gamma_{ijt}, \upsilon_{ijt}, \varepsilon_i, \alpha_{ijt}, \beta_{ijt}, \gamma_j$. Put $S_1 = a_1 W_1 \oplus \cdots \oplus a_n W_1$ and $S_2 = b_1 W_1 \oplus \cdots \oplus b_n W_1$ which are both Azumaya algebras over W_1 such that $S_1 \supset S_2$ and $S_1 = S_2$ by Proposition 1.6.3. Furthermore $RW_1 \supset S_2$ and $Z(RW_1) = W_1$. Hence $S_2 = RW_1$ and $S_2 \supset R$. □

The set of all elements in Q which are integral over V is not necessarily a subring of Q. In case Q is a skewfield there is a close connection between this property and total valuation rings.

1.6.8 Theorem

Let D be a skewfield with finite dimension over its center K, V be a valuation ring of K and T be the set of all elements in Q that are integral over V. Then T is a ring if and only if there is a total valuation ring R of D with $Z(R) = V$. In this case $T = \cap\{R, R \text{ is a total valuation ring of } D \text{ with } Z(R) = V\}$.

Proof. Suppose there is a total valuation ring of D whose center is V. It suffices to prove that $T = \cap\{R, R \text{ is a total valuation ring of } D \text{ with } Z(R) = V\}$. Let $t \in T$ and $t^m + v_{m-1} t^{m-1} + \cdots + v_1 t + v_0 = 0$ for some $v_i \in V$ and $m \in \mathbb{N}$. Assume that there is a total valuation ring R of D with $Z(R) = V$ and $t \notin R$. Then $t^{-1} \in J(R)$

and $1 = -(v_{m-1}t^{-1} + \cdots + v_0 t^{-m}) \in J(R)$, a contradiction. Hence $T \subseteq S = \bigcap\{R, R$ is a total valuation ring of D with $Z(R) = V\}$. To show the converse inclusion, let $s \in S$ and $f(X)$ be the monic irreducible polynomial of s in $K[X]$. By Wedderburn's Theorem (cf. [41, Theorem A.20]), $f(X) = (X-a_1)\cdots(X-a_n)$ for $a_i = d_i s d_i^{-1}$, $0 \neq d_i \in D$. However S is invariant under inner automorphism of D so that every $a_i \in S$. Thus the coefficients of $f(X)$ are in $S \cap K = V$ and hence s is integral over V.

Conversely suppose that T is a ring. It is clear that T is invariant under inner automorphisms of D. Since $J(V)T$ is a proper ideal of T, there is a maximal ideal M of T containing $J(V)T$ and $M \cap V = J(V)$. The set $T - M$ is a regular Ore set since T is invariant. We show that $T_M \cap K = V$. Otherwise there is an element $u \in K - V$ such that $u = ts^{-1}$, where $t \in T$, $s \in T - M$ and $s = u^{-1}t \in J(V)T$, a contradiction. It remains to prove that T_M is a total valuation ring. Let $x \in D - T_M$ and $K(x)$ is the commutative subfield of D. Then $A = T \cap K(x)$ is the integral closure of V in $K(x)$ and hence it is a Prüfer domain (cf. [22, Theorem 13.4]). Since $B = T_M \cap K(x)$ is a local overring of A, it is a valuation ring (cf. [22, Corollary 11.6]). Hence $x^{-1} \in B \subset T_M$ and T_M is a total valuation ring. □

In Proposition 1.3.2.16 we considered valuation extensions. In case the base ring is a commutative valuatin ring, we have a more concrete result. We denote by $\mathrm{Nr}(q)$ the reduced norm of $q \in Q$ (cf. [57] for reduced norm).

1.6.9 Proposition

Let D be a skewfield with finite dimension over its center K and let v be a valuation on K with the value group Γ_V. Then v is extended to a valuation on D if and only if v has a unique extension to each field F with $K \subset F \subset D$. In particular, under these conditions, if R is a valuation ring of D with $Z(R) = V$, then R is integral over V, where V is the valuation ring of K with valuation v.

Proof. Suppose that v uniquely extends to any field F with $K \subset F \subset D$. Define a function $\omega : U(D) \longrightarrow \Delta = \Gamma_V \otimes_{\mathbb{Z}} \mathbb{Q}$, $\omega(d) = \frac{1}{n}v(\mathrm{Nr}(d))$, where $d \in U(D)$ and $n = \sqrt{[D:K]} \in \mathbb{Z}$. Clearly the restriction $\omega\mid_K$ of ω to K coincides with v. We will show that ω is a valuation on D. We will prove first that for every maximal subfield F of D, $\omega\mid_F$ is the valuation on F extending v. Let N be the normal closure of F over K and let $\mu : U(N) \longrightarrow \Delta$ be any valuation on N extending v on K. For any $a \in U(F)$, $\mathrm{Nr}(a) = N_{F|K}(a) = a_1 \cdots a_n$, where $N_{F|K}$ is the norm from F to K and each $a_i \in N$ is a conjugate of a over K. Thus there is a K-automorphism σ_i of N with $\sigma_i(a) = a_i$. Since $\mu\mid_F$ and $(\mu\sigma_i\mid_F)$ are valuations on F extending v, by hypothesis, they must coincide and $\mu(a_i) = \mu\sigma_i(a) = \mu(a)$. Hence $\omega(a) = \frac{1}{n}v(\mathrm{Nr}(a)) = \frac{1}{n}\mu(a_1\cdots a_n) = \frac{1}{n}(\mu(a_1) + \cdots + \mu(a_n)) = \mu(a)$. Therefore $\omega|_F = \mu|_F$, which is the valuation on F extending v. To prove that ω is a valuation on D, let $a, b \in U(D)$. Because Nr is multiplicative, we have $\omega(ab) = \omega(a) + \omega(b)$.

Assume that $b \neq -a$ and let F be any maximal subfield of D containing ab^{-1}. Since $\omega|_F$ is a valuation, $\omega(1 + a^{-1}b) \geq \min\{\omega(1), \omega(a^{-1}b)\}$. Thus $\omega(a + b) = \omega(a) + \omega(1 + a^{-1}b) \geq \omega(a) + \min\{\omega(1), \omega(a^{-1}b)\} = \min\{\omega(a), \omega(b)\}$. Hence ω is a valuation on D.

Conversely, suppose that v extends to a valuation ω on D. Let $R = \{a, a \in U(D), \omega(a) \geq 0\} \cup \{0\}$, the valuation ring of ω. Note that Γ_R is abelian by Proposition 1.6.5. Thus $\omega(aba^{-1}) = \omega(b)$ for all $a, b \in U(D)$. For any $a \in R$, let $f(X) = X^k + c_{k-1}X^{k-1} + \cdots + c_0 \in K[X]$ be the minimal polynomial of a over K. Then $f(X) = (X - a_1)\cdots(X - a_k)$ as before, where $a_i = b_i a b_i^{-1}$ for some $b_i \in U(D)$. Thus $\omega(a_i) = \omega(a) \geq 0$ and hence the coefficients of f lie in $R \cap K = V$, i.e. every $a \in R$ is integral over V. Let F be any field with $K \subset F \subset D$. Then $R_0 = R \cap F$ is a valuation ring of F and integral over V. Hence it is the integral closure of V in F and so the valuation of R_0, i.e. $\omega|_F$ is the unique extension of v to F (cf. [22, Corollary 13.5]). □

From the properties of Henselian valuation rings (cf. [22, (16.4) and (16.6)]) and Proposition 1.6.9, we have the following.

1.6.10 Corollary

Let D, K and v be as in Proposition 1.6.9. If V is a Henselian valuation ring of K, then there is a valuation ring R of D such that $Z(R) = V$ and $R = \{a \in D, a$ is integral over $V\}$. □

1.6.11 Lemma

For a total valuation ring R of D with $Z(R) = V$, the following are equivalent:

1. R is a valuation ring.
2. R is integral over V.
3. R is the only Dubrovin valuation ring of D with its center V.

Proof. 1. ⇒ 2. If R is a valuation ring, then $\Gamma_V \subset \Gamma_R$ naturally and so the valuation on K is extended to a valuation on D. Thus R is integral over V by Proposition 1.6.9.

2. ⇒ 3. Note that R is a valuation ring by Theorem 1.6.8. Let S be a Dubrovin valuation ring of D with $Z(S) = V$. We will first claim that $J(S) \subset J(R)$. Let $0 \neq s \in J(S)$ and $f(X) = X^n + a_{n-1}X^{n-1} + \cdots + a_0$ be the minimal polynomial of s over K. Then $g(X) = X^n + a_0^{-1}a_1X^{n-1} + \cdots + a_0^{-1}a_{n-1}X + a_0^{-1}$ is the minimal polynomial of s^{-1} over K. Assume that $s^{-1} \in R$. Then $a_0^{-1}a_i \in V$ for all i, $0 \leqq i \leqq n$ with $a_n = 1$. It follows that $-1 = a_0^{-1}s^n + a_0^{-1}a_{n-1}s^{n-1} + \cdots + a_0^{-1}a_1 s \in J(S)$, a contradiction. Hence $s^{-1} \notin R$, i.e. $s \in J(R)$ and

$J(S) \subset J(R)$. Next assume that $S \neq R$. Then $S \not\subseteq R$ by Proposition 1.6.3 and let $s \in S - R$. There are $k \in K$ and $m \in \mathbb{N}$ such that $s^m R = kR$. Since $s \notin R$, we have $s^m \notin R$, $k \notin R$ and $k^{-1}s^m \notin J(R)$ and $k^{-1} \in J(R) \cap K = J(V)$. But $k^{-1}s^m \in J(V)S \subset J(R)$, a contradiction and hence $S = R$.

3. $\Rightarrow$ 1. This follows from Theorem 1.6.8. □

1.6.12 Lemma

Suppose that there is a total valuation ring of D with its center V. There is a valuation ring S of D such that $S \neq D$ and $S \supset R$ for all total valuation rings R with $Z(R) = V$.

Proof. First suppose that V has a minimal prime ideal p. Then $W = V_p$ is a valuation ring with rank one. Let $\widehat{K}$ be the completion of K with respect to the valuation defined by W, i.e. with respect to the topology defined by ideals of W and we denote by $\widehat{W}$ the corresponding completion of W. Then $\widehat{D} = D \otimes_K \widehat{K}$ is a skewfield with $Z(\widehat{D}) = \widehat{K}$. There is a valuation ring $\widehat{S}$ of $\widehat{D}$ with $Z(\widehat{S}) = \widehat{W}$ by Corollary 1.6.10, because any complete rank one valuation ring is Henselian (cf. [22, (17.18)]). It follows that $S = \widehat{S} \cap D$ is a valuation ring of D with $Z(S) = W$. Now let R be any total valuation ring of D with $Z(R) = V$. Then R_p is a total valuation ring of D with $Z(R_p) = W$ and $R_p = S$ by Lemma 1.6.11. Hence $S \supset R$ for any total valuation ring R of D with $Z(R) = V$. Next suppose that V has no minimal prime ideals. For a total valuation ring R_0 with $Z(R_0) = V$, there is a Dubrovin valuation ring S which is a finitely generated over its center $W \supset V$ such that $S \supset R_0$ by Lemma 1.6.7. Thus S is a valuation ring of D by Lemma 1.6.11, because any overring of a total valuation ring is again total. Hence, as in the first case, $S \supset R$ for all total valuation ring R of D with $Z(R) = V$.

1.6.13 Lemma

Suppose that there are, at least two, total valuation rings of D with center V. Let S be the minimal valuation ring of D with property: $S \supset R$ for all total valuation rings R of D with $Z(R) = V$ (the existence S is guaranteed by Lemma 1.6.12). Then $Z(\overline{S}) \supsetneqq \overline{W}$, where $W = Z(S)$ and $\overline{S} = S/J(S)$.

Proof. Assume that $Z(\overline{S}) = \overline{W}$. Then $\{\widetilde{R} = R/J(S), R$ runs over all total valuation rings R of D with $Z(R) = V\}$ is the set of all total valuation rings of $\overline{S}$ with $Z(\widetilde{R}) = V/J(W) = \widetilde{V}$ by Theorem 1.4.9 and Proposition 1.4.17. By Lemma 1.6.12 there is a valuation ring $\widetilde{S_1} \neq \overline{S}$ of $\overline{S}$ containing all $\widetilde{R}$. Thus $S_1 = \{s \in S, s + J(S) \in \widetilde{S_1}\}$ is a valuation ring of D with $S \supsetneqq S_1 \supset R$ for all R, contradicting the minimality of S. □

1.6.14 Proposition

Let R be a valuation ring of D with $Z(R) = V$. Then $Z(\overline{R})$ is a normal extension of $\overline{V}$ and any $\overline{V}$-automorphism of $Z(\overline{R})$ is induced by an inner automorphism i_d of D, where $d \in D$, i.e. $i_d(a) = dad^{-1}$ for any $a \in D$.

Proof. Let $\overline{r} \in Z(\overline{R})$, where $r \in R$ and $f(X)$ be the monic minimal polynomial of r in $K[X]$. Since r is integral over V, $f(X) \in V[X]$. Write $f(X) = (X - a_1)\cdots(X - a_t) \in D[X]$ with $r = a_1$ and $a_i = d_i r d_i^{-1} = i_{d_i}(r)$ for some $d_i \in D$, $i = 2, \cdots, t$. It follows that $a_i \in R$ for all i since R is invariant. Note that, for any $d \in U(D)$, i_d naturally induces an automorphism $\overline{i_d}$ of $\overline{R}$, because $i_d(J(R)) = J(R)$. Thus all elements $\overline{a}_i$ are in $Z(\overline{R})$ and hence $Z(\overline{R})$ is a normal extension of $\overline{V}$, because $\overline{f(X)} = (X - \overline{a}_1)\cdots(X - \overline{a}_t) \in \overline{V}[X]$. Let C be the maximal separable extension of $\overline{V}$ in $Z(\overline{R})$. We can choose $r \in R$ above such that $C = \overline{V}(\overline{r})$, because C is a finite separable extension. Let $g(X)$ be the irreducible separable polynomial of $\overline{r}$ over $\overline{V}$. Then $g(X) = (X - \overline{a_{k_1}})\cdots(X - \overline{a_{k_m}})$ for certain $a_{k_1}, \cdots, a_{k_m}$ in $\{a_1, \cdots, a_t\}$ above. We reorder the $\overline{a_i}$ if necessary and write $\overline{g(x)} = (x - \overline{a_1})\cdots(x - \overline{a_m})$. Each of the $\overline{a_i}$, $i = 1, \cdots, m$ is in C and this shows that C is a Galois extension of $\overline{V}$ with Galois group $\overline{G} = \{\overline{i_{d_j}}, j = 1, \cdots, m\}$. Since $\mathrm{Gal}(Z(\overline{R})/\overline{V}) = \mathrm{Gal}(C/\overline{V})$(cf. [82, Corollary 4, p. 75]), any element in $\mathrm{Gal}(Z(\overline{R})/\overline{V})$ is induced by an inner automorphism of D. □

1.6.15 Proposition

Suppose that there is a total valuation ring of a skewfield D with center V. Then any Dubrovin valuation ring of D with center V is total and any two total valuation rings are conjugate in D.

Proof. We will prove the proposition by induction on $n = [D : K]$. In the case $n = 1$, there is nothing to do and if R is the unique total valuation ring of D, then it is a valuation ring of D by Theorem 1.6.8 and is the unique Dubrovin valuation ring of D by Lemma 1.6.11. Thus we may assume that there are at least two total valuation rings of R, R_1 of D with $Z(R) = V = Z(R_1)$. Let R' be any Dubrovin valuation ring of D with $Z(R') = V$, S be the minimal valuation ring of D with the property in Lemma 1.6.13 and $W = Z(S)$. Then $Z(\overline{S}) \supsetneqq \overline{W}$ by Lemma 1.6.13 and $R_p{}'$ is a Dubrovin valuation ring of D with $Z(R'_p) = W = V_p$, where $p = J(W)$. Hence $S = R'_p \supset R'$ follows by Lemma 1.6.11. Put $\widetilde{R} = R/J(S)$ and $\widetilde{R}_1 = R_1/J(S)$. Since $Z(\overline{S})$ is a normal extension of $\overline{W}$, there are $\sigma, \tau \in \mathrm{Gal}(Z(\overline{S})/\overline{W})$ with $\sigma(Z(\widetilde{R_1})) = Z(\widetilde{R})$ and $\tau(Z(\widetilde{R'})) = Z(\widetilde{R})$ (cf. [22, (14.1)]). By Proposition 1.6.14, σ and τ are induced by inner automorphisms i_d, i_e of D such that $i_d(S) = S = i_e(S)$, where $d, e \in U(D)$. Thus we may assume that $Z(\widetilde{R_1}) = Z(\widetilde{R}) = Z(\widetilde{R'})$ and since $[D : K] \gneqq [\overline{S} : Z(\overline{S})]$, by induction hypothesis,

$\widetilde{R}'$ is a total valuation ring and so is R', which proves the first statement. Again, by induction, $\widetilde{R}$ and $\widetilde{R_1}$ are conjugate in $\overline{S}$. Hence R and R_1 are conjugate. □

1.6.16 Lemma

If rank $V = 1$, then there is a Dubrovin valuation ring R of Q such that $R \cap K = V$ and it is integral over V.

Proof. Put $Q = M_n(D)$, where D is a skewfield. As in Lemma 1.6.12, let $\widehat{K}$ be the completion of K with respect to the valuation defined by V, and $\widehat{V}$ be the valuation ring of $\widehat{K}$ with $\widehat{V} \cap K$=V, corresponding completion of V. Furthermore $\widehat{D} = D \otimes_K \widehat{K}$ is a skewfield with $Z(\widehat{D}) = \widehat{K}$ and $[D : K] = [\widehat{D} : \widehat{K}]$, and $\widehat{Q} = Q \otimes_K \widehat{K} = M_n(\widehat{D})$. By Corollary 1.6.10, there is a valuation ring $\widehat{S}$ of $\widehat{D}$ with $\widehat{S} \cap \widehat{K} = \widehat{V}$. Note that every complete valuation ring is Henselian. It follows that $\widehat{R} = M_n(\widehat{S})$ is a Dubrovin valuation ring of $\widehat{Q}$ with $Z(\widehat{R}) = \widehat{V}$. We denote $\widehat{R} \cap Q$ by R and $J(\widehat{R}) \cap Q$ by M, and we will prove that R is a Dubrovin valuation ring of Q with $Z(R) = V$. To claim $\widehat{R} = R + J(\widehat{R})$ first, let $\widehat{r} \in \widehat{R}$ and $\widehat{r} = q_1k_1 + \cdots + q_lk_l$ for some $q_i \in Q$ and $k_i \in \widehat{K}$ (we identify $Q \otimes \widehat{K} = Q \cdot \widehat{K}$). There is $0 \neq a \in \widehat{V}$ with $q_ia \in J(\widehat{R})$ since $\widehat{K} \cdot \widehat{R} = \widehat{Q}$ and $J(\widehat{V})\widehat{R} \subset J(\widehat{R})$. Since K is dense in $\widehat{K}$, there is an $x_i \in K$ with $\widehat{v}(k_i - x_i) \geq \widehat{v}(a)$ for each i, where $\widehat{v}$ is the valuation corresponding $\widehat{V}$. We obtain $\widehat{r} = \sum q_ix_i + \sum q_i(k_i - x_i)$, where $\sum q_i(k_i - x_i) \in J(\widehat{R})$, $\sum q_ix_i \in Q$ and so in R, which proves the claim. It follows that

$$\frac{R}{M} = \frac{Q \cap \widehat{R}}{Q \cap J(\widehat{R})} \cong \frac{(Q \cap \widehat{R}) + J(\widehat{R})}{J(\widehat{R})} = \frac{\widehat{R}}{J(\widehat{R})},$$

which is simple Artinian. Let $q \in Q - R$ and so $q \in \widehat{Q} - \widehat{R}$. Thus there is an $\widehat{r} \in \widehat{R}$ with $q\widehat{r} \in \widehat{R} - J(\widehat{R})$. As before, $\widehat{r} = \sum q_ik_i$, where $q_i \in Q$ and $k_i \in \widehat{K}$. Then there is $0 \neq a \in \widehat{V}$ with $qq_ia, q_ia \in J(\widehat{R})$ for all i. Again $\widehat{v}(k_i - x_i) \geq \widehat{v}(a)$ for some $x_i \in K$ and $\widehat{r} = \sum q_ix_i + \sum q_i(k_i - x_i)$, where the second sum is in $J(\widehat{R})$ and the first sum is in $Q \cap \widehat{R} = R$. Hence $q(\sum q_ix_i) = q\widehat{r} - \sum qq_i(k_i - x_i) \in (Q \cap \widehat{R}) - (Q \cap J(\widehat{R}))$, since the last sum is in $J(\widehat{R})$. Hence R is a Dubrovin valuation ring of Q with $R \cap K = V$. $\widehat{R} = M_n(\widehat{S})$ is integral over $\widehat{V}$ since so is $\widehat{S}$ by Corollary 1.6.10 and hence R is integral over V. □

1.6.17 Lemma

There is a Dubrovin valuation ring of Q whose center contains V.

Proof. If V has a minimal prime ideal p, then $W = V_p$ is a valuation ring of K with rank one. Hence there is a Dubrovin valuation ring R of Q with $Z(R) = W$. If V has no minimal prime ideals, then the lemma follows from Lemma 1.6.7. □

1.6.18 Lemma

Let R_i be total valuation rings of D with $Z(R_i) = V$ and $R_1 \neq R_2$, $i = 1, 2$. Then $[\overline{R_1} : \overline{V}] + [\overline{R_2} : \overline{V}] \leq [D : K]$.

Proof. Since $R_1 \not\subseteq R_2$, there is an $x_1 \in R_1 - R_2$. We may assume that $x_1 \notin J(R_1)$, otherwise replace x_1 by $1 + x_1$. The element $y_1 = x_1^{-1} \in (R_1 - J(R_1)) \cap J(R_2)$. Let $\overline{a_1}, \cdots, \overline{a_r}$ be a $\overline{V}$-basis of $\overline{R_1}$, where $a_i \in R_1$ and we may assume that $a_1 R_2 \supset a_i R_2$ for all i, $i = 2, \cdots, r$. After replacing a_i by $y_1 a_1^{-1} a_i$, we may assume that the a_i are all in $J(R_2)$. Similarly we can find a $\overline{V}$-basis $\overline{b_j}$ of $\overline{R_2}$ with $b_j \in R_2 \cap J(R_1)$, $j = 1, \cdots, s$. If the set $\{a_1, \cdots, a_r, b_1, \cdots, b_s\}$ is linearly dependent over K, there is an equation $\sum a_i v_i + \sum b_j v_j{}' = 0$ for some $v_i, v_j{}' \in V$, not all in $J(V)$. This leads to a contradiction. □

Total valuation rings of D whose center is V do not necessarily exist (cf. [68]). However if we extend the category of non-commutative valuation rings to Dubrovin valuation rings, then we have the following existence theorem.

1.6.19 Theorem (the Existence Theorem)

Let Q be a simple Artinian ring with finite dimension over its center K and V be a valuation ring of K. Then there is a Dubrovin valuation ring of Q such that the center is V.

Proof. We will prove the theorem by induction on $n = [Q : K]$. We may assume that $n > 1$ and let $\mathcal{R}$ be the set of Dubrovin valuation rings R of Q with $R \neq Q$ and $R \cap K \supset V$, which is non-empty by Lemma 1.6.17. We first assume that there is an $S \in \mathcal{R}$ with $\overline{S} \cong M_m(D_0)$, D_0 is a skewfield and $m \gneqq 1$. It follows from Lemma 1.6.1 that $[D_0 : Z(D_0)] \leq [D_0 : \overline{W}] \lneqq [\overline{S} : \overline{W}] \leq n$, where $W = S \cap K$ and that $\widetilde{V} = V/J(W)$ is a valuation ring of $\overline{W}$. Let $\widetilde{V_0}$ be an extension of $\widetilde{V}$ to $Z(D_0)$. By induction, there is a Dubrovin valuation ring $\widetilde{R_0}$ of D_0 with $\widetilde{R_0} \cap Z(D_0) = \widetilde{V_0}$ and so $\widetilde{R_0} \cap \overline{W} = \widetilde{V}$. Thus $M_m(\widetilde{R_0})$ is a Dubrovin valuation ring of $\overline{S}$ with $M_m(\widetilde{R_0}) \cap \overline{W} = \widetilde{V}$. It follows from Proposition 1.4.17 that $R = \{r \in S, r + J(S) \in M_m(\widetilde{R_0})\}$ is a Dubrovin valuation ring of Q with $R \cap K = V$. Next we must be the case where $J(S)$ is completely prime of S for every $S \in \mathcal{R}$, i.e. $\overline{S}$ is a skewfield. It follows from Theorem 1.4.7 that S is a 1-chain ring of Q and so is Q. Thus $Q = D$ is a skewfield and S is a total valuation ring. If there is an $S \in \mathcal{R}$ with $[\overline{S} : \overline{W}] \lneqq n$, where $W = Z(S)$, then we can use the induction to obtain a Dubrovin valuation ring $\widetilde{R}$ of $\overline{S}$ with $\widetilde{R} \cap \overline{W} = \widetilde{V}$ and there is a Dubrovin valuation ring R of Q with $R \cap K = V$ as before. We are left the case that all R in $\mathcal{R}$ are total valuation rings and $[\overline{R} : \overline{W}] = n$, where $W = Z(R)$. In this case, we will prove below that either $R_1 \subset R_2$ or $R_2 \subset R_1$ for any $R_1, R_2 \in \mathcal{R}$. Assuming this fact, let R_0 be the intersection of all $R_i \in \mathcal{R}$. Then R_0 is the minimal element in $\mathcal{R}$. If $V_0 = Z(R_0) \supsetneqq V$, then $\overline{R_0} \supset \overline{V_0} \supsetneqq \widetilde{V}$ and there is a Dubrovin valuation ring $\widetilde{R}$ of $\overline{R_0}$ with $Z(\widetilde{R}) \supset \widetilde{V}$ by Lemma 1.6.17. Then $R = \{r \in R_0, r + J(R_0) \in \widetilde{R}\}$ is a Dubrovin valuation ring of Q with $R_0 \supsetneqq R$ and $R \cap K \supset V$, a contradiction. Hence $Z(R_0) = V$. To prove the above claim, let $R_1, R_2 \in \mathcal{R}$ and we may assume that $V_1 = Z(R_1) \subset V_2 = Z(R_2)$, because these rings are overrings of V. Let $S_1 = R_1 V_2$ be an element in $\mathcal{R}$ with $Z(S_1) = V_2$. If $S_1 \neq R_2$, then, by Lemma 1.6.18 and the assumption, we have the following contradiction: $n \geqslant [\overline{S_1} : \overline{V_2}] + [\overline{R_2} : \overline{V_2}] = 2n$. Hence $R_2 = S_1 \supset R_1$. □

1.6.20 Lemma

Let R and R_1 be any Dubrovin valuation rings of Q with $Z(R) = V = Z(R_1)$. If either rank $V = 1$ or $R = a_1 V \oplus \cdots \oplus a_n V$ for some $a_i \in R$, then R and R_1 are conjugate in Q. Furthermore, in the case rank $V = 1$, $Z(\overline{R})$ is a normal extension of $\overline{V}$ and any element in $\mathrm{Gal}(Z(\overline{R})/\overline{V})$ is induced by an inner automorphism of Q.

Proof. Suppose that rank $V = 1$. Let $Q = M_n(D)$, where D is a skewfield with $Z(D) = K$. It follows from the proof of lemma 1.6.12 that there is a valuation ring B of D with $Z(B) = V$. Hence R and R_1 are conjugate in Q by Lemmas 1.4.12 and 1.6.11. In order to prove the second statement, we may assume that $R = M_n(B)$. Then $Z(\overline{R}) = Z(\overline{B})$, $Z(\overline{B})$ is a normal extension of $\overline{V}$ and any element of $\mathrm{Gal}(Z(\overline{R})/\overline{V})$ is induced by an inner automorphism of D by Proposition 1.6.14. Suppose that $R = a_1 V \oplus \cdots \oplus a_n V$ for some $a_i \in R$. There is a $q \in Q$ with $a_1 R_1 + \cdots + a_n R_1 = q R_1$. Since $R \subset a_1 R_1 + \cdots + a_n R_1$, it is clear that $q \in U(Q)$. It remains to prove that $R = q R_1 q^{-1}$. Let $b_1, \cdots, b_n \in R_1$ with $q = a_1 b_1 + \cdots + a_n b_n$, i.e. $1 = a_1 b_1 q^{-1} + \cdots + a_n b_n q^{-1}$. Since $a_i a_j \in R \subset a_1 R_1 + \ldots + a_n R_1$ we have $a_i a_j b_j q^{-1} \in (a_1 R_1 + \cdots + a_n R_1) q^{-1} = q R_1 q^{-1}$. Thus $a_i = a_i a_1 b_1 q^{-1} + \cdots + a_i a_n b_n q^{-1} \in q R_1 q^{-1}$ and $R \subset q R_1 q^{-1}$. $Z(R) = V = Z(R_1) = Z(q R_1 q^{-1})$ implies that $R = q R_1 q^{-1}$ by Proposition 1.6.3. □

1.6.21 Lemma

Let R and R_1 be Dubrovin valuation rings of a skewfield D with $Z(D) = K$ and $Z(R) = V = Z(R_1)$. If P and P_1 are the maximal completely prime ideals of R and R_1 respectively, then $P \cap K = P_1 \cap K$ and $P_1 = q P q^{-1}$ for some $q \in U(D)$.

Proof. Without restriction, we may assume that $p_1 = P_1 \cap K \subset p = P \cap K$. The completely primeness of P implies that R_p is a total valuation ring, because R_p/P_p is a skewfield. Thus R_{1p} is also a total valuation ring by Proposition 1.6.15. Hence $J(R_{1p}) = P_{1p}$ is completely prime. Since $R_{1p_1} \supset R_{1p}$, $P_1 = J(R_{1p_1}) \subset J(R_{1p})$ and, by the choice of P_1, $P_1 = J(R_{1p})$ follows. Thus $P_1 \cap K = J(R_{1p}) \cap K = p = J(R_p) \cap K = P \cap K$. By Proposition 1.6.15, $R_{1p} = qR_pq^{-1}$ for some $q \in U(D)$ and $P_1 = qPq^{-1}$ follows. □

1.6.22 Theorem (the Conjugate theorem)

Let R and R_1 be Dubrovin valuation rings of a simple Artinian ring Q with finite dimension over its center K and $Z(R) = V = Z(R_1)$. Then there is a $q \in U(Q)$ such that $R = qR_1q^{-1}$.

Proof. We will prove the theorem by induction on $n = [Q : K]$ and $n > 1$.

Case 1. $Q = M_l(D)$ for a skewfield D with $Z(D) = K$ and $l > 1$. There are Dubrovin valuation rings B, B_1 of D with $Z(B) = V = Z(B_1)$, $r, s \in U(D)$ such that $rRr^{-1} = M_l(B)$ and $sRs^{-1} = M_l(B_1)$ by Lemma 1.4.13. By induction hypothesis, B and B_1 are conjugate in D and so are R and R_1.

Case 2. Q is a skewfield and (0) is the only compltely prime ideal of R. By Lemma 1.6.21, (0) is the only completely prime ideal of R_1.

(i) If V has a minimal prime ideal p, then R_p and R_{1p} are Dubrovin valuation rings of Q with $Z(R_p) = V_p = Z(R_{1p})$, rank one. By Lemma 1.6.20, R_p and R_{1p} are conjugate. Thus we may assume that $R_p = R_{1p}$, and $\widetilde{R} = R/J(R_p)$ and $\widetilde{R_1} = R_1/J(R_p)$ are Dubrovin valuation rings of $\overline{R_p}$ with $\widetilde{R} \cap \overline{V_p} = \widetilde{R_1} \cap \overline{V_p}$. By Lemma 1.6.20, we may even assume that $\widetilde{R} \cap Z(\overline{R_p}) = \widetilde{R_1} \cap Z(\overline{R_{1p}})$. Since $J(R_p)$ is not completely prime, $\overline{R_p}$ is not a skewfield. Thus $\widetilde{R}$ and $\widetilde{R_1}$ are conjugate by case 1. So it follows that R and R_1 are conjugate.

(ii) If V has no minimal prime ideals, then there is a valuation ring W of K, $W \neq K$ containing V and K-basis $\{a_1, \cdots, a_n\}$ of Q such that $S = a_1W \oplus \cdots \oplus a_nW$ is a Dubrovin valuation ring of Q with $Z(S) = W$ and $S \supset R$ by Lemma 1.6.7. It follows from Lemma 1.6.20 that S and $S_1 = R_1W$ are conjugate. Since $\overline{S}$ is not a skewfield and $Z(\overline{S}) = \overline{W}$, we have R and R_1 are conjugate in Q which is proved in a similar way in (i).

Case 3. Q is a skewfield, and the maximal completely prime ideal P (resp. P_1) of R (resp. R_1) is non-zero. By Lemma 1.6.21 and its proof, we may assume that $P = P_1$ and R_P is a total valuation ring of Q containing R and R_1. Let S be a minimal valuation ring of Q containing R_P and $W = Z(S)$. Then $\overline{W} \cap \widetilde{R} = \widetilde{V} = \overline{W} \cap \widetilde{R_1}$ and, by Proposition 1.6.14, we may assume that $\widetilde{R} \cap Z(\overline{S}) = \widetilde{R_1} \cap Z(\overline{S})$. If R_P is not a valuation ring, then $[Q : K] \geq [\overline{S} : \overline{W}] \gneqq [\overline{S} : Z(\overline{S})]$

by Lemma 1.6.13. Hence $\widetilde{R}$ and $\widetilde{R_1}$ are conjugate by induction. If $S = R_P$, then, by case 2, $\widetilde{R}$ and $\widetilde{R_1}$ are conjugate since $\widetilde{(0)}$ is the only completely prime ideal of $\widetilde{R}$. Hence, in both cases, R is conjugate to R_1. □

1.7 Gauss Extensions

Throughout this section, V is a total valuation ring of a skewfield K, Γ is a group with a pure cone P, i.e. P is a subsemi-group of Γ such that $P \cup P^{-1} = \Gamma$ and $P \cap P^{-1} = \{e\}$, where $P^{-1} = \{\gamma^{-1}, \gamma \in P\}$ and e is the identity element. We define a left order $\leq_l$ on Γ as follows; for any $\gamma, \delta \in \Gamma$, $\gamma \leq_l \delta$ if and only if $\gamma^{-1}\delta \in P$ and similarly we define a right order. Note that Γ is a left and right totally ordered group (the simplest example is: $\Gamma = \mathbb{Z}$ and $P = \mathbb{N}_0$ in which the operation is additive).

Let $K * \Gamma$ be a **crossed product** of Γ over K, i.e. it is a ring with a left K-basis $\{\overline{\gamma}, \gamma \in \Gamma\}$, a copy of Γ. Thus each element of $K * \Gamma$ is uniquely a finite sum $\Sigma_{\gamma\in\Gamma} a_\gamma \overline{\gamma}$ with $a_\gamma \in K$. The multiplication is determined by the following; for any $\gamma, \delta \in \Gamma$, $\overline{\gamma}\overline{\delta} = c(\gamma, \delta)\overline{\gamma\delta}$, where $c : \Gamma \times \Gamma \longrightarrow U(K)$ and $\overline{\gamma}a = a^{\sigma(\gamma)}\overline{\gamma}$ for any $a \in K$, where $\sigma : \Gamma \longrightarrow \mathrm{Aut}(K)$, a mapping. The mappings σ and c have to satisfy the following two conditions:

$$c(\gamma, \delta)c(\gamma\delta, \eta) = c(\delta, \eta)^{\sigma(\gamma)}c(\gamma, \delta\eta)$$

for any $\gamma, \delta, \eta \in \Gamma$, since $(\overline{\gamma}\overline{\delta})\overline{\eta} = \overline{\gamma}(\overline{\delta}\overline{\eta})$, and

$$c(\gamma, \delta)a^{\sigma(\gamma\delta)} = (a^{\sigma(\delta)})^{\sigma(\gamma)}c(\gamma, \delta)$$

for any $\gamma, \delta \in \Gamma$ and $a \in K$, since $(\overline{\gamma}\overline{\delta})a = \overline{\gamma}(\overline{\delta}a)$.

The associativity of the multiplication in $K * \Gamma$ follows from the conditions above, and we refer the reader to the books [56] for some elementary properties of crossed products. It is easy to see that $K * \Gamma$ is a domain, because Γ is a left and right ordered group and that $K * \Gamma$ is a graded division ring, i.e. every non-zero homogeneous element is invertible (see [52, p. 38]).

We assume in this section that $D = Q(K * \Gamma)$, a *skewfield of left quotients* of $K * \Gamma$ exists, or equivalently that $K * \Gamma$ ia a left Ore domain.

If Γ is an infinite cyclic group, then $D = Q(K * \Gamma)$ exists and see [8] for such examples. Furthermore, in order to study graded subrings of $K * \Gamma$ lying over V, we consider the following condition:

$$c(\gamma, \delta) \in U(V^{\sigma(\gamma\delta)}) \text{ for any } \gamma, \delta \in \Gamma.$$

If $K * \Gamma$ satisfies the condition, then it is called a V**-crossed product** of Γ over K.

In this section, we always assume that $K * \Gamma$ is a V-*crossed product* and that c is **normarized**, i.e. $c(\gamma, e) = 1 = c(e, \gamma)$ for any $\gamma \in \Gamma$ so that $\overline{e}$ is the identity of $K * \Gamma$; (see [57, Exercise 1, p. 255] or [56, Exercise 2, p. 9]) for normalizing factor sets, and note that $K * \Gamma$ is still a V-crossed product after normalizing if $K * \Gamma$ is a V-crossed product).

A graded subring $A = \oplus_{\gamma \in \Gamma} A_\gamma \overline{\gamma}$ of $K * \Gamma$ is called a **graded total valuation** ring of $K * G$ if for any non-zero homogeneous element $a\overline{\gamma}$, either $a\overline{\gamma} \in A$ or $(a\overline{\gamma})^{-1} \in A$. A graded total valuation ring $A = \oplus_{\gamma \in \Gamma} A_\gamma \overline{\gamma}$ is said to be a **graded extension** of V in $K * \Gamma$ if $A_e = V$.

A **Gauss extension** R of V in D is a total valuation ring of D with $R \cap K = V$ and the property that for any $\alpha = a_1\overline{\gamma}_1 + \cdots + a_n\overline{\gamma}_n \in K * \Gamma$,

$$R\alpha = Ra_i\overline{\gamma}_i$$

for some i with $Ra_i\overline{\gamma}_i \supset Ra_j\overline{\gamma}_j$ for any j, $1 \leq j \leq n$.

This concept is motivated by commutative valuation rings of function fields. Let v be a valuation on a field F with value group Γ_0 and Γ_1 be a totally ordered abelian group containing Γ_0. For a fixed $\gamma \in \Gamma_1$ and any $f(X) = a_nX^n + \cdots + a_0 \in F[X]$, the mapping $v' : F[X] \longrightarrow \Gamma_1$ defined by $v'(f(X)) = \inf\{v(a_i) + i\gamma\}$ determines a valuation of the function field $F(X)$ which is an extension of v (cf. [25, (18.4)]). $O_{v'}$, the valuation ring of $F(X)$ associated to v' is called a Gauss extension of O_v in $F(X)$. Note that $O_{v'}f(X) = O_{v'}a_iX^i$ for some i with $O_{v'}a_iX^i \supset O_{v'}a_jX^j$ for all $j, 0 \leq j \leq n$.

We will use graded extensions of V in $K * \Gamma$ to determine all Gauss extensions of V in $Q(K * \Gamma)$ and study the ideal theory of Gauss extensions and residue skewfields. Furthermore we will provide some standard Gauss extensions.

The following lemma is easy and is frequently used in this section without reference.

1.7.1 Lemma

Let $A = \oplus_{\gamma \in \Gamma} A_\gamma \overline{\gamma}$ be a graded subring of $K * \Gamma$ with $A_e = V$. Then A_γ is a left V and right $V^{\sigma(\gamma)}$-submodule of K.

For any non-zero homogeneous elements $a\overline{\gamma}$ and $b\overline{\delta}$, where $\gamma, \delta \in \Gamma$ and $a, b \in K$, it follows that

$$(a\overline{\gamma})^{-1} = \overline{\gamma}^{-1}a^{-1} = c(\gamma^{-1}, \gamma)^{-1}\overline{\gamma^{-1}}a^{-1} = c(\gamma^{-1}, \gamma)^{-1}(a^{-1})^{\sigma(\gamma^{-1})}\overline{\gamma^{-1}}$$

and

$$(a\overline{\gamma})(b\overline{\delta}) = ab^{\sigma(\gamma)}c(\gamma, \delta)\overline{\gamma\delta}$$

Thus, the set of non-zero homogeneous elements of $K * \Gamma$ is a group $\mathbb{P}$ under multiplication. We write $\mathbb{P}(A) = \mathbb{P} \cap A$, the set of non-zero homogeneous elements of $A = \oplus_{\gamma \in \Gamma} A_\gamma \overline{\gamma}$.

1.7.2 Lemma

Let $A = \oplus_{\gamma \in \Gamma} A_\gamma \overline{\gamma}$ be a graded subring of $K * \Gamma$ with $A_e = V$. Then the following are equivalent:

1. A is a graded extension of V in $K * \Gamma$.
2. The set of graded left ideals of A is linearly ordered by inclusion and $K * \Gamma$ is the gr-quotient ring of A, i.e. $K * \Gamma = \{\beta^{-1}\alpha, \alpha \in A \text{ and } \beta \in \mathbb{P}(A)\} = \{\alpha\beta^{-1}, \alpha \in A \text{ and } \beta \in \mathbb{P}(A)\}$.
3. For any $\gamma \in \Gamma$, $K = A_\gamma \cup (A^{-1}_{\gamma^{-1}})^{\sigma(\gamma^{-1})^{-1}}$, where $A^{-1}_{\gamma^{-1}} = \{a^{-1}, 0 \neq a \in A_{\gamma^{-1}}\}$.

Proof 1. $\Rightarrow$ 2. For any $\alpha, \beta \in \mathbb{P}$, we have either $\alpha\beta^{-1} \in A$ or $\beta\alpha^{-1} = (\alpha\beta^{-1})^{-1} \in A$ and so $\{A\alpha, \alpha \in \mathbb{P}\}$ is linearly ordered by inclusion. Let I and J be graded left ideals of A. Suppose $I \not\supseteq J$ and let $\alpha = \alpha_1 + \cdots + \alpha_n$, $\alpha_i \in J \cap \mathbb{P}$, be an element in J but not in I. We may assume that $A\alpha_1 \supset A\alpha_i$ for all i, $1 \leqslant i \leqslant n$ and hence $A\alpha \subset A\alpha_1$. For any element $\beta = \beta_1 + \cdots + \beta_l$ in I with $\beta_i \in I \cap \mathbb{P}$, we may assume that $A\beta \subset A\beta_1$ as before. Since $A\alpha_1 \supsetneq A\beta_1$, we have $\beta \in A\alpha_1 \subset J$. Hence $I \subset J$. Let $\alpha = \alpha_1 + \cdots + \alpha_m \in K * \Gamma$ with $\alpha_i \in \mathbb{P}$. By the right version of the before, we may assume that $\alpha_1 A \supset \alpha_i A$ for all i, $1 \leqslant i \leqslant m$, i.e. $A \ni \alpha_1^{-1}\alpha_i = \beta_i$. If $\alpha_1 \in A$, then $\alpha \in A$. If $\alpha_1 \notin A$, then $\alpha_1^{-1} \in A$ and hence $\alpha = (\alpha_1^{-1})^{-1}(\beta_1 + \cdots + \beta_m)$ with $\alpha_1^{-1} \in \mathbb{P}(A)$ and $\beta_1 + \cdots + \beta_m \in A$. Similarly we have $K * \Gamma = \{\alpha\beta^{-1}, \alpha \in A \text{ and } \beta \in \mathbb{P}(A)\}$. Hence $K * \Gamma$ is the gr-quotient ring of A.
2. $\Rightarrow$ 3. Let $\gamma \in \Gamma$ and $a \in K$. If $a\overline{\gamma} \in A$, then $a \in A_\gamma$. If $a\overline{\gamma} \notin A$, then $A \ni (a\overline{\gamma})^{-1} = \overline{\gamma}^{-1}a^{-1} = c(\gamma^{-1}, \gamma)^{-1}\overline{\gamma^{-1}}a^{-1} = c(\gamma^{-1}, \gamma)^{-1}(a^{-1})^{\sigma(\gamma^{-1})}\overline{\gamma^{-1}}$ and so $c(\gamma^{-1}, \gamma)^{-1}(a^{-1})^{\sigma(\gamma^{-1})} \in A_{\gamma^{-1}}$. Since $c(\gamma^{-1}, \gamma) \in U(V)$, we have $(a^{-1})^{\sigma(\gamma^{-1})} = b$ for some $b \in A_{\gamma^{-1}}$. Hence $a = (b^{-1})^{\sigma(\gamma^{-1})^{-1}} \in (A^{-1}_{\gamma^{-1}})^{\sigma(\gamma^{-1})^{-1}}$.
3. $\Rightarrow$ 1. Let $a\overline{\gamma} \in \mathbb{P}$. If $a \in A_\gamma$, then $a\overline{\gamma} \in A$. If $a \notin A_\gamma$, then $a = (b^{-1})^{\sigma(\gamma^{-1})^{-1}}$ for some $b \in A_{\gamma^{-1}}$ and so $(a^{-1})^{\sigma(\gamma^{-1})} = b$ follows. Thus, $(a\overline{\gamma})^{-1} = c(\gamma^{-1}, \gamma)^{-1}(a^{-1})^{\sigma(\gamma^{-1})}\overline{\gamma^{-1}} \in A_{\gamma^{-1}}\overline{\gamma^{-1}} \subset A$. Hence A is a graded extension of V in $K * \Gamma$. □

1.7.3 Lemma

1. For any $a \in K$ and $\gamma, \delta \in \Gamma$, $(a^{\sigma(\delta)})^{\sigma(\gamma)} = c(\gamma, \delta)a^{\sigma(\gamma\delta)}c(\gamma, \delta)^{-1}$.
2. Let W be a left V and right V-module in K. Then $(W^{\sigma(\delta)})^{\sigma(\gamma)} = W^{\sigma(\gamma\delta)}$ for any $\gamma, \delta \in \Gamma$. In particular, if $W^{\sigma(\gamma)} \supset W$, then $W \supset W^{\sigma(\gamma^{-1})}$.

Proof. 1. From the associativity, we have $c(\gamma, \delta)a^{\sigma(\gamma\delta)} = (a^{\sigma(\delta)})^{\sigma(\gamma)}c(\gamma, \delta)$.
2. It follows from (1) that $(W^{\sigma(\delta)})^{\sigma(\gamma)} = c(\gamma, \delta)W^{\sigma(\gamma\delta)}c(\gamma, \delta)^{-1} = W^{\sigma(\gamma\delta)}$, because $c(\gamma, \delta) \in U(V^{\sigma(\gamma\delta)})$ and $W^{\sigma(\gamma\delta)}$ is a $V^{\sigma(\gamma\delta)}$-bimodule in K. If $W^{\sigma(\gamma)} \supset W$, then $W^{\sigma(\gamma^{-1})} \subset (W^{\sigma(\gamma)})^{\sigma(\gamma^{-1})} = W$. □

1.7.4 Lemma

Let $A = \oplus_{\gamma\in\Gamma} A_\gamma\overline{\gamma}$ be a graded extension of V in $K * \Gamma$. Then:

1. Either $A_\gamma \supset V$ or $A_{\gamma^{-1}} \supset V$ for any $\gamma \in \Gamma$.
2. If $A_\gamma \supset V$, then $A_{\gamma^{-1}} \subset V$ for any $\gamma \in \Gamma$. In particular, if $A_\gamma \supsetneqq V$, then $A_{\gamma^{-1}} \subset J(V)$.
3. If $A_\gamma = V = A_{\gamma^{-1}}$, then $V^{\sigma(\gamma)} = V$.
4. $A_\gamma = K$ if and only if $A_{\gamma^{-1}} = (0)$.

Proof. 1. Since A_γ is a left V-submodule of K, we have either $A_\gamma \supset V$ or $A_\gamma \subsetneqq V$ for any $\gamma \in \Gamma$. Suppose $A_\gamma \subsetneqq V$ and $A_{\gamma^{-1}} \subsetneqq V$ for some $\gamma \in \Gamma$. Then $A_\gamma \subset J(V)$ and $A_{\gamma^{-1}} \subset J(V)$. We may assume that $A_\gamma \supset A_{\gamma^{-1}}$ and let $a \in V - A_\gamma$. Then $a^{-1} \notin A_\gamma$ and there are b, c in $A_{\gamma^{-1}}$ such that $a = (b^{-1})^{\sigma(\gamma^{-1})^{-1}}$ and $a^{-1} = (c^{-1})^{\sigma(\gamma^{-1})^{-1}}$ by Lemma 1.7.2. Thus $1 = a^{-1}a = ((cb)^{-1})^{\sigma(\gamma^{-1})^{-1}}$ and $cb = 1$, a contradiction. Hence either $A_\gamma \supset V$ or $A_{\gamma^{-1}} \supset V$ for any $\gamma \in \Gamma$.

2. It follows that $A_\gamma\overline{\gamma}A_{\gamma^{-1}}\overline{\gamma^{-1}} = A_\gamma(A_{\gamma^{-1}})^{\sigma(\gamma)}c(\gamma,\gamma^{-1}) \subset V$ and $A_{\gamma^{-1}}\overline{\gamma^{-1}}A_\gamma\overline{\gamma} = A_{\gamma^{-1}}(A_\gamma)^{\sigma(\gamma^{-1})}c(\gamma^{-1},\gamma) \subset V$. Thus we have

$$A_\gamma(A_{\gamma^{-1}})^{\sigma(\gamma)} \subset V \text{ and } A_{\gamma^{-1}}A_\gamma^{\sigma(\gamma^{-1})} \subset V. \qquad (*)$$

If $A_\gamma \supset V$, then $A_{\gamma^{-1}} \subset V$ by $(*)$ since $1 \in A_\gamma^{\sigma(\gamma^{-1})}$. Suppose that $A_\gamma \supsetneqq V$ and $A_{\gamma^{-1}} \not\subseteq J(V)$. Then $A_{\gamma^{-1}} = V$ and $A_\gamma \subset V$, a contradiction. Hence $A_{\gamma^{-1}} \subset J(V)$.

3. It follows from $(*)$ that $V^{\sigma(\gamma)} \subset V$ and $V^{\sigma(\gamma^{-1})} \subset V$. Thus $V = (V^{\sigma(\gamma^{-1})})^{\sigma(\gamma)} \subset V^{\sigma(\gamma)} \subset V$ by Lemma 1.7.3 and hence $V = V^{\sigma(\gamma)}$.

4. This follows from $(*)$ and Lemma 1.7.2. □

We will write $J_g(A)$ for the **graded Jacobson radical** of A, the intersection of all graded maximal left ideals of A. It turns out that $J_g(A)$ is the intersection of all graded maximal right ideals of A (cf. [52, (1.7.4)]). If A is a graded extension of V in $K * \Gamma$, then we have the following about the graded Jacobson radical $J_g(A)$:

1.7.5 Lemma

Let $A = \oplus_{\gamma\in\Gamma} A_\gamma\overline{\gamma}$ be a graded extension of V in $K * \Gamma$. Then

1. $J_g(A) = \{\alpha = \Sigma a_\gamma\overline{\gamma},\ a_\gamma\overline{\gamma} \notin U(A)$ for all $\gamma \in \mathrm{Supp}(\alpha)\}$, where $\mathrm{Supp}(\alpha) = \{\gamma \in \Gamma, a_\gamma \neq 0\}$.
2. $J_g(A)$ is a completely prime ideal of A with $J_g(A) \cap K = J(V)$.

Proof. 1. Let $M = \{\alpha = \Sigma a_\gamma\overline{\gamma} \in A,\ a_\gamma\overline{\gamma} \notin U(A)$ for all $\gamma \in \mathrm{Supp}(\alpha)\}$. Let $a_\gamma\overline{\gamma}, a_\delta\overline{\delta} \in \mathbb{P}(A) \cap M$ and we may assume that $Aa_\gamma\overline{\gamma} \ni a_\delta\overline{\delta}$. Then $a_\gamma\overline{\gamma} + a_\delta\overline{\delta} \in$

$Aa_\gamma\overline{\gamma} \subsetneqq A$, which shows M is an additive subgroup of A. Let $a_\gamma\overline{\gamma} \in M$ and $a_\delta\overline{\delta} \in \mathbb{P}(A)$. Then $a_\gamma\overline{\gamma}a_\delta\overline{\delta}$ and $a_\delta\overline{\delta}a_\gamma\overline{\gamma}$ are both not units in A, because $a_\gamma\overline{\gamma} \notin U(A)$. Hence M is an ideal. Furthermore, it is easy to show from the definition of M that M is a unique graded maximal right ideal and hence $M = J_g(A)$ follows.

2. Let $\alpha = a_1\overline{\gamma}_1 + \cdots + a_n\overline{\gamma}_n$ and $\beta = b_1\overline{\delta}_1 + \cdots + b_m\overline{\delta}_m \in A - J_g(A)$. Suppose $\gamma_1 <_l \gamma_2 <_l \cdots <_l \gamma_n$ and $\delta_1 <_r \delta_2 <_r \cdots <_r \delta_m$. Since $J_g(A)$ is a graded ideal, there are a minimal index i with $a_i\overline{\gamma}_i \in U(A)$ and a minimal index j with $b_j\overline{\delta}_j \in U(A)$. We claim that $\gamma_i\delta_j$-term in the product $\alpha\beta$ is a unit in A and hence $\alpha\beta \notin J_g(A)$. The product $a_i\overline{\gamma}_i b_j\overline{\delta}_j$ is a unit in A. Since $\gamma_{i'}\delta_{j'} \neq \gamma_i\delta_j$ for $i' \geq i$, $j' \geq j$ and $(i,j) \neq (i',j')$, there are no contributions to the $\gamma_i\delta_j$-term of $\alpha\beta$ from the products $a_{i'}\overline{\gamma}_{i'}b_{j'}\overline{\delta}_{j'}$ with $i' \geq i$, $j' \geq j$ and $(i,j) \neq (i',j')$.
Since the remaining product $a_{i''}\overline{\gamma}_{i''}b_{j''}\overline{\delta}_{j''}$ with either $i'' < i$ or $j'' < j$ are all contained in $J_g(A)$, the $\gamma_i\delta_j$-term of the product $\alpha\beta$ has the form $c\overline{\gamma_i\delta_j} = a_i\overline{\gamma}_i b_j\overline{\delta}_j + \eta$ with $c \in K$ and $\eta \in J_g(A)$. Since $J_g(A)$ is an ideal and $a_i\overline{\gamma}_i b_j\overline{\delta}_j \notin J_g(A)$, it follows that $c\overline{\gamma_i\delta_j} \notin J_g(A)$, which proves our claim. Therefore $J_g(A)$ is a completely prime ideal. □

1.7.6 Theorem

Let $A = \oplus_{\gamma\in\Gamma}A_\gamma\overline{\gamma}$ be a graded extension of V in $K * G$. Then $A - J_g(A)$ is a left Ore set and $R = A_{J_g(A)}$ is a Gauss extension of V in D.

Proof. Let α and η be in A with $\eta \notin J_g(A)$. Then $\alpha\eta^{-1} = \lambda^{-1}\beta$ for some $\beta, \lambda \in K * G$, i.e. $\lambda\alpha = \beta\eta$. Write $\beta = \beta_1 + \cdots + \beta_n$ and $\lambda = \lambda_1 + \cdots + \lambda_m$, where $\beta_i, \lambda_j \in \mathbb{P}$. We may assume that $\beta_1 A \supset \beta_i A$ for any i, $1 \leq i \leq n$, and $\lambda_1 A \supset \lambda_j A$ for any j, $1 \leq j \leq m$ by Lemma 1.7.2. If $\beta_1 A \supset \lambda_1 A$, then $\beta_1^{-1}\lambda\alpha = \beta_1^{-1}\beta\eta$ and $\beta_1^{-1}\lambda$, $\beta_1^{-1}\beta \in A$ but $\beta_1^{-1}\beta \notin J_g(A)$. Thus $\beta_1^{-1}\lambda \notin J_g(A)$ since $J_g(A)$ is completely prime. If $\beta_1 A \subset \lambda_1 A$, then $(\lambda_1^{-1}\lambda)\alpha = (\lambda_1^{-1}\beta)\eta$ and $\lambda_1^{-1}\lambda$, $\lambda_1^{-1}\beta \in A$ but $\lambda_1^{-1}\lambda \notin J_g(A)$. Hence $A - J_g(A)$ is a left Ore set. Let $\alpha = \alpha_1 + \cdots + \alpha_n \in K * G$, where $\alpha_i \in \mathbb{P}$. Suppose $A\alpha_1 \supset A\alpha_i$ for all i, $1 \leq i \leq n$. Then $\alpha\alpha_1^{-1} \in A - J_g(A)$ and $A_{J_g(A)} = R = R\alpha\alpha_1^{-1}$, i.e. $R\alpha_1 = R\alpha$. Similarly we have $\alpha R = \alpha_i R$ for some i with $\alpha_i A \supset \alpha_j A$ for all j, $1 \leq j \leq n$. With these properties, we will prove that R is a total valuation ring of D. Let $\alpha, \beta \in K * \Gamma$. Then $\beta^{-1}\alpha R = \beta^{-1}\alpha_1 R$ for some $\alpha_1 \in \mathbb{P}$ and $R\alpha_1^{-1}\beta = R\eta_1$ for some $\eta_1 \in \mathbb{P}$, because $\alpha_1^{-1}\beta \in K * \Gamma$. Since A is a graded extension of V in $K * G$, it follows that either $\eta_1 \in A$ or $\eta_1^{-1} \in A$. Suppose $\eta_1^{-1} \in A$. Then $\beta^{-1}\alpha R = \beta^{-1}\alpha_1 R = (R : R\alpha_1^{-1}\beta)_r = (R : R\eta_1)_r = \eta_1^{-1}R \subset R$. Thus $\beta^{-1}\alpha \in R$ follows. Suppose $\eta_1 \in A$. Then $R\alpha^{-1}\beta = (R : \beta^{-1}\alpha R)_l = (R : \beta^{-1}\alpha_1 R)_l = R\alpha_1^{-1}\beta = R\eta_1 \subset R$ and $R \ni \alpha^{-1}\beta = (\beta^{-1}\alpha)^{-1}$. Hence R is a total valuation ring of D. it is clear that $R \cap K * \Gamma \supset A$. To prove the converse inclusion, let $\beta = \eta^{-1}\alpha \in K * \Gamma$, where $\alpha, \eta \in A$ and $\eta \notin J_g(A)$. Write $\beta = \beta_1 + \cdots + \beta_n$, where $\beta_i \in \mathbb{P}$ with $A\beta_1 \supset A\beta_i$ for all i, $1 \leq i \leq n$. Then we have $R \supset R\alpha = R\eta\beta = R\beta_1$ and $\beta_1 \in R \cap K * \Gamma$. Again write $\beta_1 = \mu^{-1}\lambda$ for some $\lambda, \mu \in A$ and $\mu \notin J_g(A)$ with $\lambda = \lambda_1 + \cdots + \lambda_l$ and $\mu = \mu_1 + \cdots + \mu_m$, where $\lambda_i, \mu_j \in \mathbb{P}(A)$ and $\mu_i \in U(A)$ for some i. Then $\mu_i\beta_1 = \lambda_j$ for some j and $\beta_1 = \mu_i^{-1}\lambda_j \in A$. Since $A\beta_1 \supset A\beta_i$ for all i, we have $\beta \in A$ and hence $R \cap K * \Gamma = A$. In particular, $R \cap K = R \cap K * \Gamma \cap K = A \cap K = V$. Therefore R is an extension of V in D, and it follows from an earlier observation that R is a Gauss extension of V in D. □

Conversely we will prove that every Gauss extension of V in D is obtained by left localization of a graded extension of V in $K * \Gamma$ at its graded Jacobson radical.

1.7.7 Theorem

Let R be a Gauss extension of V in D. Then $A = R \cap K * \Gamma$ is a graded extension of V in $K * \Gamma$ with $J_g(A) = J(R) \cap K * \Gamma$ and $R = A_{J_g(A)}$.

Proof. For any $\gamma \in \Gamma$, put $A_\gamma\overline{\gamma} = R \cap K\overline{\gamma}$. It is clear that $A \supset \oplus_{\gamma\in\Gamma} A_\gamma\overline{\gamma}$. To prove the converse inclusion, let $\alpha = \alpha_1 + \cdots + \alpha_n$ be any element in A, where $\alpha_i \in K\overline{\gamma_i}$ for some $\gamma_i \in \Gamma$. Suppose that $\alpha_1 R \supset \alpha_i R$ for any i, $1 \le i \le n$. Then $R \supseteq \alpha R = \alpha_1 R$ and $\alpha_i \in R$ for all i. Thus $\alpha_i \in K\overline{\gamma_i} \cap R = A_{\gamma_i}\overline{\gamma_i}$ which proves $A \subset \oplus_{\gamma\in G} A_\gamma\overline{\gamma}$. Hence $A = \oplus_{\gamma\in\Gamma} A_\gamma\overline{\gamma}$, a graded subring of $K * \Gamma$. Similarly, $J(R) \cap K * \Gamma$ is a graded ideal of A, i.e. $J(R) \cap K * \Gamma = \oplus_{\gamma\in\Gamma} J_\gamma\overline{\gamma}$, where $J_\gamma\overline{\gamma} = J(R) \cap K\overline{\gamma}$. To prove that $J_g(A) = J(R) \cap K * \Gamma$, it suffices to prove that $I = A$ for any graded left ideal I of A with $I \nsubseteq J(R) \cap K * \Gamma$. Let $\beta = \beta_1 + \cdots + \beta_m \in I - (J(R) \cap K * \Gamma)$, where $\beta_i \in \mathbb{P}$ and suppose $R\beta_1 \supset R\beta_i$ for any i, $1 \le i \le m$. Then $R = R\beta = R\beta_1$ and $\beta_1^{-1} \in R \cap K * \Gamma = A$, i.e. $\beta_1 \in U(A)$. Since $\beta_1 \in I$, we have $I = A$ as desired.

To prove that A is a graded extension of V in $K * \Gamma$, let $a\overline{\gamma} \in \mathbb{P}$. Then either $a\overline{\gamma} \in R$ or $(a\overline{\gamma})^{-1} \in R$ implies either $a\overline{\gamma} \in A$ or $(a\overline{\gamma})^{-1} \in A$. Furthermore $V = R \cap A = A \cap K$ shows that $A_e = V$. Hence A is a graded extension of V in $K * \Gamma$.

Since $J_g(A) = J(R) \cap K * \Gamma$, we have $A - J_g(A) \subset R - J(R) = U(R)$ and so $A_{J_g(A)} \subset R$. To prove the converse inclusion, let $\alpha = \eta^{-1}\beta \in R$. We may assume that $\beta, \eta \in A$, because $Q(K * \Gamma)$ is the quotient ring of $A_{J_g(A)}$ by Theorem 1.7.6. Write $\eta = \eta_1 + \cdots + \eta_n$, where $\eta_i \in \mathbb{P}(A)$ with $\eta_1 A \supset \eta_i A$ for any i, $1 \le i \le n$. Then $\eta_1^{-1}\eta \in A$ and $\eta_1^{-1}\eta \notin J_g(A)$. So $\eta_1^{-1}\eta \in U(A_{J_g(A)})$. Hence since $\alpha = \eta^{-1}\beta = (\eta^{-1}\eta_1)(\eta_1^{-1}\beta)$, we have $R \supset A_{J_g(A)}\alpha = A_{J_g(A)}(\eta^{-1}\eta_1)(\eta_1^{-1}\beta) = A_{J_g(A)}\eta_1^{-1}\beta \ni \eta_1^{-1}\beta$, i.e. $\eta_1^{-1}\beta \in R \cap K * G = A$. Thus $\alpha \in A_{J_g(A)}\alpha = A_{J_g(A)}\eta_1^{-1}\beta \subset A_{J_g(A)}$. Therefore $R = A_{J_g(A)}$ follows. □

Let R be a Gauss extension of V in D and let $A = R \cap K * \Gamma$. Then A is called a **graded extension of** V in $K * \Gamma$ **corresponding** to R. From Theorems 1.7.6 and 1.7.7, we have the following.

1.7.8 Corollary

There is a one-to-one correspondence between the set of all Gauss extensions of V in D and the set of all graded extensions of V in $K * \Gamma$, which is given by $A \longrightarrow \Phi(A) = A_{J_g(A)}$ and $R \longrightarrow \Psi(R) = R \cap K * \Gamma$, where A is a graded extension of V in $K * \Gamma$ and R is a Gauss extension of V in D.

Proof. It follows from Theorem 1.7.7 that $\Phi(\Psi(R)) = R$ for any Gauss extension R of V. Let A be a graded extension of V. Then $\Phi(A) = A_{J_g(A)}$ is a Gauss extension of V by Theorem 1.7.6 and $\Psi(\Phi(A)) = \widetilde{A}$ is a graded extension of V containing A. Let $\alpha \in \widetilde{A}$. To prove $\alpha \in A$, we may assume that α is a homogeneous element since $\widetilde{A}$ is a graded extension of V and that $\alpha = \eta^{-1}\beta$ for $\beta \in A$ and $\eta \in A - J_g(A)$. Then $\eta\alpha = \beta \in A$ implies $\alpha \in A$, since the least component of η is a unit in A. Hence $\Psi(\Phi(A)) = A$. □

1.7.9 Theorem

Let R be a Gauss extension of V in D and $A = \oplus_{\gamma\in\Gamma} A_\gamma\overline{\gamma}$ be the graded extension of V in $K * \Gamma$ corresponding to R. Then:

1. There is a one-to-one correspondence between the set of all left ideals of R and the set of all graded left ideals of A, which is given by $I \longrightarrow I_g = I \cap K * \Gamma$, where I is a left ideal of R and $I_g \longrightarrow RI_g$ for a graded left ideal I_g of A.
2. I is an ideal of R if and only if I_g is a graded ideal of A.
3. I is a completely prime ideal of R if and only if I_g is a graded completely prime ideal of A, i.e. a graded and completely prime ideal of A.
4. I is a prime ideal of R if and only if I_g is a graded prime ideal of A, i.e. a graded and prime ideal of A.

Proof. 1. Let I be a left ideal of R and $I_g = I \cap K * \Gamma$. It is clear that I_g is a left ideal of A. To prove that I_g is graded, let $\alpha = \alpha_1 + \cdots + \alpha_n \in I_g$, where $\alpha_i \in \mathbb{P}(A)$ with $A\alpha_1 \supset A\alpha_i$ for all i, $1 \le i \le n$. Then, since R is a Gauss extension, it follows that $R\alpha_1 = R\alpha \subset I$ so that $\alpha_1 \in I \cap K * \Gamma = I_g$, and hence $\alpha_i \in I_g$ for all i. Hence I_g is graded. To prove that $RI_g = I$, let $\alpha = \eta^{-1}\beta \in I$, where $\beta, \eta \in A$ and $\eta \notin J_g(A)$. Then $I \supset R\alpha = R\eta^{-1}\beta = R\beta$ and so $\beta \in I_g$. Hence $\alpha \in RI_g$ and $I = RI_g$ follows.
Conversely, let I_g be a graded left ideal of A. Then we will prove that $RI_g \cap K * \Gamma = I_g$. Let $\alpha = \eta^{-1}\beta \in RI_g \cap K * \Gamma$, where $\beta \in I_g$ and $\eta \in A - J_g(A)$. Then $\eta\alpha = \beta \in I_g$. Write $\beta = \beta_1 + \cdots + \beta_n$, where $\beta_i \in \mathbb{P}(A)$ and $A\beta_1 \supset A\beta_i$ for all i, $1 \le i \le n$. It follows that $R\alpha = R\eta\alpha = R\beta = R\beta_1$ and so $\alpha\beta_1^{-1} \in R \cap K * \Gamma = A$. Hence $\alpha \in A\beta_1 \subset I_g$, because I_g is graded, proving that $RI_g \cap K * \Gamma = I_g$.

2. If I is an ideal of R, then I_g is an ideal of A. Conversely suppose I_g is a graded ideal of A. In order to prove that RI_g is an ideal of R, it suffices to prove that $\alpha\eta^{-1} \in RI_g$ for any $\alpha \in I_g$ and $\eta \in A - J_g(A)$. There exist $\beta, \lambda \in A$ with $\lambda \notin J_g(A)$ such that $\alpha\eta^{-1} = \lambda^{-1}\beta$, i.e. $\lambda\alpha = \beta\eta$. Write $\beta = \beta_1 + \cdots + \beta_m$ and $\eta = \eta_1 + \cdots + \eta_n$, where $\beta_i, \eta_j \in \mathbb{P}(A)$ with $A\beta_1 \supset A\beta_i$ for all i, $1 \leq i \leq m$ and $A\eta_1 \supset A\eta_j$ for all j, $1 \leq j \leq n$. Then $RI_g \supset R\lambda\alpha = R\beta\eta = R\beta_1\eta$ and so $I_g = RI_g \cap K * \Gamma \ni \beta_1\eta$ by (1). Since I_g is graded, it follows that $\beta_1\eta_j \in I_g$ and in particular, $\beta_1\eta_1 \in I_g$ so that $\beta_1 \in I_g$ because $\eta_1 \in U(A)$ by Lemma 1.7.5. Hence $\beta \in I_g$ follows and thus $\alpha\eta^{-1} = \lambda^{-1}\beta \in RI_g$, as desired.
3. Suppose I is a completely prime ideal of R. Then I_g is completely prime, because $A/I_g \subset R/I$, a domain. Conversely, suppose that I_g is a graded completely prime ideal of A with $A \supsetneqq I_g$ and that $\eta^{-1}\alpha \notin I$, $\lambda^{-1}\beta \notin I$, where $\alpha, \beta, \eta, \lambda \in A$ and η, λ not in $J_g(A)$. Note that $\alpha \notin I_g$ and $\beta \notin I_g$ by (1). There are $\mu, \kappa \in A$ with $\mu \notin J_g(A)$ such that $\beta\eta^{-1} = \mu^{-1}\kappa$, i.e. $\mu\beta = \kappa\eta \notin I_g$, because $J_g(A) \supset I_g$ and so $\kappa \notin I_g$ follows. Thus we have $R\lambda^{-1}\beta\eta^{-1}\alpha = R\beta\eta^{-1}\alpha = R\mu^{-1}\kappa\alpha = R\kappa\alpha$ which is not contained in I, because $\kappa\alpha \notin I_g$. Hence I is a completely prime ideal of R.
4. Suppose I_g is prime. It is easy to see that I is prime by (1). Conversely, suppose that I is prime and that $\alpha A\beta \subset I_g$ with $\alpha \notin I_g$, where $\alpha, \beta \in A$. Then $R\alpha_1 A\beta = R\alpha A\beta \subset RI_g = I$ for some $\alpha_1 \in \mathbb{P}(A)$ with $R\alpha = R\alpha_1$. Since $A\alpha_1 A$ is a graded ideal of A, we have $RA\alpha_1 A$ is an ideal of R by (2). Thus $\beta \in I$, because $\alpha_1 \notin I$, $\beta \in I \cap A = I_g$ and hence I_g is prime. □

Let R be a Gauss extension of V in D. $\overline{R} = R/J(R)$ is called the **residue skew field** of R by $J(R)$. We will show that $\overline{R}$ is isomorphic to $Q(\overline{V} * E)$ for some subgroup E of Γ, where $Q(\overline{V} * E)$ is the quotient ring of a crossed product algebra $\overline{V} * E$ of E over $\overline{V} = V/J(V)$.

1.7.10 Lemma

Let $A = \oplus_{\gamma\in\Gamma} A_\gamma\overline{\gamma}$ be a graded extension of V in $K * \Gamma$ and $J_g(A) = \oplus_{\gamma\in\Gamma} J_\gamma\overline{\gamma}$ be the graded Jacobson radical of A. Then:

1. $E = \{\gamma \in \Gamma \mid A_\gamma \supsetneqq J_\gamma\}$ is a subgroup of Γ.
2. For any $\gamma \in E$ and any $c_\gamma \in A_\gamma - J_\gamma$, $A_\gamma = Vc_\gamma$ and $J_\gamma = J(V)c_\gamma$.
3. If $a\overline{\gamma}$ and $b\overline{\gamma}$ are both units in A, where $\gamma \in E$ and a, $b \in A_\gamma$, then $Va = Vb$.

Proof. 1. Let γ and δ be in E. Then there are $c_\gamma \in A_\gamma - J_\gamma$ and $c_\delta \in A_\delta - J_\delta$. Since $J_g(A)$ is completely prime, we have $c_\gamma\overline{\gamma}c_\delta\overline{\delta} = c_\gamma c_\delta^{\sigma(\gamma)}c(\gamma,\delta)\overline{\gamma\delta} \notin J_g(A)$, i.e. $c_\gamma c_\gamma^{\sigma(\gamma)}c(\gamma,\delta) \in A_{\gamma\delta} - J_{\gamma\delta}$ and $\gamma\delta \in E$ follows. Furthermore, since $c_\gamma\overline{\gamma} \in U(A)$ by Lemma 1.7.5, $(c_\gamma\overline{\gamma})^{-1} = c(\gamma^{-1},\gamma)^{-1}(c_\gamma^{-1})^{\sigma(\gamma^{-1})}\overline{\gamma^{-1}} \in A_{\gamma^{-1}}\overline{\gamma^{-1}}$, but not in $J_{\gamma^{-1}}\overline{\gamma^{-1}}$ and so $\gamma^{-1} \in E$. Hence E is a subgroup of Γ.

2. Since J_γ is a left V-module in K and $c_\gamma \notin J_\gamma$, it follows that $Vc_\gamma \supsetneqq J_\gamma$. Let $a \in A_\gamma - J_\gamma$. Then $a\overline{\gamma} \in U(A)$, i.e. $Aa\overline{\gamma} = A = Ac_\gamma\overline{\gamma}$ and so $Va = Vc_\gamma$ follows. Thus $A_\gamma = Vc_\gamma$. To prove that $J_\gamma = J(V)c_\gamma$, let $b \in J(V)$ and suppose that $bc_\gamma \notin J_\gamma$. Then $b \in U(V)$, because $c_\gamma\overline{\gamma} \in U(A)$, a contradiction. Hence $bc_\gamma \in J_\gamma$. Conversely, let $a \in J_\gamma$. Then $a = dc_\gamma$ for some $d \in V$. If $d \notin J(V)$, then $dc_\gamma\overline{\gamma} \in U(A)$, a contradiction. So $a = dc_\gamma \in J(V)c_\gamma$ and hence $J_\gamma = J(V)c_\gamma$.
3. $Aa\overline{\gamma} = A = Ab\overline{\gamma}$ implies $Aa = Ab$ and so $Va = Vb$ follows. □

1.7.11 Proposition

Let R be a Gauss extension of V in D and $A = \oplus_{\gamma\in\Gamma} A_\gamma\overline{\gamma}$ be the graded extension of V in $K * \Gamma$ corresponding to R with $J_g(A) = \oplus_{\gamma\in\Gamma} J_\gamma\overline{\gamma}$. Then $\overline{A} = A/J_g(A)$ is isomorphic to $\overline{V} * E$, a crossed product algebra of E over $\overline{V} = V/J(V)$, where $E = \{\gamma \in \Gamma \mid A_\gamma \supsetneqq J_\gamma\}$ and $\overline{R} = R/J(R) \cong Q(\overline{V} * E)$.

Proof. By Lemma 1.7.10, E is a subgroup of Γ. For $\gamma \in E$, we can write $A_\gamma = Vc_\gamma$ for a fixed $c_\gamma \in A_\gamma - J_\gamma$. Since A_γ is a right $V^{\sigma(\gamma)}$-module, it follows from the right version of Lemma 1.7.10 that $A_\gamma = c_\gamma V^{\sigma(\gamma)}$ and $J_\gamma = c_\gamma J(V)^{\sigma(\gamma)}$. Thus we have $V = c_\gamma V^{\sigma(\gamma)} c_\gamma^{-1}$ and $J(V) = c_\gamma J(V)^{\sigma(\gamma)} c_\gamma^{-1}$. Hence, for any $v \in V$, the mapping $\rho(\gamma)$ defined by

$$v^{\rho(\gamma)} = c_\gamma v^{\sigma(\gamma)} c_\gamma^{-1}$$

is an automorphism of V and $\rho(\gamma)$ naturally induces an automorphism $\overline{\rho}(\gamma)$ of $\overline{V}$ by $\overline{u}^{\overline{\rho}(\gamma)} = u^{\rho(\gamma)} + J(V)$. Furthermore, for any $\gamma, \delta \in E$, $c_\gamma\overline{\gamma}c_\delta\overline{\delta} = c_\gamma c_\delta^{\sigma(\gamma)} c(\gamma,\delta)\overline{\gamma\delta} \in U(A)$. So $Vc_{\gamma\delta} = Vc_\gamma c_\delta^{\sigma(\gamma)} c(\gamma,\delta)$ by Lemma 1.7.10, i.e. $c_\gamma c_\delta^{\sigma(\gamma)} c(\gamma,\delta) c_{\gamma\delta}^{-1} \in U(V)$. Put $\widetilde{\gamma} = c_\gamma\overline{\gamma}$ and $\widetilde{\delta} = c_\delta\overline{\delta}$. Then

$$\widetilde{\gamma}\widetilde{\delta} = c_\gamma\overline{\gamma}c_\delta\overline{\delta} = c_\gamma c_\delta^{\sigma(\gamma)} c(\gamma,\delta) c_{\gamma\delta}^{-1}\widetilde{\gamma\delta} = f(\gamma,\delta)\widetilde{\gamma\delta}$$

where $f(\gamma,\delta) = c_\gamma c_\delta^{\sigma(\gamma)} c(\gamma,\delta) c_{\gamma\delta}^{-1} \in U(V)$. For any $u \in U(V)$ and $\gamma, \delta \in E$, we have

$$\widetilde{\gamma}u = c_\gamma\overline{\gamma}u = c_\gamma u^{\sigma(\gamma)} c_\gamma^{-1} c_\gamma\overline{\gamma} = c_\gamma u^{\sigma(\gamma)} c_\gamma^{-1}\widetilde{\gamma} = u^{\rho(\gamma)}\widetilde{\gamma},$$

$$(\widetilde{\gamma}\widetilde{\delta})u = f(\gamma,\delta)\widetilde{\gamma\delta}u = f(\gamma,\delta)u^{\rho(\gamma\delta)}\widetilde{\gamma\delta},$$

$$\widetilde{\gamma}(\widetilde{\delta}u) = \widetilde{\gamma}(u^{\rho(\delta)}\widetilde{\delta}) = (u^{\rho(\delta)})^{\rho(\gamma)} f(\gamma,\delta)\widetilde{\gamma\delta}.$$

Thus $u^{\rho(\gamma\delta)} = f(\gamma,\delta)^{-1}(u^{\rho(\delta)})^{\rho(\gamma)} f(\gamma,\delta) \in U(V)$. Similarly $\widetilde{\gamma}(\widetilde{\delta}\widetilde{\eta}) = (\widetilde{\gamma}\widetilde{\delta})\widetilde{\eta}$ implies $f(\gamma,\delta)f(\gamma\delta,\eta) = f(\delta,\eta)^{\rho(\gamma)} f(\gamma,\delta\eta) \in U(V)$. Hence, for $\overline{\rho} : E \longrightarrow \mathrm{Aut}(\overline{V})$ and $\overline{f} : E \times E \longrightarrow \overline{V} - \{0\}$ with $\overline{f}(\gamma,\delta) = f(\gamma,\delta) + J(V) \in \overline{V}$, $\gamma, \delta \in E$, we have a crossed product algebra $\overline{V} * E$ with basis $\{\widetilde{\gamma},\ \gamma \in E\}$ over

$\overline{V}$, $\widetilde{\gamma}\overline{u} = \overline{u}^{\overline{\rho}(\gamma)}\widetilde{\gamma}$ and $\widetilde{\gamma}\widetilde{\delta} = \overline{f}(\gamma,\delta)\widetilde{\gamma\delta}$ for $\overline{u} \in \overline{V}$, $\gamma, \delta \in E$. Then $\overline{A} = A/J_g(A) = \oplus_{\gamma\in E} Vc_\gamma\overline{\gamma}/J(V)c_\gamma\overline{\gamma} \cong \oplus_{\gamma\in E}\overline{V}\widetilde{\gamma} = \overline{V} * E$, and $\overline{R} = R/J(R) = Q(\overline{V} * E)$, the skewfield of left quotients of $\overline{V} * E$. □

In the remainder of this section, we will give some standard Gauss extensions of V in D.

A subset H of Γ is called a **cone** of Γ if $H \cdot H \subset H$ and $\Gamma = H \cup H^{-1}$. The cone H is **proper** if $H \neq \Gamma$. We write $U(H) = H \cap H^{-1}$ for the group of units of H. We can define (one-sided) ideals, (completely) prime ideals and the Jacobson radical of a cone in the same way as in rings. Cones and valuation rings have similar properties, see for example [5]. We collect some of these properties in the following two lemmas:

1.7.12 Lemma

Let H be a proper cone of Γ. Then:

1. $J(H) = H - U(H)$ is the maximal left and the maximal right ideal of H; it is the Jacobson radical of H and a completely prime ideal of H.
2. Let $\gamma \in \Gamma$. Then $\gamma \in J(H)$ if and only if $\gamma^{-1} \in \Gamma - H$.
3. $\Gamma - H$ is a semigroup.

Proof. 1. Let $\gamma \in J(H)$ and $\delta \in H$. If $\delta\gamma \notin J(H)$, i.e. $\delta\gamma \in U(H)$, then $\gamma^{-1} = (\delta\gamma)^{-1}\delta \in H$ and $\gamma \in U(H)$ follows. Hence $\delta\gamma \in J(H)$ and so it is a left ideal. It is clear that $J(H)$ is the maximal left ideal which is the Jacobson radical of H. Similarly $J(H)$ is the maximal right ideal of H.

2. If $\gamma \in J(H)$, then $\gamma^{-1} \notin H$, i.e. $\gamma^{-1} \in \Gamma - H$. If $\gamma^{-1} \in \Gamma - H$, then $\gamma^{-1} \in H^{-1}$, i.e. $\gamma \in H$. If $\gamma \notin J(H)$, then it is in $U(H)$, a contradiction. Hence $\gamma \in J(H)$.
3. Let γ and $\delta \in \Gamma - H$. Then $(\gamma\delta)^{-1} = \delta^{-1}\gamma^{-1} \in J(H)$ by (1) and (2) and so $\gamma\delta \in \Gamma - H$ by (2) again.

1.7.13 Lemma

1. Let H be a proper cone of Γ. Then there is a one-to-one correspondence between the set of all proper cones H_0 of Γ with $H_0 \supset H$ and the set of all completely prime ideals P_0 of H, which is obtained by: $H_0 \longrightarrow J(H_0)$ and $H_0 = H \cup (H - P_0)^{-1}$.
2. Let P_0 be a completely prime ideal of H and let $H_0 = H \cup (H - P_0)^{-1}$. Then $H - P_0$ is a cone of $U(H_0)$ with $J(H - P_0) = J(H) - P_0$ and $U(H - P_0) = U(H)$.

Proof. 1. Let H_0 be a proper cone of Γ containing H and $\gamma \in J(H_0)$. Then $\gamma^{-1} \in \Gamma - H_0 \subset \Gamma - H$ and so $\gamma \in J(H)$ by Lemma 1.7.12. Hence $J(H) \supset J(H_0)$

and $H_0 = H \cup (H - J(H_0))^{-1}$. Since $J(H_0)$ is a completely prime ideal of H_0, it is a completely prime ideal of H.
Conversely, let P_0 be a completely prime ideal of H and let $H_0 = H \cup (H - P_0)^{-1}$. It is clear that $\Gamma = H_0 \cup H_0^{-1}$. To prove that H_0 is a semi-group, let $\gamma, \delta \in H_0$. It is clear that $\gamma\delta \in H_0$ if either $\gamma, \delta \in H$ or $\gamma_1 = \gamma^{-1}, \delta_1 = \delta^{-1} \in H - P_0$. We may assume that $\gamma \in H$ and $\delta \in (H - P_0)^{-1}$, i.e. $\delta = \delta_1^{-1}$ for some $\delta_1 \in H - P_0$. Suppose that $\gamma\delta_1^{-1} \notin H$. Then $\gamma\delta_1^{-1} = \eta^{-1}$ for some $\eta \in H$, $\eta\gamma = \delta_1$ and so $\eta \notin P_0$, i.e. $\eta^{-1} \in H_0$. Hence $\gamma\delta_1^{-1} \in H_0$. Therefore H_0 is a cone of Γ. To prove $J(H_0) = P_0$, let $\delta \in H - P_0$. Then $H\delta \supsetneqq P_0$ and $H \supsetneqq P_0\delta^{-1}$. Thus $P_0 = P_0\delta^{-1}\delta$ implies $P_0\delta^{-1} \subset P_0$ and P_0 is a right ideal of H_0. Similarly it is a left ideal of H_0. It is now clear that $J(H_0) = P_0$.
2. This is obvious, because $U(H_0) = H_0 - P_0$. □

1.7.14 Proposition

Let H be a proper cone of Γ with $H \supset P$. Then:

1. $A = \oplus_{\gamma \in U(H)} V\overline{\gamma} \oplus (\oplus_{\gamma \in J(H)} K\overline{\gamma})$ is a graded extension of V in $K * \Gamma$ if and only if $V^{\sigma(\gamma)} = V$ for all $\gamma \in U(H)$. The Gauss extension $R = A_{J_g(A)}$, with $J_g(A) = \oplus_{\gamma \in U(H)} J(V)\overline{\gamma} \oplus (\oplus_{\gamma \in J(H)} K\overline{\gamma})$, satisfies
   (a) $R = R\overline{\gamma}$ for any $\gamma \in U(H)$.
   (b) $R\overline{\gamma} \subsetneqq Rk$ for any $\gamma \in J(H), 0 \neq k \in K$.
2. Let S be a Gauss extension of V in D satisfies *(a)* and *(b)*. Then $S \cap K * G = A$, where $A = \oplus_{\gamma \in U(H)} V\overline{\gamma} \oplus (\oplus_{\gamma \in J(H)} K\overline{\gamma})$.

Proof. 1. If $A = \oplus_{\gamma \in U(H)} V\overline{\gamma} \oplus (\oplus_{\gamma \in J(H)} K\overline{\gamma})$ is a graded extension and $\gamma \in U(H)$, $v \in V$, then $\overline{\gamma}v = v^{\sigma(\gamma)}\overline{\gamma}$ and $V^{\sigma(\gamma)} \subset V$. Since, $V^{\sigma(\gamma^{-1})} \subset V$, it follows from Lemma 1.7.3 that $V^{\sigma(\gamma)} = V$. Conversely, if $V^{\sigma(\gamma)} = V$ for all $\gamma \in U(H)$, then A is multiplicatively closed and hence it is a graded subring of $K * \Gamma$. Therefore, by Lemma 1.7.2, A is a graded extension of V with $J_g(A) = \oplus_{\gamma \in U(H)} J(V)\overline{\gamma} \oplus (\oplus_{\gamma \in J(H)} K\overline{\gamma})$. Furthermore $R = R\overline{\gamma}$, since $\overline{\gamma} \in U(A)$ for any $\gamma \in U(H)$. Also, for any $\gamma \in J(H)$ and any $0 \neq k \in K$, we have $\overline{\gamma}k^{-1} = (k^{-1})^{\sigma(\gamma)}\overline{\gamma} \in J_g(A) \subset J(R) \subsetneqq R$ and so $R\overline{\gamma} \subsetneqq Rk$ follows.
2. Let $B = S \cap K * \Gamma = \oplus_{\gamma \in \Gamma} B_\gamma\overline{\gamma}$, a graded extension of V in $K * \Gamma$ by Theorem 1.7.7. By *(a)* $\overline{\gamma} \in B$ for any $\gamma \in U(H)$ and so $B_\gamma \supset V$. Furthermore, $B_\gamma\overline{\gamma}B_{\gamma^{-1}}\overline{\gamma^{-1}} = B_\gamma B_{\gamma^{-1}}^{\sigma(\gamma)}c(\gamma, \gamma^{-1}) \subset V$ implies $B_\gamma \subset V$, because $1 \in B_{\gamma^{-1}}^{\sigma(\gamma)}$. Hence $B_\gamma = V$ and $V^{\sigma(\gamma)} = V$ for any $\gamma \in U(H)$ by Lemma 1.7.4. For any $\gamma \in J(H)$ and $0 \neq k \in K$, it follows from *(b)* that $k\overline{\gamma^{-1}} = kc(\gamma^{-1}, \gamma)\overline{\gamma}^{-1} \notin S$. Hence $k \notin B_{\gamma^{-1}}$ and so $B_{\gamma^{-1}} = (0)$ for any $\gamma \in J(H)$. It follows from Lemma 1.7.4 that $B_\gamma = K$ and hence $B = A$ follows. □

The mapping $\sigma : \Gamma \longrightarrow \mathrm{Aut}(K)$ is called **compatible** with V if $V = V^{\sigma(\gamma)}$ for all $\gamma \in \Gamma$. If σ is compatible with V, then the V-crossed product algebra of Γ over K naturally induces a crossed product algebra $V * \Gamma$ over V.

1.7.15 Proposition

1. If σ is compatible with V, then $A = V * \Gamma$ is a graded extension of V in $K * \Gamma$ with $J_g(A) = J(V) * \Gamma$ and $\overline{\gamma} \in U(A)$ for all $\gamma \in \Gamma$. Conversely,
2. If $B = \oplus_{\gamma \in \Gamma} B_\gamma \overline{\gamma}$ is a graded extension of V in $K * \Gamma$ with $\overline{\gamma} \in U(B)$ for all $\gamma \in \Gamma$, then σ is compatible with V and $B = V * \Gamma$.

Proof. 1. This is clear from Lemmas 1.7.2, 1.7.5 and the definition of crossed product algebras.

2. Since $\overline{\gamma} a = a^{\sigma(\gamma)} \overline{\gamma}$ for any $a \in V$ and $\gamma \in \Gamma$, we have $a^{\sigma(\gamma)} = \overline{\gamma} a \overline{\gamma}^{-1} \in B \cap K = V$, which implies $V^{\sigma(\gamma)} \subset V$ for any $\gamma \in \Gamma$. Thus it follows from Lemma 1.7.3 that $V = V^{\sigma(\gamma)}$, which shows σ is compatible with V. Furthermore, $B \supset V\overline{\gamma}$ and $B \supset V\overline{\gamma^{-1}}$ imply $B_\gamma \supset V$ and $B_{\gamma^{-1}} \supset V$. So $B_\gamma = V$ for any $\gamma \in \Gamma$ by Lemma 1.7.4. Hence $B = V * \Gamma$ follows. □

In case $A = V * \Gamma$, we have a precise information about the relation between the (one-sided) ideals of $R = A_{J_g(A)}$ and the (one-sided) ideals of V. An ideal $\mathfrak{a}$ of V is called σ-invariant if $\mathfrak{a}^{\sigma(\gamma)} = \mathfrak{a}$ for any $\gamma \in \Gamma$. A σ-invariant ideal $\mathfrak{p}$ of V is called σ-prime if $\mathfrak{a}\mathfrak{b} \subset \mathfrak{p}$, where $\mathfrak{a}$ and $\mathfrak{b}$ are σ-invariant ideals of V, implies either $\mathfrak{a} \subset \mathfrak{p}$ or $\mathfrak{b} \subset \mathfrak{p}$.

1.7.16 Proposition

Suppose that σ is compatible with V. Let $A = V * \Gamma$ be the crossed product algebra of Γ over V and let $R = A_{J_g(A)}$. Then

1. There is a one-to-one correspondence between the set of all left ideals of R and the set of all left ideals of V, which is given by $I \longrightarrow I_e = I \cap K$ and $I_e \longrightarrow RI_e$, where I is a left ideal of R and I_e is a left ideal of V.
2. I is an ideal of R if and only if I_e is a σ-invariant ideal of V.
3. I is a completely prime ideal of R if and only if I_e is a σ-invariant completely prime ideal of V, i.e. a σ-invariant and completely prime ideal of V.
4. I is a prime ideal of R if and only if I_e is a σ-invariant prime ideal of V.

Proof. We recall that there is a one-to-one correspondence between the set of left ideals of R and the set of the graded left ideals of A by Proposition 1.7.9.

1. We will show that $I_g \longrightarrow I_e = I \cap K$ and $I_e \longrightarrow AI_e$ establishes a one-to-one correspondence between the set of graded left ideals I_g of A and the set of left ideals I_e of V.
   For a left ideal I_e of V, clearly AI_e is a graded left ideal of A and $AI_e \cap K = VI_e = I_e$. It remains to show that $AI_e = I_g$ for a graded left ideal I_g of A with $I_e = I_g \cap K$. By construction, $AI_e \subset I_g$. Conversely, let $\alpha = \alpha_1 + \cdots + \alpha_n$ be an element in I_g with $\alpha_i = a_i\overline{\gamma_i} \in I_g$, for $\gamma_i \in \Gamma$, $a_i \in V$, since I_g is a graded left ideal. To prove that $\alpha_i \in AI_e$, it is sufficient to find an element $b_i \in I_e$ such that $a_i\overline{\gamma_i} = \overline{\gamma_i}b_i$, i.e. $a_i = b_i^{\sigma(\gamma_i)}$. By Lemma 1.7.3, we have $b_i = b_i^{\sigma(\gamma_i^{-1}\gamma_i)} = c(\gamma_i^{-1}, \gamma_i)^{-1}(b_i^{\sigma(\gamma_i)})^{\sigma(\gamma_i^{-1})}c(\gamma_i^{-1}, \gamma_i)$. Thus, $b_i = c(\gamma_i^{-1}, \gamma_i)^{-1}(a_i)^{\sigma(\gamma_i^{-1})}c(\gamma_i^{-1}, \gamma_i)$. Since $\overline{\gamma_i^{-1}} \in A$, $\alpha_i = a_i\overline{\gamma_i} \in I_g$, $c(\gamma_i^{-1}, \gamma_i) \in U(V)$, it follows that $\overline{\gamma_i^{-1}}\alpha_i = \overline{\gamma_i^{-1}}a_i\overline{\gamma_i} = a_i^{\sigma(\gamma_i^{-1})}c(\gamma_i^{-1}, \gamma_i)$ is in $I_g \cap K = I_e$ and hence $b_i \in I_e$. Therefore, $\alpha_i = a_i\overline{\gamma_i} \in AI_e$ for all i and $I_g = AI_e$.
2. Suppose that I is an ideal of R. Let $a \in I_e$ and $\gamma \in \Gamma$. Then $\overline{\gamma}a(\overline{\gamma})^{-1} = a^{\sigma(\gamma)} \in I_e$, because $\overline{\gamma} \in U(A)$ and so I_e is a σ-invariant ideal of V by Lemma1.7.3. Conversely, let I_e be a σ-invariant ideal of V. Then for any $\gamma \in \Gamma$ and $a \in I_e$, we have $a\overline{\gamma} = \overline{\gamma}b$, where $b = a^{\sigma(\gamma^{-1})} \in I_e$ and so $a\overline{\gamma} \in AI_e$. Now, we apply Theorem 1.7.9 to conclude that $I = RI_e$ is an ideal of R since AI_e is a graded ideal of A.
3. Suppose that I is a completely prime ideal of R. Then I_e is completely prime, because $R/I \supset V/I_e$. Conversely, suppose that I_e is σ-invariant and completely prime. Then $I = RI_e$ is an ideal of R by (2). It suffices to prove that $I_g = I \cap K * \Gamma$ is completely prime by Theorem 1.7.9. Assume that $\alpha\beta \in I_g$ and $\alpha \notin I_g$, where α, $\beta \in A$. Then $R\alpha = R\alpha_1$ for some $\alpha_1 \in \mathbb{P}(A)$ and $\alpha_1\beta \in R\alpha\beta \cap K * \Gamma \subset I \cap K * G = I_g$ with $\alpha_1 \notin I_g$. Let $\alpha_1 = a\overline{\gamma}$ for some $a \in V$ and $\gamma \in \Gamma$. Then $\overline{\gamma}^{-1}\alpha_1 = \overline{\gamma}^{-1}a\overline{\gamma} \notin I_g \cap K = I_e$ and $\overline{\gamma}^{-1}\alpha_1\beta \in I_g$. Thus we may assume that $\alpha_1 \in V - I_e$ with $\alpha_1\beta \in I_g$. Write $\beta = \beta_1 + \cdots + \beta_n$, where $\beta_i = b_i\overline{\gamma_i}$ for some $b_i \in V$ and $\gamma_i \in \Gamma$. Since I_g is graded, it follows that $\alpha_1\beta_i \in I_g$ for any i, $1 \le i \le n$ and that $\alpha_1\beta_i\overline{\gamma_i}^{-1} = \alpha_1 b_i \in I_e$. Thus $b_i \in I_e$ for any i and so $\beta_i \in I_eA \subset I_g$. Hence I_g is completely prime.
4. Suppose that I_e is a σ-invariant prime ideal of V and let $I = RI_e$, an ideal of R. Assume that $BC \subset I$, where B and C are ideals of R. By (1) and (2), $B = RB_e$ and $C = RC_e$ for some σ-invariant ideals B_e and C_e of V. Thus $B_eC_e \subset I_e$ implies that either $B_e \subset I_e$ or $C_e \subset I_e$. Hence we have either $B \subset I$ or $C \subset I$ so that I is a prime ideals of R.
   Conversely suppose that I is a prime ideal of R. Then I_e is a σ-invariant ideal of V by (2). Assume that I_e is not a prime ideal of V. Then there exists an ideal $\mathfrak{a}$ of V such that $\mathfrak{a} \supsetneqq I_e \supset \mathfrak{a}^2$. Let $\mathfrak{b} = \bigcup\{\mathfrak{a}, \mathfrak{a}$ is an ideal of V with $\mathfrak{a} \supsetneqq I_e \supset \mathfrak{a}^2\}$. Since the set of ideals of V is linearly ordered by inclusion, $\mathfrak{b}$ is an ideal of V and $\mathfrak{b} \supsetneqq I_e \supset \mathfrak{b}^2$. Hence $\mathfrak{b}^{\sigma(\gamma)} \supsetneqq I_e^{\sigma(\gamma)} \supset (\mathfrak{b}^{\sigma(\gamma)})^2$ and $\mathfrak{b}^{\sigma(\gamma)} \subset \mathfrak{b}$ for all $\gamma \in \Gamma$. It follows from Lemma 1.7.3 that $\mathfrak{b}^{\sigma(\gamma)} = \mathfrak{b}$ for all $\gamma \in \Gamma$, i.e. $\mathfrak{b}$ is a σ-invariant

ideal of V. Thus $R\mathfrak{b} \supsetneq RI_e = I \supset R\mathfrak{b}^2 = R\mathfrak{b}R\mathfrak{b} = (R\mathfrak{b})^2$ by (1) and (2), a contradiction. This completes the proof of (4). □

Any graded ideal of $A = V * \Gamma$ is of the form $\mathfrak{a} * \Gamma = \oplus_{\gamma\in\Gamma}\mathfrak{a}\overline{\gamma}$ for some σ-invariant ideal $\mathfrak{a}$ of V. So we have the following result.

1.7.17 Corollary

Suppose that σ is compatible with V. Let $A = V * \Gamma$ be the crossed product algebra of Γ over V and let $\mathfrak{p}$ be a σ-invariant ideal of V. Then

1. $\mathfrak{p} * \Gamma = \oplus_{\gamma\in\Gamma}\mathfrak{p}\overline{\gamma}$ is a graded completely prime ideal of A if and only $\mathfrak{p}$ is a completely prime ideal of V.
2. $\mathfrak{p} * \Gamma$ is a graded prime ideal of A but not completely prime if and only if $\mathfrak{p}$ is a prime ideal of V but not completely prime.

Finally we provide examples of graded extensions of V in $K * \Gamma$ which are related with overrings of V and cones of Γ.

1.7.18 Proposition

Let W be an overring of V and H be a proper cone of Γ with $H \supset P$. Set $A = \oplus_{\gamma\in\Gamma-H}J(W)\overline{\gamma}\oplus(\oplus_{\gamma\in U(H)}V\overline{\gamma})\oplus(\oplus_{\gamma\in J(H)}W^{\sigma(\gamma)}\overline{\gamma})$. Then A is a graded extension of V in $K * \Gamma$ if and only if

1. $V^{\sigma(\gamma)} = V$ and $W^{\sigma(\gamma)} = W$ for any $\gamma \in U(H)$
2. $W^{\sigma(\gamma)} \supset W$ for any $\gamma \in J(H)$

Proof. Suppose that A is a graded extension of V in $K * \Gamma$. To prove the first statement (1), let $\gamma \in U(H)$ and $\delta \in J(H)$. Then $\overline{\gamma}V = V^{\sigma(\gamma)}\overline{\gamma}$ and $\overline{\gamma^{-1}}V = V^{\sigma(\gamma^{-1})}\overline{\gamma^{-1}}$ imply $V^{\sigma(\gamma)} \subset V$ and $V^{\sigma(\gamma^{-1})} \subset V$. Thus $V^{\sigma(\gamma)} = V$ follows as before. Furthermore

$$W^{\sigma(\delta)}\overline{\delta}V\overline{\gamma} = W^{\sigma(\delta)}V^{\sigma(\delta)}c(\delta,\gamma)\overline{\delta\gamma} = W^{\sigma(\delta)}c(\delta,\gamma)\overline{\delta\gamma},$$

which shows $W^{\sigma(\delta)}c(\delta,\gamma) \subset W^{\sigma(\delta\gamma)}$ since $\delta\gamma \in J(H)$ and so $W^{\sigma(\delta)} \subset W^{\sigma(\delta\gamma)} = (W^{\sigma(\gamma)})^{\sigma(\delta)}$. Thus $W \subset W^{\sigma(\gamma)}$ follows. Similarly, we have $W^{\sigma(\gamma^{-1})} \supset W$ since $\gamma \in U(H)$ and hence $W = W^{\sigma(\gamma)}$ by Lemma 1.7.3.

To prove the second statement (2), let $\gamma \in J(H)$. Then we have

$$W^{\sigma(\gamma)}\overline{\gamma}J(W)\overline{\gamma^{-2}} = J(W)^{\sigma(\gamma)}c(\gamma,\gamma^{-2})\overline{\gamma^{-1}}$$

and so $J(W)^{\sigma(\gamma)}c(\gamma,\gamma^{-2}) \subset A_{\gamma^{-1}} = J(W)$. Since $A_{\gamma^{-1}}$ is a right $V^{\sigma(\gamma^{-1})}$-module and $c(\gamma,\gamma^{-2}) \in U(V^{\sigma(\gamma^{-1})})$, it follows that $J(W)^{\sigma(\gamma)} \subset J(W)$ and so $W^{\sigma(\gamma)} \supset W$ follows for any $\gamma \in J(H)$.

Conversely, suppose that (1) and (2) hold. To prove that A is a ring, it suffices to prove that A is closed under multiplication. This will be done in the following four cases. Let $\gamma,\ \delta \in \Gamma$.

Case 1. $\gamma, \delta \in H$. If γ and δ are both in $U(H)$, then $\gamma\delta \in U(H)$ and $V\overline{\gamma}V\overline{\delta} = Vc(\gamma,\delta)\overline{\gamma\delta} = V\overline{\gamma\delta}$. If $\gamma \in U(H)$ and $\delta \in J(H)$, then $\gamma\delta \in J(H)$ and so

$$V\overline{\gamma}W^{\sigma(\delta)}\overline{\gamma} = V(W^{\sigma(\delta)})^{\sigma(\gamma)}c(\gamma,\delta)\overline{\gamma\delta} = W^{\sigma(\gamma\delta)}\overline{\gamma\delta}.$$

If $\gamma \in J(H)$ and $\delta \in U(H)$, then $\gamma\delta \in J(H)$ and so

$$\begin{aligned} W^{\sigma(\gamma)}\overline{\gamma}V\overline{\delta} &= W^{\sigma(\gamma)}V^{\sigma(\gamma)}c(\gamma,\delta)\overline{\gamma\delta} = W^{\sigma(\gamma)}c(\gamma,\delta)\overline{\gamma\delta} \\ &= (W^{\sigma(\delta)})^{\sigma(\gamma)}c(\gamma,\delta)\overline{\gamma\delta} = W^{\sigma(\gamma\delta)}\overline{\gamma\delta}. \end{aligned}$$

If γ, δ are both in $J(H)$, then $\gamma\delta \in J(H)$ and

$$W^{\sigma(\gamma)}\overline{\gamma}W^{\sigma(\delta)}\overline{\delta} = W^{\sigma(\gamma)}(W^{\sigma(\delta)})^{\sigma(\gamma)}c(\gamma,\delta)\overline{\gamma\delta} = W^{\sigma(\delta\gamma)}\overline{\gamma\delta}$$

by (2).

Case 2. $\gamma \in H$ and $\delta \in \Gamma - H$. Note that $W \supset W^{\sigma(\delta)}$ and $W^{\sigma(\gamma)} \supset W$ by Lemmas 1.7.3, 1.7.12 and the assumption, and so $J(W) \subset J(W)^{\sigma(\delta)}$ and $J(W)^{\sigma(\gamma)} \subset J(W)$.

If $\gamma \in U(H)$, then $\gamma\delta \in \Gamma - H$ and so

$$V\overline{\gamma}J(W)\overline{\delta} = VJ(W)^{\sigma(\gamma)}c(\gamma,\delta)\overline{\gamma\delta} = J(W)c(\gamma,\delta)\overline{\gamma\delta} = J(W)\overline{\gamma\delta},$$

because $W = W^{\sigma(\gamma)} \supset (W^{\sigma(\delta)})^{\sigma(\gamma)} = W^{\sigma(\gamma\delta)}$ and $c(\gamma,\delta) \in U(W)$. If $\gamma \in J(H)$, then $W^{\sigma(\gamma)}\overline{\gamma}J(W)\overline{\delta} = J(W)^{\sigma(\gamma)}c(\gamma,\delta)\overline{\gamma\delta}$, which is contained in $J(W)\overline{\gamma\delta}$ if $\gamma\delta \in \Gamma - H$ as before. If $\gamma\delta \in U(H)$, then $J(W)^{\sigma(\gamma)}c(\gamma,\delta)\overline{\gamma\delta} \subset V\overline{\gamma\delta}$, because $W^{\sigma(\gamma)} \supset W^{\sigma(\gamma\delta)}$ and $V \supset J(W)^{\sigma(\gamma)}$. If $\gamma\delta \in J(H)$, then $J(W)^{\sigma(\gamma)}c(\gamma,\delta)\overline{\gamma\delta} \subset W^{\sigma(\gamma\delta)}\overline{\gamma\delta}$.

Case 3. $\gamma \in \Gamma - H$ and $\delta \in H$. If $\delta \in U(H)$, then $\gamma\delta \in \Gamma - H$ and thus

$$J(W)\overline{\gamma}V\overline{\delta} = J(W)Vc(\gamma,\delta)\overline{\gamma\delta} = J(W)\overline{\gamma\delta},$$

because $c(\gamma,\delta) \in U(V^{\sigma(\gamma\delta)}) \subset U(W^{\sigma(\gamma\delta)}) \subset U(W)$. If $\delta \in J(H)$, then

$$J(W)\overline{\gamma}W^{\sigma(\delta)}\overline{\delta} = J(W)W^{\sigma(\gamma\delta)}c(\gamma,\delta)\overline{\gamma\delta}. \qquad (**)$$

If $\gamma\delta \in U(H)$, then $V = A_{\gamma\delta}$, the $\gamma\delta$-component of A. If $\gamma\delta \in J(H)$, then $A_{\gamma\delta} = W^{\sigma(\gamma\delta)} \supset W$ and if $\gamma\delta \in \Gamma - H$, then $A_{\gamma\delta} = J(W)$ and $W^{\sigma(\gamma\delta)} \subset W$. Hence in any case, $(**)$ implies that multiplication is closed.

Case 4. $\gamma,\ \delta \in \Gamma - H$, then $\gamma\delta \in \Gamma - H$ by Lemma 1.7.12. So

$$J(W)\overline{\gamma}J(W)\overline{\delta} = J(W)J(W)^{\sigma(\gamma)}c(\gamma,\delta)\overline{\gamma\delta} = J(W)\overline{\gamma\delta},$$

because $c(\gamma,\delta) \in U(V^{\sigma(\gamma\delta)}) \subset U(W^{\sigma(\gamma\delta)}) \subset U(W)$. We have proved that A is a ring and it is easy to see that A is a graded extension of $K * \Gamma$ by Lemma 1.7.2. □

1.7.19 Proposition

Under the notation as in Proposition 1.7.18, suppose that $A = \oplus_{\gamma\in\Gamma-H} J(W)\overline{\gamma} \oplus (\oplus_{\gamma\in U(H)} V\overline{\gamma}) \oplus (\oplus_{\gamma\in J(H)} W^{\sigma(\gamma)}\overline{\gamma})$ is a graded extension of V in $K * \Gamma$. Then

1. $J_g(A) = \oplus_{\gamma\in\Gamma-H} J(W)\overline{\gamma} \oplus (\oplus_{\gamma\in U(H)} J(V)\overline{\gamma}) \oplus (\oplus_{\gamma\in J(H)} W^{\sigma(\gamma)}\overline{\gamma})$.
2. $R = A_{J_g(A)}$ satisfies the following conditions:
   (a) $RJ(W) \subsetneq R\overline{\gamma} \subset RW^{\sigma(\gamma)}\overline{\gamma}) \subsetneq R$ for any $\gamma \in J(H)$.
   (b) $\overline{\gamma} \in U(R)$ for any $\gamma \in U(H)$.

Proof. 1. Set $J_g(A) = \oplus_{\gamma\in\Gamma} J_\gamma\overline{\gamma}$ as before. For any $\gamma \in J(H)$ and $0 \neq a \in W^{\sigma(\gamma)}$, we have $(a\overline{\gamma})^{-1} = (c(\gamma^{-1},\gamma))^{-1}(a^{-1})^{\sigma(\gamma^{-1})}\overline{\gamma^{-1}}$. Assume that $a\overline{\gamma} \in U(A)$, then $(c(\gamma^{-1},\gamma))^{-1}(a^{-1})^{\sigma(\gamma^{-1})} \in J(W)$ and so $(a^{-1})^{\sigma(\gamma^{-1})} \in J(W)$ and $((a^{-1})^{\sigma(\gamma^{-1})})^{\sigma(\gamma)} \in J(W)^{\sigma(\gamma)}$. It follows from Lemma 1.7.3 that

$$\begin{aligned} a^{-1} &= (c(\gamma,\gamma^{-1}))^{-1}((a^{-1})^{\sigma(\gamma^{-1})})^{\sigma(\gamma)}c(\gamma,\gamma^{-1}) \\ &\in c(\gamma,\gamma^{-1})J(W)^{\sigma(\gamma)}c(\gamma,\gamma^{-1}), \end{aligned}$$

which is equal to $J(W)^{\sigma(\gamma)}$, because $c(\gamma,\gamma^{-1}) \in U(W) \subset U(W^{\sigma(\gamma)})$. Thus $1 = a^{-1}a \in J(W)^{\sigma(\gamma)}W^{\sigma(\gamma)} = J(W)^{\sigma(\gamma)}$, a contradiction and hence $J_\gamma = W^{\sigma(\gamma)}$. Similarly $J_\gamma = J(W)$ for any $\gamma \in \Gamma - H$. For any $\gamma \in U(H)$ and $a \in V$, $a\overline{\gamma} \in U(A)$ if and only if $(c(\gamma^{-1},\gamma))^{-1}(a^{-1})^{\sigma(\gamma^{-1})} \in V$, which is equivalent to $a^{-1} \in V$, i.e. $a \in U(V)$. Hence $J_\gamma = J(V)$ for any $\gamma \in U(H)$.

2. For any $\gamma \in J(H)$, we have

$$J(W)\overline{\gamma}^{-1} = J(W)(c(\gamma^{-1},\gamma))^{-1}\overline{\gamma^{-1}} = J(W)\overline{\gamma^{-1}} \subset J_g(A) \subset J(R) \subsetneq R,$$

which implies $RJ(W) \subsetneqq R\overline{\gamma}$. Furthermore, $W^{\sigma(\gamma)}\overline{\gamma} \subset J_g(A)$ entails $R\overline{\gamma} \subset RW^{\sigma(\gamma)}\overline{\gamma} \subset RJ_g(A) \subset J(R) \subsetneqq R$. It is clear that $\overline{\gamma} \in U(R)$ for any $\gamma \in U(H)$ since $\overline{\gamma} \in U(A)$.

The converse of Proposition 1.7.19 is true as it will be shown in the following:

1.7.20 Proposition

Let H be a proper cone of Γ with $H \supset P$, W be an overring of V and let S be a Gauss extension of V in D. Suppose that S satisfies

(a) $SJ(W) \subsetneqq S\overline{\gamma} \subset SW^{\sigma(\gamma)}\overline{\gamma} \subsetneqq S$ for any $\gamma \in J(H)$.
(b) $\overline{\gamma} \in U(S)$ for any $\gamma \in U(H)$.

Then:

$$A = S \cap (K * \Gamma) = \oplus_{\gamma \in \Gamma - H} J(W)\overline{\gamma} \oplus (\oplus_{\gamma \in U(H)} V\overline{\gamma}) \oplus (\oplus_{\gamma \in J(H)} W^{\sigma(\gamma)}\overline{\gamma})$$

is a graded extension of V in $K * \Gamma$.

Proof. Let $A = S \cap (K * \Gamma) = \oplus_{\gamma \in \Gamma} A_\gamma \overline{\gamma}$, a graded extension of V in $K * \Gamma$ by Theorem 1.7.7. First, we prove $A_\gamma = J(W)$ for any $\gamma \in \Gamma - H$. Let $\gamma \in \Gamma - H$, equivalently, $\gamma = \delta^{-1}$ for some $\delta \in J(H)$. By (a), $S \supset J(W)(\overline{\delta})^{-1} = J(W)(c(\delta^{-1}, \delta))^{-1}\overline{\delta^{-1}} = J(W)\overline{\delta^{-1}}$ and so $J(W) \subset A_{\delta^{-1}} = A_\gamma$. Since $V \supset A_{\delta^{-1}}\overline{\delta^{-1}}W^{\sigma(\delta)}\overline{\delta} = A_{\delta^{-1}}Wc(\delta^{-1}, \delta)$, we have $A_\gamma = A_{\delta^{-1}} \subset (V : W)_l$. If $W \supsetneqq V$, then $(V : W)_l = J(W)$ by Proposition 1.5.32. Thus $A_\gamma = J(W)$ follows. If $W = V$, then $(V : W)_l = V$ and so we have either $A_\gamma = J(V)$ or $A_\gamma = V$. Suppose that $A_\gamma = V$, then $\overline{\gamma} \in S$ and $c(\gamma, \gamma^{-1}) = \overline{\gamma}\overline{\gamma^{-1}} \in S$, which is a unit in V. Thus $\overline{\gamma^{-1}} \in U(S)$, contradicting the fact $\overline{\gamma^{-1}} \in J(S)$ by (a). Hence $A_\gamma = J(V)$ follows and thus $A_\gamma = J(W)$ for any $\gamma \in \Gamma - H$.

Secondly we prove that $A_\gamma = V$ for any $\gamma \in U(H)$. Since $\overline{\gamma} \in U(S)$, we have $A_\gamma \overline{\gamma}\overline{\gamma^{-1}} = A_\gamma c(\gamma, \gamma^{-1}) \subset V$ and $A_\gamma \subset V$. On the other hand, $V\overline{\gamma} \subset S$ implies $V \subset A_\gamma$. Hence $A_\gamma = V$ for any $\gamma \in U(H)$.

Finally we prove that $A_\gamma = W^{\sigma(\gamma)}$ for any $\gamma \in J(H)$. It is clear from (a) that $W^{\sigma(\gamma)} \subset A_\gamma$. To prove the converse inclusion, we claim first that $W^{\sigma(\gamma)} \supset W$. To prove this we consider:

$$W^{\sigma(\gamma)}\overline{\gamma}J(W)\overline{\gamma^{-2}} = J(W)^{\sigma(\gamma)}c(\gamma, \gamma^{-2})\overline{\gamma^{-1}} \subset A_{\gamma^{-1}}\overline{\gamma^{-1}} = J(W)\overline{\gamma^{-1}}.$$

Thus $J(W)^{\sigma(\gamma)}c(\gamma, \gamma^{-2}) \subset J(W)$ and so $J(W)^{\sigma(\gamma)} \subset J(W)$ since $A_{\gamma^{-1}} = J(W)$ is a right $V^{\sigma(\gamma^{-1})}$-module. Hence we have $W^{\sigma(\gamma)} \supset W$, as desired.

To prove that $W^{\sigma(\gamma)} \supset A_\gamma$, consider:

$$J(W)\overline{\gamma^{-1}}A_\gamma\overline{\gamma} = J(W)A_\gamma^{\sigma(\gamma^{-1})}c(\gamma^{-1},\gamma) \subset V \text{ and } A_\gamma^{\sigma(\gamma^{-1})} \subset (V : J(W))_r.$$

If $W \supsetneqq V$, then $(V : J(W))_r = W$ by Proposition 1.5.32 Thus:

$$A_\gamma = c(\gamma,\gamma^{-1})^{-1}(A_\gamma^{\sigma(\gamma^{-1})})^{\sigma(\gamma)}c(\gamma,\gamma^{-1}) \subset c(\gamma,\gamma^{-1})^{-1}W^{\sigma(\gamma)}c(\gamma,\gamma^{-1}) = W^{\sigma(\gamma)},$$

since $c(\gamma,\gamma^{-1}) \in U(V) \subset U(W^{\sigma(\gamma)})$.

If $W = V$ and $J(V)^2 = J(V)$, then $(V : J(W))_r = V$ and so $A_\gamma \subset V^{\sigma(\gamma)}$ follows similarly. If $W = V$ and $J(V)$ is principal, say, $J(V) = bV = Vb$ for some $b \in J(V)$, then $(V : J(W))_r = (V : J(V))_r$. On the other hand, $V^{\sigma(\gamma)} \subset A_\gamma$ implies $V \subset A_\gamma^{\sigma(\gamma^{-1})}$ and so $V \subset A_\gamma^{\sigma(\gamma^{-1})} \subset b^{-1}V = Vb^{-1}$. Thus we have either $V = A_\gamma^{\sigma(\gamma^{-1})}$ or $A_\gamma^{\sigma(\gamma^{-1})} = Vb^{-1}$.

In the former case, we have $V^{\sigma(\gamma)} = (A_\gamma^{\sigma(\gamma^{-1})})^{\sigma(\gamma)} = c(\gamma,\gamma^{-1})A_\gamma c(\gamma,\gamma^{-1})^{-1}$, which shows $V^{\sigma(\gamma)} = A_\gamma$, because $U(V) \subset U(V^{\sigma(\gamma)})$.

In the latter case, it follows that:

$$(Vb^{-1})^{\sigma(\gamma)} = (A_\gamma^{\sigma(\gamma^{-1})})^{\sigma(\gamma)} = c(\gamma,\gamma^{-1})A_\gamma c(\gamma,\gamma^{-1})^{-1}$$

which implies $A_\gamma = V^{\sigma(\gamma)}(b^{-1})^{\sigma(\gamma)}$, because $U(V) \subset U(V^{\sigma(\gamma)})$, and so $(b^{-1})^{\sigma(\gamma)} \in A_\gamma$. Furthermore, since $A_\gamma\overline{\gamma} \subset S\overline{\gamma} \subsetneqq S$ by (a), we have $(b^{-1})^{\sigma(\gamma)}\overline{\gamma} \in A_\gamma\overline{\gamma} \subset J(S) \cap A = J_g(A)$ and so $c(\gamma,\gamma^{-1}) = \overline{\gamma}b^{-1}b\overline{\gamma^{-1}} = (b^{-1})^{\sigma(\gamma)}\overline{\gamma}b\overline{\gamma^{-1}} \in A_\gamma\overline{\gamma}J(V)\overline{\gamma^{-1}} \subset A$. Thus $(b^{-1})^{\sigma(\gamma)}\overline{\gamma} \in U(A)$, because $c(\gamma,\gamma^{-1}) \in U(V)$. This is a contradiction since $S\overline{\gamma} \subsetneqq J(S)$. Thus $A_\gamma = V^{\sigma(\gamma)}$ follows. Hence in any cases, $W^{\sigma(\gamma)} \supset A_\gamma$ and so $W^{\sigma(\gamma)} = A_\gamma$.

Therefore: $S \cap K * \Gamma = \oplus_{\gamma\in\Gamma - H} J(W)\overline{\gamma} \oplus (\oplus_{\gamma\in U(H)} V\overline{\gamma}) \oplus (\oplus_{\gamma\in J(H)} W^{\sigma(\gamma)}\overline{\gamma})$.

1.7.21 Remark

1. See [8] for complete descriptions of prime ideals in standard Gauss extensions of V in D and for some concrete examples of graded extensions of V in $K * \Gamma$.
2. If $\Gamma = \langle X \rangle$ is an infinite cyclic group, then the crossed product $K * \Gamma$ is a skew Laurent polynomial ring ([47, (1.5.11)]), i.e. $K * \Gamma = K[X, X^{-1}, \sigma]$ for some automorphism σ of K. A complete classification of graded extensions $A = \oplus_{i\in\mathbb{Z}} A_i X^i$ of V in $K[X, X^{-1}, \sigma]$ in terms of A_1, A_{-1} and $O_l(A_1)$ is given in [43, 44].
3. Let O_v be a commutative valuation ring of a field F and let K be a finite Galois extension of F with Galois group Γ. Then Dubrovin valuation rings in $Q(K * \Gamma)$ lying over O_v were studied in [28, 32].

1.8 Filtrations by Totally Ordered Groups

In Definition 1.3.2.23 a Γ-filtration of a ring A, where Γ is a totally ordered group, has been defined. Separated filtrations were defined in Definition 1.3.2.25. A Γ-filtration on a ring A is said to be a **strong filtration** if for all $\sigma, \tau \in \Gamma$ we have: $F_\sigma A F_\tau A = F_{\sigma+\tau} A$. It is clear that FA is a strong filtration if and only if for every $\sigma \in \Gamma$, $F_\sigma A F_{-\sigma} A = F_0 A$. Strong filtrations on simple Artinian rings may be thought of as being generalizations of Dubrovin valuations, this is motivated by the following results concerning value functions of Dubrovin valuation rings.

Consider a simple Artinian ring Q and let Γ be a totally ordered group with additively written operation, even if Γ is not assumed to be abelian.

1.8.1 Definition

A surjective map $v : Q \to \Gamma \cup \{\infty\}$ is a value function of Q if the following condition hold:

V.1. $v(q) = \infty$ if and only if $q = 0$. We have $v(-1) = 0$.
V.2. $v(q + q') \geq \min\{v(q), v(q')\}$, for $q, q' \in Q$.
V.3. $v(qq') \geq v(q) + v(q')$, for $q, q' \in Q$.
V.4. Define $\mathrm{st}(v) = \{q \in U(Q), v(q^{-1}) = -v(q)\}$, then we have $\Gamma = v(st(v))$.

1.8.2 Properties

For a value function v on Q the following properties hold:

1. If $q \in st(v)$ and $q' \in Q$ then $v(qq') = v(q) + v(q')$, $v(q'q) = v(q') + v(q)$.
2. The set $\mathrm{st}(v)$ is a subgroup of $U(Q)$ and $v : \mathrm{st}(v) \to \Gamma$ is an epimorphism.
3. If $v(q) \neq v(q')$ then $v(q + q') = \min\{v(q), v(q')\}$.
4. $R_v = \{q \in Q, v(q) \geq 0\}$ is a ring and $M_v = \{q \in Q, v(q) > 0\}$ is an ideal of R_v.

Proof. 1. For $q \in \mathrm{st}(v), q' \in Q$ we calculate: $v(q') = v(q^{-1}qq') \geq v(q^{-1}) + v(qq') = -v(q) + v(qq') \geq -v(q) + v(q) + v(q') = v(q')$, yielding $v(q') = -v(q) + v(qq')$, hence the statement (1) follows.

2. Look at $q, q' \in st(v)$. Then $v((q^{-1})^{-1}) = v(q) = -v(q^{-1})$ yields $q^{-1} \in \mathrm{st}(v)$. Using (1) we arrive at:

$$\begin{aligned} v((qq')^{-1}) &= v(q'^{-1}q^{-1}) = v(q'^{-1}) + v(q^{-1}) = \\ &= -v(q') - v(q) = -(v(q) + v(q')) = -v(qq') \end{aligned}$$

hence $qq' \in \mathrm{st}(v)$. From $v((-1)^{-1}) = v(-1) = 0 = -v(-1)$ it follows that $-1 \in \mathrm{st}(v)$, hence $v(1) = v((-1)(-1)) = v(-1) + v(-1) = 0 = -v(1)$, i.e. $1 \in \mathrm{st}(v)$. That $v : \mathrm{st}(v) \to \Gamma$ is an epimorphism is clear from (1).

3. For all $q \in Q$ we have $v(-q) = v(-1) + v(q) = v(q)$. If $v(q) > v(q')$, then $v(q') = v((-q) + (q + q')) \geq \min\{v(q), v(q + q')\}$. From $v(q) > v(q')$ it follows: $v(q') \geq v(q + q') \geq \min\{v(q), v(q')\} = v(q')$, establishing $v(q + q') = \min\{v(q), v(q')$.
4. In view of V.2, V.3, R_v is closed under addition and multiplication; from $v(q) = v(-q)$ and $1 \in R_v$ it follows that R_v is a ring. It is easy to see that M_v is a twosided ideal of R_v, as claimed. □

To a value function v on Q we define a chain consisting of $F_\gamma Q, \gamma \in \Gamma$, in Q defined by putting $F_\gamma Q = \{q \in Q, v(q) \geq -\gamma\}$.

1.8.3 Lemma

If $v : Q \to \Gamma \cup \{\infty\}$ is a value function on Q then FQ is a strong filtration of Q.

Proof. If $q, q' \in F_\gamma Q$ then V.2 yields $q + q' \in F_\gamma Q$; moreover $-q \in F_\gamma Q$ follows from V.3 and V.1. From $v(-q) \geq v(q)$ and $v(q) = v(-(-q)) \geq v(-q)$, $v(q) = v(-q)$ follows. It is clear that $FQ = \{F_\gamma Q, \gamma \in \Gamma\}$ forms an ascending chain of additive subgroups of Q with $1 \in F_0 Q = R$.

From V.3 it follows that $F_\gamma Q F_\delta Q \subset F_{\delta+\gamma} Q$ for γ, δ in Γ (using also V.2 as every element of $F_\gamma Q F_\delta Q$ is a sum of products of elements in $F_\gamma Q$ and $F_\delta Q$). Since v is surjective FQ is exhaustive, i.e. $\cup_\gamma F_\gamma Q = Q$. To see that FQ is strong, consider $F_\gamma Q$ then there is a $q_\gamma \in \mathrm{st}(v)$ such that $v(q_\gamma) = -\gamma$, thus $q_\gamma \in F_\gamma Q$. Since $v(q_\gamma^{-1}) = -v(q_\gamma)$, $q_\gamma^{-1} \in F_{-\gamma} Q$, hence we have $1 \in F_\gamma Q F_{-\gamma} Q$, and since the latter is an ideal of $F_0 R$, $F_\gamma Q F_{-\gamma} Q = F_0 Q$ for all $\gamma \in \Gamma$. □

Consider a Dubrovin valuation ring R of Q and define the stabilizer of R in Q as $\mathrm{st}(R) = \{q \in Q : qR = Rq\}$. Observe that a $q \neq 0$ in $\mathrm{st}(R)$ is regular, e.g. $aq = 0$ yields $aRq = 0$ but R is a prime ring, hence $\mathrm{st}(R) = \{q \in Q, qRq^{-1} = R\}$. Put $\Gamma_R = \mathrm{st}(R)/U(R)$. We define an ordering on Γ_R by putting $\overline{\alpha} \geq \overline{\beta}$, for $\alpha, \beta \in \mathrm{st}(R)$, if $\alpha R \subset \beta R$. Since R-ideals of Q are linearly ordered it is obvious that Γ_R is totally ordered. Observe that Γ_R is a group because $U(R)$ is a normal subgroup of $\mathrm{st}(R)$, so Γ_R is a totally ordered group.

1.8.4 Theorem

1. Let R be a Dubrovin valuation ring of Q and suppose that for all $q \neq 0$ in Q there is an $x \in \mathrm{st}(R)$ such that $RqR = xR$. Then there exists a value function v_R on Q with $R = \{q \in Q, v_R(q) \geq 0\}$, $M = \{q \in Q, v_R(q) > 0\}$. For any such

v, $\mathrm{st}(v) = \mathrm{st}(R), U(R) = \{q \in \mathrm{st}(R), v(q) = 0\}$ and v induces an isomorphism $\Gamma_R \cong \mathrm{Im}v$.

2. Suppose v is a value function on Q with $R_v = \{q \in Q, v(q) \geq 0\}, M_v = \{q \in Q, v(q) > 0\}$. Suppose that R_v/M_v is simple Artinian, then R_v is a Dubrovin valuation ring of Q with $J(R_v) = M_v$, moreover for $q \neq 0$ in Q there exists $x \in \mathrm{st}(R_v)$ such that $R_v q R_v = x R_v$.

Proof. 1. Define $v_R : Q \to \Gamma_R \cup \{\infty\}$ by $v_R(0) = \infty$ and for $q \neq 0$ in Q, $v_R(q) = \alpha \bmod U(R)$ where $\alpha \in \mathrm{st}(R)$ is such that $RqR = \alpha R$. Since $\alpha R = \beta R$ if and only if $\alpha U(R) = \beta U(R)$ in Γ_R, the map v_R is well-defined. It is clear that $RqR \subset Rq'R$ if and only if $v_R(q) \geq v_R(q')$. Condition V.1 is clear.
In order to check V.2, let $v_R(q) \geq v_R(q')$, thus $RqR \subset Rq'R$ and $q+q' \in Rq'R$. Hence $R(q+q')R \subset Rq'R$, yielding $v_R(q+q') \geq v_R(q') = \min\{v_R(q), v_R(q')\}$. To check V.3 look at $qq' \in (RqR)Rq'R), Rqq'R \subset (RqR)(Rq'R)$, and thus $v_R(qq') \geq v_R(q) + v_R(q')$ follows from the definition.
For V.4 it is clear that $\mathrm{Im}v_R = v_R(\mathrm{st}(R))$. If $\alpha \in \mathrm{st}(R)$ then $v_R(\alpha^{-1}) = -v_R(\alpha)$ since $v_R(\alpha^{-1}) + v_R(\alpha)$ corresponds to $\alpha^{-1}R\alpha R = \alpha^{-1}\alpha R = R$, thus $\mathrm{st}(R) \subset \mathrm{st}(v_R)$ and $\mathrm{Im}v_R \subset v_R(\mathrm{st}(v_R))$. Since the converse inclusion is obvious, $\mathrm{Im}v_R = v_R(\mathrm{st}(v_R))$ follows. Now $RqR \subset R$, resp. $RqR = R$, if and only if $v_R(q) \geq v_R(1) = 0$, resp. $v_R(q) = 0$. Since $RqR = R$ if and only if $q \in R - J(R)$ it follows that $R = \{q \in Q, v_R(q) \geq 0\}, J(R) = \{q \in Q, v_R(q) > 0\}$. Finally consider any value function v on Q with $R = R_v$ and $M = J(R_v)$. Take $b \in U(R)$, then $b^{-1} \in R, v(b^{-1}) \geq 0$. From $0 = v(1) = v(bb^{-1}) \geq v(b) \neq v(b^{-1})$ it follows that $0 \geq -v(b^{-1}) \geq v(b) \geq 0$, thus $v(b) = 0, b \in \{q \in \mathrm{st}(v), v(q) = 0\}$. Conversely, if $q \in \mathrm{st}(v)$ has $v(q) = 0$, then $v(q^{-1}) = 0$ and thus $q, q^{-1} \in R$, or $q \in U(R)$. Consequently $U(R) = \{q \in \mathrm{st}(v), v(q) = 0\}$. To establish $\mathrm{st}(v) = \mathrm{st}(R)$ look at $q \in \mathrm{st}(v)$ then: $v(qq'q^{-1}) = v(q')$. Thus for any $q' \in Q$ we have $v(qq'q^{-1}) \geq 0$ if and only if $v(q') \geq 0$. Therefore $qRq^{-1} = R$ and thus $q \in \mathrm{st}(R)$. Conversely, if $q \in \mathrm{st}(R)$, then there is a $q_0 \in \mathrm{st}(v)$ such that $v(q) = v(q_0)$ in view of V.4. Then $v(qq_0^{-1}) = 0$ (foregoing properties) and $qq_0^{-1} \in (R - J(R)) \cap \mathrm{st}(R) = U(R) \subset \mathrm{st}(v)$. Thus $q \in \mathrm{st}(v)$ and $\mathrm{st}(v) = \mathrm{st}(R)$ follows. Clearly $v : \mathrm{st}(R) \to \mathrm{Im}v$ is a homomorphism and $\Gamma_R = \mathrm{st}(R)/U(R) = \mathrm{Im}v$.

2. We already know that R_v/M_v is simple Artinian. Pick $q \in Q - R_v$, hence $v(q) < 0$. Condition V.4 yields that there is a $q_0 \in \mathrm{st}(v)$ with $v(q) = v(q_0)$. Hence, $v(q_0^{-1}) = -v(q_0) > 0$, or $q_0^{-1} \in R_v$. Thus $v(qq_0^{-1}) = v(q) - v(q_0) = 0$. Similarly, $v(q_0^{-1}q) = 0$. Thus $qq_0^{-1}, q_0^{-1}q \in R_v - M_v$. Hence R_v is a Dubrovin valuation ring. Consider $q \neq 0$ in Q, as before there exists $q_0 \in \mathrm{st}(v)$ such that $v(qq_0^{-1}) = 0$, hence $qq_0^{-1} \in R_v - M_v$. Thus $R_v qq_0^{-1} R_v = R_v$. The argument used in 1 yields $\mathrm{st}(v) \subset \mathrm{st}(R_v)$ and thus $R_v q R_v = q_0 R_v$. □

We have defined a Γ-filtration on any ring A in Sect. 1.3.2 and a separated Γ-filtration in Definition 1.3.1.25. We obtain from the principal symbol σ a map ord $\sigma : A \to \Gamma \cup \{\infty\}, \sigma(a) = \gamma$ if $\gamma \in \Gamma$ is such that $a \in F_\gamma A, a \notin \sum_{\delta<\gamma} F_\delta A = F_\gamma^0 A$ (corresponding to the principal symbol map $\sigma : A \to G(A)$, where $G_F(A)$

is the associated graded ring of FA, see hereafter). Put $v_F = -\text{ord}\sigma$. Then v_F satisfies, for a separated filtration FQ on a simple Artinian Q, the properties V.1, V.2, V.3 of Definition 1.8.1, but not necessarily V.4. Conversely to any given map $w: A \to \Gamma \cup \{\infty\}$ satisfying V.1, V.2, V.3, there corresponds a separated filtration of A defined by $F_\gamma(A) = \{a \in A, w(a) \geq -\gamma\}$.

The fact that we consider ascending filtrations forced us to write $-\gamma$ in the definition of $F_\gamma(A)$ above and if we insist on $F_\sigma(A)F_\tau(A) \subset F_{\sigma\tau}(A)$ we have to consider the filtration by the opposite group Γ^0 for a value function to Γ. Usually we deal with abelian Γ (value functions extending a central valuation, or value functions on Weyl fields etc...) so this then does not matter. For noncommutative Γ when the total order is respected both by left and right multiplication (as we always assumed here) the switch from Γ to Γ^0 is harmless. To any prime of A there corresponds a filtration in view of Lemma 1.3.2.22 albeit that we then consider only partially ordered value groups Γ. For an arbitrary Dubrovin valuation R in Q there is a value function with partial ordered value group as well as a filtration with respect to the totally ordered (sub)group $\Gamma_R = \text{st}(R)/U(R)$; the latter follows from the proof of Theorem 1.8.4.1. where V.1,...,V.3 are checked without using the extra condition on the Dubrovin valuation ring assumed in the theorem.

To a filtration FA of type Γ we associate a Γ-graded ring $G_F(A) = \oplus_{\gamma\in\Gamma} G_F(A)_\gamma$ where $G_F(A)_\gamma = F_\gamma A/F_\gamma^0 A$ and multiplication is defined by the following, for $\overline{a} \in G_F(A)_\gamma$, $\overline{b} \in G_F(A)_\delta$, for γ and $\delta \in \Gamma$, we define $\overline{a}.\overline{b}$ by taking $\overline{a}\overline{b} = (ab)\text{mod}F_{\gamma+\delta}^0 A$ where $a \in F_\gamma A, b \in F_\delta A$ are any representatives of $\overline{a}$, resp. $\overline{b}$. We can also follow the Γ-graded **Rees ring** (or blow up ring) $\widetilde{A} = \oplus_{\gamma\in\Gamma} F_\gamma A$ that may be identified to the subring of the groupring $A\Gamma$ given as $\widetilde{A} = \sum_{\gamma\in\Gamma} F_\gamma A.\gamma$. Since $1 \in F_\delta A$ for $\delta \geq 0$ it follows that $\Gamma \subset \widetilde{A}$ and as Γ commutes with A in $A\Gamma$, $\widetilde{A}\Gamma^+$ is a graded ideal of $\widetilde{A}$, where $\Gamma^+ = \Gamma_{>0} = \{\gamma \in \Gamma, \gamma > 0\}$. We write Γ_+ for $0 \cup \Gamma^+$ in Γ, $\Gamma_+ = \{\gamma \in \Gamma, \gamma \geq 0\}$. Note that the unit of the groupring is written as $1, 1 = 1_A.0$ for $0 \in \Gamma$, 1_A the unit element of A.

1.8.5 *Proposition*

Let us write Γ multiplicatively. The multiplicative set Γ_+ is a left and right Ore set in $\widetilde{A}$ such that $(\Gamma_+)^{-1}\widetilde{A} = A\Gamma$.

Proof. Look at $\delta \in \Gamma_+$ and $a_\gamma\gamma \in \widetilde{A}_\gamma$, then we see that $(\gamma\delta\gamma^{-1})a_\gamma\gamma = \gamma a_\gamma\delta$ yields the left Ore condition (the right Ore condition follows in the symmetric way). Now look at $\sum_\gamma a_\gamma\gamma$ and we reason by induction on the length of this sum, the case where only one $a_\gamma \neq 0$ has been dealt with above. Suppose the claim has been established for elements of length strictly less than n. Consider $\sum_{i=1}^n a_{\gamma_i}\gamma_i \in \widetilde{A}$ and $\tau \in \Gamma_+$, assuming without loss of generality that $\gamma_1 > \gamma_i$ for $i = 2, \ldots, n$. Then we have: $(\gamma_1\tau\gamma_1^{-1})(\sum_{i=1}^n a_{\gamma_i}\gamma_i) = a_{\gamma_1}\gamma_1\tau + \sum_{i=2}^n a_{\gamma_i}\gamma_1\tau\gamma_1^{-1}\gamma_i$.

The induction hypothesis yields the existence of $\delta \in \Gamma_+$ such that:

$$\delta(\sum_{i=2}^{n} a_{\gamma_i}\gamma_1\tau\gamma_1^{-1}\gamma_i) = a'\tau \text{ with } a' \in \widetilde{A}$$

Look at $\delta\gamma_n\tau\gamma_1^{-1} \in \Gamma_+$, then we obtain:

$$\delta(\gamma_1\tau\gamma_1^{-1})(\sum_{i=1}^{n} a_{\gamma_1}\delta\gamma_1\tau) = a_{\gamma_1}\delta\gamma_1\tau + a'\tau = (\delta a_{\gamma_1}\gamma_1 + a')\tau$$

with $\delta a_{\gamma_1}\gamma_1 + a'$ in $\widetilde{A}$ since $\delta \in \widetilde{A}$. For the final statement observe that Γ_+ consists of regular elements in $\widetilde{A}$ because they are invertible in $A\Gamma$. Hence $\Gamma_+^{-1}\widetilde{A} \subset A\Gamma$. If $\gamma \in \Gamma$ then either $\gamma \in \Gamma_+$ or $\gamma^{-1} \in \Gamma_+$ hence $\Gamma_+^{-1}\widetilde{A} = A\Gamma$, we always assume FA to be exhaustive. □

1.8.6 Lemma

Let FA be a Γ-filtration on A, then:

1. $\widetilde{A}/\widetilde{A}\Gamma^+ \cong G_F(A)$ as graded rings.
2. Let I be the ideal of $\widetilde{A}$ generated by the elements $\{\gamma - 1, \gamma \in \Gamma^+\}$, then $\widetilde{A}/I = A$ and FA is obtained from the gradation filtration of $\widetilde{A}$.

Proof. 1. For $\delta \in \Gamma, (\widetilde{A\Gamma^+})_\delta = \sum_{\gamma<\delta}(F_\gamma A.\gamma)(-\gamma^{-1}\delta) = F_\delta^o A \cdot \delta$ hence we obtain: $\widetilde{A}/\widetilde{A\Gamma^+} \cong \oplus_{\delta\in\Gamma} F_\delta A/F_\delta^0 A \cong G_F(A)$ that the latter is indeed a ring isomorphism (graded) is easily verified.
2. Note that the left ideal I of $\widetilde{A}$ generated by the $\{1-\gamma, \gamma \in \Gamma^+\}$ is actually a two-sided ideal even if Γ is not abelian; this follows from $a_\tau\tau(\gamma - 1) = (\tau\gamma\tau^{-1} - 1)\tau a_\tau$, for $\gamma \in \Gamma^+, a_\tau \in F_\tau A$, where $\tau\gamma\tau^{-1} \in \Gamma^+$ too.
3. If $\pi : \widetilde{A} \to \widetilde{A}/I$ is the canonical morphism then it extends naturally to $\pi^e : \Gamma_+^{-1}\widetilde{A} \to \widetilde{A}/I$, i.e. $A\Gamma \to \widetilde{A}/I$ and $\pi^e(A\Gamma) = A$; the image of $\widetilde{A}_\gamma = F_\gamma A.\gamma$ is identified with $F_\gamma A$, hence the filtration of A is determined by the gradation of $\widetilde{A}$. □

1.8.7 Lemma

If FA is a strong filtration then the ring $G_F A$ is strongly graded i.e.

$$G_F(A)_\gamma G_F(A)_\delta = G_F(A)_{\gamma+\delta} \textit{ for all } \gamma, \delta \in \Gamma$$

Proof. It is clear that $\widetilde{A}$ is strongly graded if FA is a strong filtration. Then any graded quotient ring of $\widetilde{A}$, in particular $G_F(A)$, is strongly graded too. □

The converse of Lemma 1.8.7 is true in case $\Gamma = \mathbb{Z}$ but we do not have a proof for general totally ordered groups. In case $F_R Q$ corresponds to a Dubrovin valuation ring R of Q it follows that $F_R Q$ is a strong filtration hence the associated graded ring $G_R(Q)$ is strongly graded with $G_R(Q)_0 = R/M$ a simple Artinian ring, hence $G_R(Q)$ is a so-called gr-simple Artinian ring. Since for $\gamma \in \Gamma_R$ there is $q_\gamma \in \mathrm{st}(v_R)$ with $v_R(q_\gamma) = \gamma$ and $v_R(q_\gamma^{-1}) = -\gamma$ it is easily checked that $F_{R,\gamma} Q = Rq_\gamma^{-1}$ from $(F_{R,\gamma} Q)q_\gamma \subset F_{R,0} Q = R$ while $Rq_\gamma^{-1} \subset F_{R,\gamma} Q$ is clear. Moreover $F_{R,\gamma}^0 Q = Mq_\gamma^{-1}$ follows for all $\gamma \in \Gamma_R$, and $G_R(Q)_\gamma = (R/M)q_\gamma^{-1} \cong R/M$.

1.8.8 Proposition

Let R be a Dubrovin valuation ring of Q then $G_{F_R}(Q) = R/M * \Gamma_R$, a crossed product of the simple Artinian ring R/M by the value group Γ_R.

Proof. Write z_γ for the image of q_γ^{-1} in $G_R(Q)_\gamma$, $z_{\gamma^{-1}}$ for the image of q_γ in $G_R(Q)_\gamma$. Clearly $z_\gamma z_{\gamma^{-1}} = 1 = z_{\gamma^{-1}}.z_\gamma$ hence $z_\gamma^{-1} = z_{\gamma^{-1}}$. Since $q_\gamma R q_\gamma^{-1} = R$, $q_\gamma M q_\gamma^{-1} = M$ it follows that q_γ induces an automorphism of R/M in fact induced by conjugation with z_γ. Since $q_\gamma q_\delta$ is $q_{\gamma\delta}$ up to some $c(\gamma, \delta) \in U(R)$, say $q_\gamma q_\delta = c(\gamma, \delta) q_{\gamma\delta}$. Associativity of Q yields that $c(\gamma, \delta)$ defines a two-cocycle $\Gamma_R \times \Gamma_R \to U(R)$. It follows that $G_{F_R}(Q)$ is a crossed product over $G_{F_R}(Q)_0 = R/M$ with basis $z_\gamma, \gamma \in \Gamma_R$. □

1.8.9 Lemma

Let A be a Γ-strongly graded ring.

1. A is a domain if and only if A_0 is a domain.
2. If A_0 is a prime ring then A is a prime ring.

Proof. 1. We only have to establish that A is a domain if A_0 is. Let $ab = 0$ for some nonzero $a, b \in A$ and let $a = a_{\gamma_1} + \ldots + a_{\gamma_d}$, $b = b_{\delta_1} + \ldots + b_{\delta_e}$ with $\gamma_1 < \ldots < \gamma_d, \delta_1 < \ldots < \delta_e$ in Γ. Then $ab = 0$ yields $a_{\gamma_d} b_{\delta_e} = 0$, yielding $A_{\gamma_d^{-1}} a_{\gamma_d} b_{\delta_e} A_{\delta_e^{-1}} = 0$ with $A_{\gamma_d^{-1}} a_{\gamma_d} \neq 0$ (otherwise $A_{\gamma_d} A_{\gamma_d^{-1}} a_{\gamma_d} = 0$ yields $a_{\gamma_d} = 0$) and also $b_{\delta_e} A_{\delta_e^{-1}} \neq 0$. This thus contradicts the assumption that A_0 is a domain.

2. If I, J are homogeneous ideals of A such that $IJ = 0$ then $I = AI_0, J = AJ_0$ because A is strongly graded, hence $I_0 A J_0 = 0$ and $I_0 J_0 = 0$. Since A_0 is prime, then either I_0 or $J_0 = 0$ hence I or $J = 0$ too. Now look at $aAb = 0$ with $a = a_{\gamma_1} + \ldots + a_{\gamma_d}, b = b_{\delta_1} + \ldots + b_{\delta_e}$ with $\gamma_1 < \ldots < \gamma_d, \delta_1 < \ldots < \delta_e$

in Γ. Then for every homogenous $x \in A, a_{\gamma_d} x b_{\delta_e} = 0$, hence for every $y \in A$, $a_{\gamma_d} y b_{\delta_e} = 0$ or $a_{\gamma_d} A b_{\delta_e} = 0$. Now the first part of (2) yields $a_{\gamma_a} = 0$ or $b_{\delta_e} = 0$, a contradiction. □

1.8.10 Proposition

Let FA be a Γ-separated filtration on A with principal symbol map $\sigma : A \to G_F(A)$.

1. For $a \neq 0$ in A, $\sigma(a)$ is (right) regular in $G_F(A)$ if and only if $v_F(ab) = v_F(a).v_F(b)$ for all $b \in A$.
2. If $G_F(A)$ is a domain then v_F satisfies $v_F(ab) = v_F(a) + v_F(b)$ and A is a domain.
3. If moreover A is Artinian in (2) then it is a skewfield and F_0A is a valuation ring of A. If $\Gamma_s = \{\gamma \in \Gamma, G_F(A)_\gamma \neq 0\}$ then Γ_s is a normal subgroup of Γ and $G_F(A)$ is strongly Γ_s-graded.

Proof. 1. If $b = 0$ then $v_F(ab) = v_F(a).v_F(b)$ holds, because $\infty = v_F(a).\infty$. If $b \neq 0$ then there is a unique $\tau \in \Gamma$ such that $b \in F_\tau A - \sum_{\gamma<\tau} F_\gamma A$; similarly $a \in F_\delta A - \sum_{\gamma<\delta} F_\gamma A$. If $\sigma(a)$ is right regular then $\sigma(a)\sigma(b) \neq 0$ yields $\sigma(a)\sigma(b) = (ab)\mathrm{mod}(\sum_{\gamma<\delta+\tau} F_\gamma A) = \sigma(ab)$. Conversely, assume $v_F(ab) = v_F(a).v_F(b)$ for all $b \in A$, then $\deg \sigma(ab) = \deg\sigma(a) \neq \deg\sigma(b)$ i.e. $\deg\sigma(ab) = \delta.\tau$ but this entails $\sigma(a)\sigma(b) \neq 0$ as $ab \notin \sum_{\gamma<\delta+\tau} F_\gamma A$.
2. That v_F satisfies $v_F(ab) = v_F(a).v_F(b)$ follows from (1). If $a \neq 0, b \neq 0$, then $v_F(a) \neq \infty, v_F(b) \neq \infty$ hence $ab \neq 0$ and thus A is a domain.
3. If A is moreover Artinian then since it is a domain it is a skewfield. If $a_\gamma \in F_\gamma A - \sum_{\delta<\gamma} F_\delta A$ then we have $a_\gamma^{-1} \in F_\tau A - \sum_{\delta<\tau} F_\delta A$ for some $\tau \in \Gamma$. From $1 = a_\gamma a_\gamma^{-1}$ and $\sigma(1) = \sigma(a_\gamma) + \sigma(a_\gamma^{-1})$, $\tau = \gamma^{-1}$ follows. So for every $z \in A$ either z or z^{-1} is in F_0A and since $v_F(uxu^{-1}) \geq 0$ if $v_F(x) \geq 0$, F_0A is invariant under inner automorphisms of A, hence a valuation ring of A. It is also clear that $G_F(A)$ is a gr-skewfield i.e. every homogeneous element of $G_F(A)$ is invertible. The statements about Γ_s are easily checked. □

Recall (cf. [55]) that a graded ring R of type G is said to have property (E) (the gradation is nondegenerate) if every nonzero graded left ideal of R intersects R_e nontrivially (e is the neutral element of G), equivalently for every $r_\sigma \neq 0$ in R_σ we have $R_{\sigma^{-1}} r_\sigma \neq 0$. In case R_e is semisimple, then property (E) is left–right symmetric. In case R is strongly graded then it satisfies condition (E) (both left and right versions).

1.8.11 Theorem

Let A be a simple Artinian ring with separated filtration FR. Assume that $G_F(A)$ satisfies condition (E) and $G_F(A)_0$ is a prime (left) Goldie ring, then:

1. $G_F(A)_0$ is a simple Artinian ring.
2. $F_0^0 A = J(F_0 A)$ and $F_0 A$ is local in the sense of Goldie.
3. The ring $F_0 A$ is a Dubrovin valuation of A.

Proof. 1. From condition (E) it follows that a left regular $\overline{s}$ in $G_F(A)_0$ is still left regular in $G_F(A)$; indeed if $z\overline{s} = 0$ for some $z \in G_F(A)$ then $z_{\gamma_d}\overline{s} = 0$, where $z = z_{\gamma_1} + \ldots + z_{\gamma_d}$ with $\gamma_1 < \ldots < \gamma_d$ is the homogeneous decomposition of z, with $z_{\gamma_d} \neq 0$ hence there is a $y_{\gamma_d^{-1}} \in G_F(A)$ such that $y_{\gamma_d^{-1}} z_{\gamma_d} \neq 0$ in $G_F(A)_0$, but $y_{\gamma_d^{-1}} z_{\gamma_d}\overline{s} = 0$ contradicts left regularity of $\overline{s}$.
Pick $s \in F_0 A$ such that $\sigma(s) = \overline{s}$ is left regular, then s is left regular in A because $as = 0$ would yield $\sigma(as) = 0$, but $\sigma(as) = \sigma(a)\sigma(s)$ follows from Proposition 1.8.10.1, hence $\sigma(a)\sigma(s) = 0$ a contradiction. Hence there is an inverse s^{-1} for s in the simple Artinian ring A and again $\sigma(s^{-1}s) = \sigma(s^{-1})\sigma(s)$ hence $\sigma(s^{-1}) = \sigma(s)^{-1}$ in $G_F(A)$, and in view of the gradation of $G_F(A), \sigma(s)^{-1} \in G_F(A)_0$. Since $G_F(A)_0$ is a prime Goldie ring its classical ring of fractions is a simple Artinian ring but since left regular elements of $G_F(A)_0$ are invertible in $G_F(A)_0$ it follows that $G_F(A)_0$ is itself simple Artinian.

2. Take $x \in F_0^0 A = \sum_{\gamma<0} F_\gamma A$ and look at $1 + ax$ for $a \in F_0 A$. Clearly, $ax \in F_\gamma A$ for some $\gamma < 0$ hence $\sigma(1 + ax) = 1$. As in the proof of (1), $1 + ax$ is regular in A, hence it has an inverse y. Then from $y(1 + ax) = 1$ it follows that $\sigma(y)\sigma(1 + ax) = 1$, or $\sigma(y) = 1$ and $y \in F_0 A$. This establishes $F_0 A^0 \subset J(F_0 A)$. Since $F_0 A/F_0^0 A$ is simple (see (1)), $F_0^0 A$ is a maximal ideal of $F_0 A$ and thus $F_0 A^0 = J(F_0 A)$ follows.

3. We already know that $G_F(A)_0$ is simple Artinian so $F_0 A$ will be a Dubrovin valuation ring of A if for all $a \in A - F_0 A$ there exist $x, y \in F_0 A$ such that $xa, ay \in F_0 A - F_0^0 A$. Fix $a \in A - F_0 A$, let $\sigma(a) \in G_F(A)_\gamma$. By condition (E) there is a $\overline{b} \in G_F(A)_{\gamma^{-1}}$ such that $\overline{b}\sigma(a) \neq 0$; since $a \notin F_0 A, \gamma > 0$ and $\gamma^{-1} < 0$. Pick $b \in F_\gamma^{-1} A$ such that $\sigma(b) = \overline{b}$; since $\overline{b}\sigma(a) \neq 0$, $ba \bmod F_0^0 A \neq 0$, hence $ba \in F_0 A - F_0^0 A$, take $x = b$. Similarly, we find y such that $ay \in F_0 A - F_0^0 A$. Hence $F_0 A$ is a Dubrovin valuation ring of A. □

1.8.12 Corollary

Let FA be a separated Γ-filtration on a simple Artinian ring A such that $G_F(A)_0$ is a prime (left) Goldie ring. Then FA is a strong filtration, $G_F(A)$ is strongly graded and in fact a crossed product $G_F(A)_0 * \Gamma$.

Proof. Follows from the fact that F_0A is a Dubrovin valuation ring of A (cf. Proposition 1.8.8). □

1.9 Reductions of Algebras and Filtrations

In this section we consider a field K, Γ an abelian totally ordered group written additively and $v : K \to \Gamma \cup \{\infty\}$ a valuation of K, which is assumed to be surjective. We write O_v for the valuation ring of K associated to v, m_v for its unique maximal ideal and $k_v = O_v/m_v$ for the residual field. To a valuation v on K we associate a Γ-filtration on K, say f^vK, defined by $f^v_\gamma K = \{x \in K, v(x) \geq -\gamma\}$. Obviously $f^v_o K = O_v$, $f^v_{<0}K = m_v$.

Consider a finite dimensional K-vector space V and M an O_v-submodule of V. Recall that M is said to be an **O_v-lattice** in V if it contains a K-basis of V and it is an O_v-submodule of a finitely generated O_v-submodule of V. We denote the k_v-vector space M/m_vM by $\overline{V}$. For an O_v-lattice M in V we have $\dim_{k_v} \overline{V} \leq \dim_K V$; when equality holds in the latter we say that M defines an unramified reduction of V.

1.9.1 Lemma

Let M be an O_v-lattice of V.

1. If M is a finitely generated O_v-module, then M is a free O_v-module of rank less than or equal to $\dim_K V$ (this happens e.g. when $\Gamma = \mathbb{Z}$ because then O_v is Noetherian).
2. If M defines an unramified reduction of V then M is a free O_v-module of rank equal to $\dim_K V$.

Proof. 1. Since the O_v-module M is contained in a K-vector space it is a torsion free O_v-module and finitely generated torsion free O_v-modules over a valuation ring are known to be free O_v-modules; the rank condition is obvious.
2. Let $\{\overline{x}_1, \ldots, \overline{x}_n\}$ be a k_v-basis of $\overline{V}$, $n = \dim_K V$. The $\{x_1, \ldots, x_n\} \subset M$ is linearly independent over O_v hence over K. For $m \in M$, there exist $\lambda_1, \ldots, \lambda_n \in K$ such that $m = \sum_{i=1}^n \lambda_i x_i$. For some $j_0 \in \{1, \ldots, n\}$ we have $\lambda_{j_0}^{-1}\lambda_i \in O_v$ for all $i \in 1, \ldots, n$.
   If not all $\lambda_i \in O_v$ (which we may assume since otherwise $m \in O_v x_1 + \ldots + O_v x_n$) then $\lambda_{j_0}^{-1} \in m_v$ and we obtain $\lambda_{j_0}^{-1} m = \sum_{i=1}^n (\lambda_{j_0}^{-1}\lambda_i) x_i$, reducing modulo m_vM to a nontrivial (because $\lambda_{j_0}^{-1}\lambda_{j_0} = 1$) relation for the $\overline{x}_1, \ldots, \overline{x}_n$, a contradiction to the choice of the $\overline{x}_i, i = 1, \ldots, n$. □

Given any O_v-lattice M in V we may define a filtration on V by putting: $F^v_\gamma V = (f^v_\gamma K)M$ for $\gamma \in \Gamma$. It is clear that F^vV makes V into an exhaustively filtered K-

vector space, in fact for $\gamma \geq 0$, $F^v_\gamma V \supset M$ and exhaustivity follows from $KM = V$. In general it is not clear whether $F^v V$ is a separated filtration.

1.9.2 Lemma

If the O_v-lattice M in V is a free O_v-module then the filtration $F^v V$ defined by M on V is Γ-separated.

Proof. Let $\{x_1, \ldots, x_r, \ldots\}$ be an O_v-basis of M, so they are also K-independent in V. Pick $v \in V$. From $KM = V$ it follows that there is an $a \in O_v - \{0\}$ such that $av \in M$. Write $av = \sum_{i=1}^r \lambda_i x_i$ with $\lambda_i \in O_v$, for some finite r. Putting $\lambda_i^* = a^{-1}\lambda_i \in K$, we have $x = \sum_{i=1}^r \lambda_i^* x_i$. For the $\lambda_i^* \neq 0$ we put $v(\lambda_i^*) = -\gamma_i \in \Gamma$. Up to a reordering we may assume $\gamma_1 < \ldots < \gamma_r = \gamma$. We claim that $x \in F^v_\gamma V$ but $x \notin \sum_{\delta<\gamma} F^v_\delta V$. It is obvious that $x \in F^v_\gamma V$. Suppose $x \in \sum_{\delta<\gamma} F^v_\delta V$, then $x \in (f^v_{\delta_0} K)M$ for some $\delta_0 < \gamma$, and we may write $x = \Sigma' \mu_j x_j$ with $\mu_j \in f^v_{\delta_0} K$ and $\delta_0 < \gamma$. Since $\{x_1, \ldots, x_r, \ldots\}$ is K-independent in V, we derive $x = \sum_{i=1}^r \lambda_i^* x_i = \sum \mu_j x_j$ and thus $\lambda_r^* = \mu_r$, but $v(\lambda_r^*) = -\gamma$ and $v(\mu_r) \geq -\delta_0$ with $\delta_0 < \gamma$, a contradiction. Hence for any $x \in V$ there exists a $\gamma \in \Gamma$ such that $x \in F^v_\gamma V$ and $x \notin \sum_{\delta<\gamma} F^v_\delta V$. □

Consider a field K and a K-algebra A with a separated $\mathbb{Z}$-filtration FA such that $K \subset F_0 A$. We assume moreover that FA is finite in the sense that $\dim_K F_n A < \infty$ for all $n \in \mathbb{Z}$; then there is an $n_0 \in \mathbb{Z}$ such that $F_n A = 0$ for all $n \leq n_0$. Without loss of generality we may assume that FA is a positive filtration. For a subring Λ of A we define the induced filtration $F\Lambda$ by putting for $n \in \mathbb{Z}$, $F_n \Lambda = F_n A \cap \Lambda$.

1.9.3 Definition

The subring Λ is called an **F-reductor** if $\Lambda \cap K = O_v$ is a valuation ring of K and $F_n \Lambda$ is an O_v-lattice in $F_n A$ for all $n \in \mathbb{Z}$. The ring $\underline{A} = \Lambda / m_v \Lambda$, where m_v is the maximal ideal of O_v, is the **reduction of A with respect to** Λ. We say that Λ is an **unramified F-reductor** in case $F_n \Lambda$ is an unramified reductor of $F_n A$ for all $n \in \mathbb{Z}$. The **valuation filtration $F^v A$ on** A (defined from Λ) is given by: $F^v_\gamma A = (f^v_\gamma K)\Lambda, \gamma \in \Gamma$, where $f^v K$ is the valuation filtration of K.

For a general reductor it is not clear whether the associated valuation filtration on A is separated, for the unramified case we have the following result.

1.9.4 Proposition

Let A be a filtered K-algebra with finite $\mathbb{Z}$-filtration FA and $\Lambda \subset A$ an unramified F-reductor, then the valuation filtration F^vA is Γ-separated.

Proof. Since $F_n\Lambda$ defines an unramified reduction of F_nA, we may apply Lemma 1.9.1.2 and conclude that $F_n\Lambda$ is a free O_v-module, for every $n \in \mathbb{Z}$. The filtration $F^v(F_nA)$ given by $F^v_\gamma(F_nA) = (f^v_\gamma K)F_n\Lambda$ is Γ-separated in view of Lemma 1.6.2, for every $n \in \mathbb{Z}$. Now we claim that for $\gamma \in \Gamma, n \in \mathbb{Z}$ we have:

$$(f^v_\gamma K)\Lambda \cap F_nA = (f^v_\gamma K)(\Lambda \cap F_nA) = (f^v_\gamma K)(F_n\Lambda) \qquad (*)$$

First the inclusion $(f^v_\gamma K)(\Lambda \cap F_nA) \subset (f^v_\gamma K)\Lambda \cap F_nA$ is obvious. For the converse pick $x \in (f^v_\gamma K)\Lambda \cap F_nA$. Then $(f^v_{-\gamma}K)x \subset \Lambda \cap F_nA$ follows and thus $x \in (f^v_\gamma K)(\Lambda \cap F_nA)$ because $(f^v_\gamma K)(f^v_{-\gamma}K) = f^v_0K \ni 1$.
Now $(*)$ means exactly that the filtration F^vA induces the valuation filtration on F_nA defined with respect to $F_n\Lambda$. Take $x \in A$, say $x \neq 0$, then $x \in F_nA$ for some n. Since the valuation filtration on F_nA is Γ-separated there exists a $\gamma \in \Gamma$ such that $x \in F^v_\gamma(F_nA) - F^v_{<\gamma}(F_nA)$ thus $x \in F^v_\gamma A - F^v_{<\gamma}A$, because if $x \in F^v_\delta A$ with $\delta < \gamma$ then $x \in F_nA \cap F^v_\delta A = F_nA \cap (f^v_\delta K)F_n\Lambda$ which by $(*)$ equals $(f^\mu_\gamma K)F_n\Lambda$ and this leads to $x \in F^v_{\delta<\gamma}(F_nA)$, a contradiction. Therefore F^vA is Γ-separated. □

Note that for $\Gamma = \mathbb{Z}$ an F-reductor necessarily defines a separated valuation filtration, that is because in this case $F_n\Lambda$ is a finitely generated O_v-module in F_nA hence it is free, making all $F^v(F_nA)$ separated and then the same proof as in Proposition 1.9.4 allows to conclude that F^vA is separated!

On $\overline{A}$ we induce a $\mathbb{Z}$-filtration $F\overline{A}$ by putting $\overline{F_nA} = F_n\Lambda/m_vF_n\Lambda$ with respect to an F-reductor Λ of A. This filtration is exhaustive and is finite if FA is finite (see remarks before Lemma 1.9.1). It is easily seen that $G_{\overline{F}}(\overline{A}) = G_F(\Lambda)/m_vG_F(\Lambda)$ which may be seen as a graded reduction of $G_F(A)$.

For any $\mathbb{Z}$-graded K-algebra $R = \sum_{n\in\mathbb{Z}} R_n$, with $K \subset R_0$, a graded subring S is a **graded reductor** for R if $S \cap K = O_v$ and $S \cap R_n$ is an O_v-lattice in R_n for all $n \in \mathbb{Z}$. The $G_F(\Lambda)$ appearing above is a graded reductor for $G_F(A)$.

In the sequel we assume that R is finite graded, that is $\dim_K R_n < \infty$ for almost all n and $K \subset R_0$. The valuation filtration F^vR is defined by $F^vR = (f^v_\gamma K)\Lambda$. A **graded filtration** FR is such that $F_\gamma R = \oplus_{n\in\mathbb{N}}(F_\gamma R \cap R_n)$ for all $\gamma \in \Gamma$.

1.9.5 Proposition

Let $\Lambda \subset A$ be an F-reductor, then:

1. The valuation filtration F^vA is exhaustive.
2. F^vA is separated if and only if $F^v(F_nA)$ is Γ-separated for all $n \in \mathbb{N}$.

3. The induced filtration of $F^v A$ on $F_n A$ is the valuation filtration on $F_n A$ and provided that it is Γ-separated we get $G_v(F_n A) = \overline{F_n A} * \Gamma$, where $\overline{F_n A} = F_n \Lambda / m_v F_n \Lambda$.
4. $\overline{F_n A}$ defines a finite filtration on $\overline{A} = \Lambda / m_v \Lambda$, denoted by $\overline{F}\,\overline{A}$, and $G_{\overline{F}}(\overline{A}) = \overline{G(A)} = G(\Lambda)/m_v G(\Lambda)$.

Proof. 1. Since $F_n \Lambda = F_n A \cap \Lambda$ is an O_v-lattice in $F_n A$, we have $K(F_n \Lambda) = F_n A$ for all $n \in \mathbb{N}$. As FA is exhaustive, we obtain $K\Lambda = A$. The valuation filtration $f^v K$ is exhaustive hence $K\Lambda = A$ entails that $F^v A$ is an exhaustive filtration.
2. From (∗) in the proof Proposition 1.9.4 we retain:

$$(f^v_\gamma K)\Lambda \cap F_n A = (f^v_\gamma K)(\Lambda \cap F_n A)$$

Fix an $x \neq 0$ in A. There exits an $n \in \mathbb{N}$ such that $x \in F_n A$. If we assume that the induced valuation filtrations on all $F_n A$ are separated, then there exists a $\gamma \in \Gamma$ such that $x \in F^v_\gamma (F_n A) - F^v_{<\gamma}(F_n A)$, hence $x \in F^v_\gamma A - F^v_{<\gamma} A$ since $F^v(F_n A)$ is induced in $F_n A$ by $F^v A$. Conversely if $F^v A$ is separated then each $F^v(F_n A)$ is separated, hence the valuation filtration of each $F_n A$ is separated.
3. We have seen in (2) that the induced filtration of $F^v A$ on $F_n A$ is the valuation filtration of $F_n A$, $F^v_\gamma(F_n A) = (f^v_\gamma K)(\Lambda \cap F_n A)$. This is obviously a strong filtration and $G_v(F_n A) \subset G_v(A)$ as $G_f(R)$-modules because the filtrations are supposed to be separated. Then $G_v(F_n A)$ is $G_{v,0}(F_n A) \otimes_k (k_v * \Gamma)$ where $k_v * \Gamma = G_v(K)$, i.e. $\overline{F_n A} * \Gamma$ where $G_{v,0}(F_n A) = \overline{F_n A} = (F_n A \cap \Lambda)/m_v(F_n A \cap \Lambda)$.
4. Since $\Lambda \cap F_n A$ is an O_v-lattice in $F_n A$ we obtain that $\dim_{k_v} \overline{F_n A} \le \dim_K F_n A$, hence $\overline{FA}$ is a finite filtration. It is easily verified that $G_{\overline{F}}(\overline{A}) = \overline{G_F(A)}$. □

In the foregoing we may verify that $G_F(\Lambda) \subset G_F(A)$ is a graded reductor for $G_F(A)$ since we have that $\Lambda \cap F_n A / \Lambda \cap F_{n-1} A$ is an O_v-lattice in $G_F(A)_n$ for every $n \in \mathbb{N}$. We have a graded version of the foregoing property.

1.9.6 Proposition

Let R be a positively graded K-algebra and $\Lambda \subset R$ a graded reductor, then:

1. The valuation filtration $F^v R$ is an exhaustive graded filtration.
2. $F^v R$ is separated if and only if $F^v R_n$ is separated for all $n \in \mathbb{N}$.
3. The induced filtration of $F^v R$ on R_n is the valuation filtration on R_n and provided it is separated we obtain: $G_v(R_n) = \overline{R}_n * \Gamma$ where $\overline{R}_n = \Lambda_n / m_v \Lambda_v$.
4. $\overline{R} = \oplus_n \overline{R}_n$ and $G_v(R) = \overline{R} * \Gamma$.

Proof. An easy modification of the proof of the foregoing proposition. □

We may now consider the lifting properties from (associated) graded rings to filtered rings.

1.9.7 *Proposition*

Let A be a K-algebra with finite positive filtration FA and consider an O_v-subring Λ of A.

1. If Λ is an F-reductor then $G_F(\Lambda) \subset G_F(A)$ and the Rees ring $\widetilde{\Lambda} \subset \widetilde{A}$ are graded reductors.
2. If $G_F(\Lambda) \subset G_F(A)$ or $\widetilde{\Lambda} \subset \widetilde{A}$ is a graded reductor then $\Lambda \subset A$ is an F-reductor.

Proof. 1. That $G_F(\Lambda) \subset G_F(A)$ is a graded reductor has been observed before, for $\widetilde{\Lambda} \subset \widetilde{A}$ it suffices to recall that $\widetilde{\Lambda}_n \cong F_n\Lambda$ i.e. $\widetilde{\Lambda} \cap \widetilde{A}_n$ is an O_v-lattice for every n.

2. If $\widetilde{\Lambda} \subset \widetilde{A}$ is a graded reductor then $\widetilde{\Lambda} \cap \widetilde{A}_0 = \widetilde{\Lambda} \cap K = O_v$ and $\widetilde{\Lambda} \cap \widetilde{A}_n \cong \Lambda \cap F_n A$ is an O_v-lattice.
3. If $\widetilde{\Lambda}$ is a graded reductor then from $\widetilde{\Lambda} \cap \widetilde{A}_n \cong \Lambda \cap F_n A$ it follows that Λ is an F-reductor. Assume $G_F(\Lambda) \subset G_F(A)$ is a graded reductor. Thus $G_F(\Lambda)_1$ is an O_v-lattice in $G_F(A)_1$, since $G_F(\Lambda)_1 = F_1\Lambda/O_v$ it follows that $F_1\Lambda$ is an O_v-lattice. The proof can now be completed by induction i.e. $G_F(\Lambda)_n$ is an O_v-lattice hence $F_n\Lambda/F_{n-1}\Lambda$ is an O_v-lattice, the induction hypothesis yields that F_{n-1}, Λ is an O_v-lattice and then it follows that $F_n\Lambda$ is an O_v-lattice. □

1.9.8 *Example*

Let g be a finite dimensional Lie algebra over K and $U_K(g)$ the enveloping algebra of g over K. Fix a K-basis for g, say $\{x_1, \dots, x_n\}$. There are structure constants $\lambda_{ij} \in K$ given by:

$$[x_i, x_j] = \sum_{k=1}^{n} \lambda_{ij}^k x_k$$

without loss of generality (up to multiplying by one $(\lambda_{ij}^k)^{-1}$) we may assume that all $\lambda_{ij}^k \in O_v$ but not all in m_v. Put $g_{O_v} = O_v x_1 + \dots + O_v x_n$, it is a Lie O_v-algebra with the induced bracket and it is an O_v-lattice in g. On $\overline{g} = g_{O_v}/m_v g_{O_v}$ we define a Lie algebra structure over k_v by putting:

$$[\overline{x}_i, \overline{x}_j] = \sum_{k=1}^{n} \overline{\lambda_{ij}^k}\, \overline{x}_k$$

where $\overline{x}_i = x_i$-mod $m_v g_{O_v}$. By assumption $\overline{g}$ is not the trivial Lie algebra (some $\lambda_{ij} \neq 0$). Obviously $\overline{g}$ depends on the choice of the K-basis $\{x_1, \dots, x_n\}$ for g. Put $\Lambda = U_{O_v}(g_{O_v})$. Consider on $A = U_K(g)$ the standard filtration (coming from the gradation filtration of the tensor algebra $T(g)$) and on Λ the induced filtration. Clearly FA is a finite filtration. We have that $G_F(\Lambda) = O_v[X_1, \dots, X_n] \subset K[X_1, \dots, X_n] = G_F(A)$ and Λ is an F-reductor, $\overline{G}(\Lambda)$ is a graded reduction for

$G_F(A)$ (use Proposition 1.9.7 for example). We also have that F^vA is Γ-separated and $G_v(R)$ is a domain isomorphic to $U_{k_v}(\overline{g})$. The latter means that we can extend v to a noncommutative valuation on the skewfield $\mathcal{D}(g)$, the total ring of fractions of $U(g)$ (see Proposition 1.8.10 (3).).

Again look at a K-algebra A as in Proposition 1.9.7. On $G_v(A)$ we define a filtration $fG_v(A)$ by: $f_n(G_v(A)) = G_v(F_nA)$, for $n \in \mathbb{N}$. When $G_F(\Lambda) \subset G_F(A)$ is a graded reductor then we also have a valuation filtration $f^vG_F(A)$ given by:

$$f^v_\gamma(G_F(A)) = (f^v_\gamma K)G_F(\Lambda), \text{ for } \gamma \in \Gamma$$

If F^vA is Γ-separated then the valuation filtration $f^vG_F(A)$ is also Γ-separated (similar to the proof of Proposition 1.9.5).

1.9.9 Proposition

Let A be as before and $\Lambda \subset A$ an F-reductor such that F^vA is Γ-separated. Then we have that: $G_v(G_F(A)) \cong \overline{G(A)} * \Gamma$ and $G_f(G_v(A) = G_v(G_F(A))$.

Proof. We have $G_v(G_F(A)) = \oplus_{\gamma\in\Gamma}(f^v_\gamma K)G_F(\Lambda)/(f^v_{<\gamma}K)G_F(\Lambda)$ and since $G_F(\Lambda)$ is O_v-flat (because torsionfree) we obtain that $G_v(G_F(A)) = \oplus_{\gamma\in\Gamma}(f^v_\gamma K)/(f^v_{<\gamma}K)\otimes_{O_v} G_F(\Lambda)$. The latter means $G_v(K)\otimes_{O_v} G_F(\Lambda) = (k_v * \Gamma)\otimes_{O_v} G_F(\Lambda) = \overline{G(A)} * \Gamma$. The fact that $G_f(G_v(A)) = G_v(G_F(A))$ may be verified directly, it follows by a general compatibility result of M. Hussein and F. Van Oystaeyen [49] (see Proposition 2.4. loc. cit).

1.9.10 Corollary

Let $\Lambda \subset A$ be an F-reductor such tat F^vA is Γ-separated. If $\overline{G(A)}$ is a domain then so are $G_F(A)$, $G_v(A)$ and A.

Proof. By Proposition 1.8.4 we get that $G_v(G_F(A))$ is a domain; then $G_F(A)$ is a domain and A is a domain. From $G_f(G_v(A)) = G_v(G_F(A))$ it follows that $G_f(G_v(A))$ is a domain and so is $G_v(A)$. □

1.10 Appendix: Global Dimension and Regularity of Reductions

This section is less self-contained so we add it as an appendix. Basic references are:

- For Krull dimension, R. Gordon, J. C. Robson [27]
- For Noetherian rings, J. McConnell, J. C. Robson [47]

- For dimensions of rings, C. Năstăsescu, F. Van Oystaeyen [53]
- For Homological algebra, J. T. Rotman [59]

We study $\mathbb{Z}$-graded D algebras having the commutative domain D for the part of degree zero, where D is usually a regular (local) domain. The properties we focus on are related to homological dimension and Auslander–Gorenstein regularity. A problem is that the maximal ideal w of the local domain D is not necessarily contained in the Jacobson radical of the D-algebra under consideration. In particular when D is a discrete valuation ring, there is a link to noncommutative valuation theory, hence the inclusion of these results in this work.

Let Λ be an algebra over a commutative domain D. Usually Λ is given by generators and finitely many relations $0 \to Q \to D < \underline{X} > \to \Lambda \to 0$, where $D < \underline{X} >$ is the free algebra over $\underline{X} = \{X_1, \ldots, X_n\}$ and Q is generated as a two-sided ideal by finitely many $p_1, \ldots, p_t \in D < \underline{X} >$. If K is the field of fractions of D then $A = K \otimes_D \Lambda$ may be given by generators and the same relations viewed in $K < \underline{X} >$. Note however that for a given set of relations $q_1, \ldots, q_s \in K < \underline{X} >$ defining A, even if the q_i have coefficients in O_v, need not yield a set of relations defining Λ! We come back to this problem when studying extensions of valuations of K to valuations of $Q(A)$ in Chap. 3. Now starting from $p_1, \ldots, p_t$ we obtain the residue algebra $\overline{\Lambda} = \Lambda / w\Lambda$ with respect to a maximal ideal w of D, by the reduced relations $\overline{p}_1, \ldots, \overline{p}_t$ over $k_w = D/w$ (not by $\overline{q}_1, \ldots, \overline{q}_s$ in general!).

A ring Δ is called a **quotient ring** if every regular element of Δ is invertible. A subring Λ of Δ is a **right (left) order** in Δ if each $q \in \Delta$ is of the form as^{-1} (resp. $s^{-1}a$) for some $a, s \in \Lambda$; by an **order** in Δ we mean a left and right order. Two orders Λ_1 and Λ_2 are **equivalent**, denoted by $\Lambda_1 \sim \Lambda_2$, if there exist units a_1, a_2, b_1, b_2 in Δ such that: $a_1 \Lambda_1 b_1 \subset \Lambda_2, a_1 \Lambda_2 b_2 \subset \Lambda_1$. An order Λ is a **maximal order** in Δ if Λ is maximal in its equivalence class. It is well known that a prime Goldie ring is an order in a simple Artinian ring. We will study maximal orders in Chap. 2. A commutative domain is a maximal order if it is completely integrally closed, in case it is a Noetherian domain this reduces to integrally closedness.

For a prime ideal p of D, let D_p be the localization of D at p. We obtain the exact sequence: $0 \to D_p \otimes_D Q \to D_p < \underline{X} > \to \Lambda_p \to 0$. Let $\mathrm{Spec} D$ be the prime spectrum of D and $\Omega(D)$ the space of maximal ideals; both $\oplus_{p \in \mathrm{Spec} D} \Lambda_p$ and $\oplus_{w \in \Omega(D)} \Lambda_w$ are left and right faithfully flat Λ-modules.

1.10.1 Proposition

Consider a domain Λ having a quotient ring Δ. If Λ_p is a maximal order for every $p \in \mathrm{Spec} D$, then Λ is a maximal order in Δ.

Proof. Since $\Lambda = \cap\{\Lambda_p, p \in \mathrm{Spec} D\}$ the statement follows easily from the faithful flatness of $\oplus_{p \in \mathrm{Spec} D} \Lambda_p$ (or by a rather direct verification). □

We write $K\text{dim}$ for the Krull dimension in the sense of Gabriel, Rentschler [24] and gldim for the global (projective) dimension [47, 53].

1.10.2 Proposition

1. $K\text{dim}\Lambda = \sup\{K\text{dim}\Lambda_p, p \in \text{Spec}D\}$.
2. If Λ is left and right Noetherian then: $\text{inj.dim}\Lambda = \sup\{\text{inj.dim}\Lambda_p, p \in \text{Spec}D\}$.
3. If Λ is left and right Noetherian and $\text{gldim}\Lambda$ is finite, then $\text{gldim}\,\Lambda = \sup\{\text{gldim}\Lambda_p, p \in \text{Spec}D\}$.

Proof. See [40].

Recall that a left and right Noetherian ring R is said to be an **Auslander–Gorenstein** (resp. **Auslander**) regular ring if R has finite injective (resp. global) dimension, say μ, and satisfies the Auslander condition, that is, for any finitely generated R-module M, any $0 \leq k \leq \mu$ and any submodule N of $\text{Ext}^k_R(M, R)$ we have $\text{Ext}^n_R(N, R) = 0$ for $n < k$.

1.10.3 Proposition

Let Λ as before be left and right Noetherian

1. Λ is Auslander–Gorenstein regular if and only if Λ_p is Auslander–Gorenstein regular for every $p \in \text{Spec}D$.
2. If Λ has finite global dimension then Λ is Auslander regular if and only if Λ_p is Auslander regular for every $p \in \text{Spec}D$.

Proof. In view of Proposition 1.10.2 we only have to establish that Λ satisfies the Auslander condition if and only if Λ_p satisfies the Auslander condition for every $p \in \text{Spec}D$. First assume that Λ satisfies the Auslander condition. Let N be any Λ_p-submodule of $\Lambda_p \otimes_\Lambda \text{Ext}^k_\Lambda(M, \Lambda) = \text{Ext}^k_{\Lambda_p}(S_p^{-1}M, \Lambda_p)$ where M is a finitely generated Λ-module, $S_p = D - p$, then there is a Λ-submodule N' of $\text{Ext}^k_\Lambda(M, \Lambda)$ such that $N = S_p^{-1}N'$. Hence for $n < k$, we have $0 = \Lambda_p \otimes_\Lambda \text{Ext}^n_\Lambda(N', \Lambda) = \text{Ext}^n_{\Lambda_p}(S_p^{-1}N', \Lambda_p) = \text{Ext}^n_{\Lambda_p}(N, \Lambda_p)$.

Conversely, suppose that Λ_p satisfies the Auslander condition for every $p \in \text{Spec}D$. Let N be any Λ-submodule of $\text{Ext}^A_\Lambda(M, \Lambda)$ where M is a finitely generated Λ-module. Since $S_p^{-1}N$ is a Λ_p-submodule of $\Lambda_p \otimes_\Lambda \text{Ext}^k_\Lambda(M, \Lambda) = \text{Ext}^k_{\Lambda_p}(S_p^{-1}M, \Lambda_p)$. For $n < k$ we obtain: $\Lambda_p \otimes_\Lambda \text{Ext}^n_\Lambda(N, \Lambda) = \text{Ext}^n_{\Lambda_p}(S_p^{-1}N, \Lambda_p) = 0$. Then from the faithful flatness of $\oplus_{p \in \text{Spec}D}\Lambda_p$ it follows that $\text{Ext}^n_\Lambda(N, \Lambda) = 0$.

In the sequel D is a commutative domain, w an ideal of D contained in the Jacobson radical $J(D)$ of D; we write $k = D/w$ and consider a $\mathbb{Z}$-graded D-algebra Λ with $\Lambda_0 = D$. We write Λ-gr for the category of graded left Λ-modules and graded Λ-morphisms of degree zero.

1.10.4 Lemma

With notation as above:

1. $w\Lambda$ is contained in the graded Jacobson radical $J^g(\Lambda)$ which is the largest proper graded ideal of Λ such that its intersection with Λ_0 is contained in the Jacobson radical of Λ_0.
2. If Λ is a free D-module of finite rank than $w\Lambda$ is contained in the Jacobson $J(\Lambda)$ of Λ.

Proof. 1. Since $\Lambda_0 = D$ it follows that $w\Lambda \cap \Lambda = w \subset J(\Lambda_0)$. Hence $w\Lambda \subset J^g(\Lambda)$ (cf. [52] for detail on the graded Jacobson radical).
2. If Λ is a free D-module of finite rank then it follows from a result of G. Bergman (cf. [52] p. 56) that $w \subset J(D) = D \cap J(\Lambda) \subset J(\Lambda)$, or $w\Lambda \subset J(\Lambda)$. □

From hereon we assume that $\mathbf{\Lambda}$ is flat over D; we write $\overline{\Lambda} = \Lambda/w\Lambda$ and for any Λ-module M, $\overline{M} = M/wM$.

1.10.5 Proposition

Suppose M is finitely generated in Λ-gr, M is flat over D and Λ is left Noetherian. If $\overline{M}$ is projective in $\overline{\Lambda}$-gr then M is projective in Λ-gr.

Proof. The proof may be reduced to the case where $\overline{M}$ is free in $\overline{\Lambda}$-gr; then let $\{\overline{m}_1, \dots, \overline{m}_n\}$ be a $\overline{\Lambda}$-basis of $\overline{M}$ consisting of homogeneous elements and let $m_i \in M$ be homogeneous preimages for the $\overline{m}_i$, $i = 1, \dots, n$. Since $w\Lambda \subset J^g(\Lambda)$, $M = \sum_{i=1}^n \Lambda m_i$ follows from the graded version of Nakayama's lemma (cf. [52]). Consider the gr-free Λ-module $L = \oplus_{i=1}^n \Lambda e_i$ and the exact sequence in Λ-gr: $0 \to K \to L \underset{\pi}{\longrightarrow} M \to 0$, π given by $\pi(e_i) = m_i$, $i = 1, \dots, n$, and $K = \operatorname{Ker} \pi$. Since M is flat over D we obtain exactness of $0 \to \overline{K} \to \overline{L} \underset{\overline{\pi}}{\longrightarrow} \overline{M} \to 0$, where $\overline{L} = D/w \otimes_D L$, $\overline{K} = D/w \otimes_D K$. Since the $\overline{\pi}(\overline{e}_i)$ are free generators for $\overline{M}$, $\overline{\pi}$ is an isomorphism, i.e. $\overline{K} = 0$. Since K is finitely generated as Λ is left Noetherian, and a graded Λ-module the graded version of Nakayama's Lemma applies again, leading to $K = 0$ and $M \cong L$ or M is free. □

1.10.6 Proposition

Suppose Λ is left Noetherian.

1. $\text{grgldim}\,\Lambda \leq \text{grgldim}\,\overline{\Lambda} + \text{pdim}_{\Lambda}\overline{\Lambda}$, where grgldim denotes the left graded global dimension.
2. If Λ is positively graded then $\text{gldim}\Lambda \leq \text{gldim}\overline{\Lambda} + \text{pdim}_{\Lambda}\overline{\Lambda}$.

Proof. Statement (2) follows from (1) since for a positively graded ring the global dimension coincides with the graded global dimension.

1. We have that $\text{pdim}_{\Lambda}\overline{\Lambda}$ equals the graded projective dimension so we can restrict to resolutions of graded modules. By the change of rings theorem (cf. [40, 59]) $\text{pdim}_{\Lambda}\overline{\Lambda}$ is introduced in the comparison of grgldim for Λ and $\overline{\Lambda}$. Applying the lifting of projectivity established in the foregoing proposition then yields the result. □

In a similar way we may establish the following.

1.10.7 Proposition

Suppose Λ is left Noetherian and suppose $w = aD$ for some $a \in D$ regular on Λ (i.e. Λ is without a-torsion), then we have:

1. If Λ is positively graded then: $\text{injdim}\,\Lambda = 1 + \text{injdim}\,\overline{\Lambda}$.
2. $\text{grgldim}\,\Lambda \leq \text{grgldim}\,\overline{\Lambda} + 1$ with equality holding in case $\text{grgldim}\Lambda < \infty$.

1.10.8 Corollary

In the situation of foregoing proposition: but assuming now that $w = (x_1, \ldots, x_d)$ where $\{x_1, \ldots, x_d\}$ is a regular sequence for the domain D which is regular on Λ. Then we have:

1. If Λ is positively graded then $\text{injdim}\Lambda = d + \text{injdim}\overline{\Lambda}$.
2. $\text{grgldim}\,\Lambda \leq \text{grgldim}\,\overline{\Lambda} + d$ with equality in case $\text{grgldim}\,\overline{\Lambda} < \infty$.

1.10.9 Proposition

Suppose Λ is Noetherian and $w = (x_1, \ldots, x_d)$ as in Corollary 1.10.8. If $\overline{\Lambda}$ is Auslander (-Gorenstein) regular then Λ is Auslander (-Gorenstein) regular.

Proof. Cf. [40]. □

1.10.10 Proposition

Let K be the field of fractions of D and A a K-algebra such that Λ is a subring over D of A such that $K\Lambda = A$. Suppose Λ is left Noetherian and $w = (x_1, \ldots, x_d)$ as before. If $\overline{\Lambda}$ is a domain then Λ is a domain.

Proof. It suffices (via an easy iteration argument) to prove the claims in case $w = Da, a = x_1$. Since $w\Lambda = a\Lambda$ is an invertible ideal and graded, the w-adic filtration on the ring $S = \cup\{a^{-1}\Lambda, n \in \mathbb{Z}\} \subset A$ defined via $F_nS = a^{-n}\Lambda, n \in \mathbb{Z}$, is a strong filtration, i.e. $F_nSF_mS = F_{n+m}S$ for all $n, m \in \mathbb{Z}$. Hence the associated graded ring $G_w(S)$ is a strongly $\mathbb{Z}$-graded ring with $G_w(S)_0 = \Lambda/w\Lambda = \overline{\Lambda}$ a domain, therefore $G_w(S)$ is a domain. The negative part $G_w(S)_- = \oplus_{n\leq 0}G_w(S)_n$ is also a domain but it is also the associated graded ring, $G\ (\Lambda)$ say, of Λ with respect to the w-adic filtration on Λ. So if we can prove that $F\Lambda$ is separated then it follows that Λ is a domain too. Now $w\Lambda = a\Lambda$ is invertible and Λ is left Noetherian, thus a Λ has the Artin–Rees property. Then, since $a\Lambda \subset J^g(\Lambda)$ it follows that $\cap_{n\geq 1}a^n\Lambda = 0$. □

1.10.11 Corollary

Let A be as in the proposition. Suppose Λ is left Noetherian and $w = aD$. If $\overline{\Lambda}$ is a domain, then the $w\Lambda$-adic filtration on Λ defines a valuation function v on Λ and v naturally extends to a discrete valuation of the total quotient ring $Q_{cl}(\Lambda)$ of Λ.

Proof. From the proposition it follows that Λ is a left Noetherian domain hence an order in a skewfield. Moreover the $w\Lambda$-adic filtration is separated (proof of foregoing proposition), hence we may apply Proposition 1.8.10 (3). □

If we suppose that the graded ring Λ has a graded ring of fractions, Δ^g say, then Λ is a gr-**maximal order** in Δ^g if for any graded subring T of Δ^g such that $T \supset \Lambda$ and $aTb \subset \Lambda$ for some homogeneous units $a, b \in \Delta^g$, we have $T = \Lambda$. It is known that Λ is a gr-maximal order in Δ^g if and only if the graded ring Λ is a maximal order in Δ, the total quotient ring of Δ^g (cf. [38], Graded Orders).

1.10.12 Proposition

Suppose Λ is a positively graded left Noetherian ring and $w = aD$ where a is regular on Λ. If $\overline{\Lambda}$ is a gr-maximal order in a gr-simple Artinian ring, then Λ is a gr-max order in a gr-simple Artinian ring.

Proof. The element a is a central regular homogeneous element of Λ with $a\Lambda \subset J^g(\Lambda)$ and the $a\Lambda$-adic filtration of Λ is separated (see proof of Proposition 1.8.10). It follows that Λ is a prime ring. An argument similar to the one used in [65] for "lifting maximal orders" finishes the proof. Note that in loc. cit. the condition that X is of positive degree is only used to guarantee the existence of a gr-simple Artinian ring of fractions of the $\mathbb{Z}$-graded ring, but in our present case this is not a problem since we assumed Λ to be positively graded (cf. [52], p. 126 Corollary 1.1.7).

Chapter 2
Maximal Orders and Primes

2.1 Maximal Orders

Throughout this section R is always an order in a simple Artinian ring Q. We study divisorial R-ideals in case R is a maximal order which was defined in Sect. 1.10. If R satisfies the ascending chain condition on divisorial ideals, then we show that $D(R)$, the set of divisorial R-ideals, is an Abelian group generated by maximal divisorial ideals. We define Dedekind and Asano orders which are both important subclasses of maximal orders. A Dedekind order is an Asano order and the converse does not necessarily hold. However in case an order is bounded, the converse is also true. We start with the ideal theoretical characterization of maximal orders.

2.1.1 Proposition

Let R be an order in Q. Then the following conditions are equivalent:

1. R is a maximal order in Q.
2. $O_l(I) = R$ for all left R-ideals I and $O_r(J) = R$ for all right R-ideals J.
3. $O_l(A) = R = O_r(A)$ for all R-ideals A.
4. $O_l(A) = R = O_r(A)$ for all non-zero ideals A of R.

Proof. 1. $\Rightarrow$ 2. If I is a left R-ideal, then there exists a regular element c in R with $Ic \subset R$. It is easy to see that $O_l(I) \supset R$ and $O_l(I)Ic \subset Ic \subset R$, which imply $O_l(I) \sim R$. Hence $O_l(I) = R$.

2. $\Rightarrow$ 3. and 3. $\Rightarrow$ 4. These are special cases.

4. $\Rightarrow$ 1. Suppose $S \supset R$ and $S \sim R$. Then there are regular elements c and d with $cSd \subset R$. Put $T = R + cS + RcS$, an overring of R and $Td \subset R$. The set $A = \{r \in R, Tr \subset R\}$ is an integral R-ideal and $O_l(A) \supset T$. Hence $T = R \subset S$ follows. Since $cS \subset T$, $B = \{r, rS \subset T = R\}$ is an ideal of R with $O_r(B) \supset S$ and hence $R = O_r(B) = S$. □

Marubayashi and Van Oystaeyen, *Prime Divisors and Noncommutative Valuation Theory*, Lecture Notes in Mathematics 2059, DOI 10.1007/978-3-642-31152-9_2,

In Chap. 1 we defined divisorial ideals in Dubrovin valuation rings. In this section, we define divisorial one-sided R-ideals as follows:

For any left R-ideal I, $^*I = (R : (R : I)_r)_l$ which is a left R-ideal containing I. I is said to be **left divisorial** if $I = {}^*I$. Similarly we define a right divisorial R-ideal. Suppose R is a maximal order. Then for any R-ideal A, $^*A = (A^{-1})^{-1} = A^*$, because $(R : A)_r = A^{-1} = (R : A)_l$.

2.1.2 Theorem

Suppose R is a maximal order in Q.

1. $D(R) = \{A, A$ is a divisorial R-ideal$\}$ is a group under the multiplication "$\circ$": $A \circ B = {}^*(AB)$ for any $A, B \in D(R)$.
2. If R satisfies the ascending chain condition on divisorial ideals of R, then
   (i) $D(R)$ is an abelian group generated by maximal divisorial ideals.
   (ii) Any maximal divisorial ideal is a minimal prime ideal.

Proof. 1. If A is a divisorial R-ideal, then it follows that $(R : A^{-1}A)_r = R$ and so $R = {}^*(A^{-1}A)$. Similarly $(A^{-1}A)^* = R$ and hence $D(R)$ is a group.

2. (i) Let M and N be maximal divisorial ideals of R. Then $M \circ N \subset M \cap N$ and $M \cap N \in D(R)$. So we have either $M = (M \cap N) \circ N^{-1}$ or $(M \cap N) \circ N^{-1} = R$. In the later case, we have $M \cap N = N$ which is a contradiction. Thus $M = (M \cap N) \circ N^{-1}$ and $M \circ N = M \cap N$ follows. Hence $M \circ N = M \cap N = N \cap M = N \circ M$. Assume that there is a divisorial ideal A which is not a finite product of maximal divisorial ideals in $D(R)$. Choose A maximal with this property. Since A is not a maximal divisorial ideal, there is a maximal divisorial ideal M with $M \supsetneqq A$. Then $R \supset M^{-1} \circ A \supsetneqq A$, hence $M^{-1} \circ A$ is a finite product of maximal divisorial ideals in $D(R)$ and therefore so is A which is a contradiction. Now let B be a divisorial R-ideal. Then there exists a regular element c in R with $Bc \subset R$. Put $C = (RcR)^*$. Then $B = (B \circ C) \circ C^{-1}$ and $B \circ C = {}^*(BRcR) \subseteq R$. Since $B \circ C$ and C are both finite products of maximal divisorial ideals, $D(R)$ is generated by maximal divisorial ideals and it is abelian, because maximal divisorial ideals commute under the product "$\circ$".
   (ii) Let M be a maximal divisorial ideal. It is clear that M is prime ideal. If P is a non-zero prime ideal with $M \supset P$, then $P^{-1}P \not\subset P$, because if $P^{-1}P \subset P$, then $P^{-1} \subset O_l(P) = R$ but $P^{-1} = (R : P)_l \supset (R : M)_l = M^{-1} \supsetneqq R$ which is a contradiction. $P^*P^{-1}P \subset P$ implies $P^* \subset P$, i.e. P is a divisorial ideal. Since P is a finite product of maximal divisorial ideals, it must be a maximal divisorial ideal. □

In case of commutative domains, maximal orders are nothing but completely integrally closed domains.

2.1.3 *Proposition*

A commutative domain D is a maximal order in its quotient field K if and only if it is completely integrally closed.

Proof. Suppose D is a maximal order in K and $aq^n \in D$ for all $n \in \mathbb{N}$, where $0 \neq a \in D$ and $q \in K$. Then $D' = D[q]$ is equivalent to D, because $aD' \subset D$ and so $q \in D$. Hence D is completely integrally closed. Conversely suppose D is completely integrally closed. Let D' be an overring of D which is equivalent to D, i.e. $aD' \subset D$ for some non-zero $a \in D$. Pick any $q \in D'$. Then $aq^n \in D$ for all $n \in \mathbb{N}$. Thus $q \in D$ and hence $D = D'$. □

2.1.4 *Proposition*

The center of a maximal order in Q is a completely integrally closed domain.

Proof. Let R be a maximal order with $D = Z(R)$. It is clear that D is a domain. Let K be a quotient field of D, which is contained in Q. Let k be an element in K which is almost integral over D, i.e. $dk^n \in D$ for some $0 \neq d \in D$ and all $n \geq 0$. Then $S = R + R[k]$ is equivalent to R and so $S = R$, i.e. $k \in D$. Hence D is completely integrally closed. □

An R-ideal A is said to be **invertible** if $(R : A)_l A = R = A(R : A)_r$. If A is invertible, then $(R : A)_l = A^{-1} = (R : A)_r$ and $O_r(A) = R = O_l(A)$. An order in Q is said to be **Asano** if each non-zero ideal is invertible, and is said to be **Dedekind** if it is Asano and hereditary.

2.1.5 *Lemma*

Let I be an essential left ideal of an order R, then I is projective if and only if

$$(R : I)_r I = O_r(I)$$

Proof. This is clear from the dual basis theorem for projective modules [59] since $(R : I)_r = \mathrm{Hom}_R(I, R)$ and $O_r(I) = \mathrm{Hom}_R(I, I)$. □

From Lemma 2.1.5, we see that an invertible ideal is projective. The converse is not necessarily true. But in case of commutative domains, the converse is also true. □

2.1.6 Proposition

Let D be a commutative domain with its quotient field K. The following conditions are equivalent:

1. D is a Dedekind domain.
2. D is an Asano order in K.
3. D is a hereditary order in K.

Proof. Suppose an ideal I is projective. Then $O_r(I) = (D : I)_r I = (D : I)_l I \subset D$ and so $O_r(I) = D$. Hence I is invertible. □

In case of noncommutative rings, the Dedekind assumption of course implies the hereditary and Asano properties. However the converse implications do not necessarily hold and there are no implications between the properties of being Asano and hereditary.

2.1.7 Example

1. $\begin{pmatrix} \mathbb{Z} & \mathbb{Z} \\ \mathbb{Z} & \mathbb{Z} \end{pmatrix}$ is a Dedekind order in $\begin{pmatrix} \mathbb{Q} & \mathbb{Q} \\ \mathbb{Q} & \mathbb{Q} \end{pmatrix}$ and for any prime number p, $\begin{pmatrix} \mathbb{Z} & p\mathbb{Z} \\ \mathbb{Z} & \mathbb{Z} \end{pmatrix}$ is hereditary but neither Dedekind nor Asano.
2. Let $A_n(K)$ be the nth Weyl algebra over a field K with $\mathrm{char}(K) = 0$. Then $A_n(K)$ is a Noetherian simple ring and so it is a trivial Asano order but not hereditary, because $\mathrm{gl.dim} A_n(K) = n$. Let $R = A_n(K)[X]$ be the polynomial ring in an indeterminate X over $A_n(K)$. Then any ideal of R is principal by the division algorithm. Thus R is a non-trivial Asano order. Furthermore, let S be a quasi-simple, i.e. a simple ring but not necessarily Artinian. Then the Ore extension $S[X; \sigma, \delta]$ is Asano, where σ is an automorphism of S and δ is a left σ-derivation of S.

The following is just an application of Theorem 2.1.2 to Asano orders.

2.1.8 Corollary

Let R be an Asano order with the ascending chain conditions on ideals. Then the set of R-ideals is the abelian group generated by maximal ideals and every maximal ideal is a minimal prime ideal.

An essential one-sided ideal of R is said to be **bounded** if it contains a non-zero ideal and R is called **bounded** if every essential one-sided ideal is bounded.

2.1.9 Lemma

Let R be a bounded Noetherian Asano order in Q. For each non-zero prime ideal P of R, R/P is simple Artinian.

Proof. Note that P is a maximal ideal by Corollary 2.1.8. By the Goldie theorem, it is enough to show that regular elements of R/P are units in R/P. Let $c \in R$ be the preimage of such a regular element. Assume that $xc = 0$ for some $x \in R$. Then $xc \in P$, therefore $x \in P$ and $P^{-1}x \subset R$. Then $0 = P^{-1}xc \subset R$ and $P^{-1}x \subset P$, hence $x \in P^2$. By induction we get $x \in \bigcap_{n=1}^{\infty} P^n$. But if $0 \neq A = \bigcap_{n=1}^{\infty} P^n$, then $AP = A$ and so $P = R$ which is a contradiction. Thus $x = 0$ and c is regular in R. Hence Rc is an essential left ideal and contains a non-zero ideal, say, B. Choose B maximal with this property. Suppose $B \subset P$. Then $Pc \supset B$ since if $b = rc \in P$ then $r \in P$ by the choice of c. Therefore $Rc \supset P^{-1}B$ and so $P^{-1}B = B$, and $P^{-1} = R$ which is a contradiction. Hence $B \not\subset P$ and $Rc + P = R$. □

2.1.10 Proposition

Bounded Noetherian Asano orders are Dedekind orders.

Proof. First note that for each non-zero ideal A, R/A is Artinian since A is a finite product of maximal ideals. In particular R/I has a composition series for every essential left ideal I. Let I be a maximal left ideal. Then I contains a maximal ideal M. This is because I contains a finite product $M_1 \cdots M_n$ of maximal ideals. If $M_n \not\subset I$ then $M_n + I = R$ and $M_1 \cdots M_{n-1} = M_1 \cdots M_{n-1}(M_n + I) \subset I$. By iteration we have the result. The short exact sequence $0 \longrightarrow M \longrightarrow R \longrightarrow R/M \longrightarrow 0$ shows that $\mathrm{pd}_R(R/M) \leq 1$ since M is invertible, where $\mathrm{pd}_R(N)$ denotes the **projective dimension** of the left R-module N. But since R/M is simple artinian $R/M \cong I/M \bigoplus R/I$. Thus $\mathrm{pd}_R(R/I) \leq 1$. So from the short exact sequence $0 \longrightarrow I \longrightarrow R \longrightarrow R/I \longrightarrow 0$ we deduce that I is projective. To prove that R is left hereditary, suppose there is an essential left ideal I which is not projective. Choose I maximal with this property and let J be a left ideal such that $J \supsetneq I$ and J/I is a simple left R-module. Since $J/I \cong R/I_0$ for some maximal left ideal I_0, we have $\mathrm{pd}_R(J/I) \leq 1$. Hence from the exact sequence $0 \longrightarrow I \longrightarrow J \longrightarrow J/I \longrightarrow 0$ we deduce that I is projective which is a contradiction and so any essential left ideal is projective. Hence R is left hereditary since any non-essential left ideal is a direct summand of an essential one. Similarly R is right hereditary and so hereditary. □

2.1.11 Proposition

Suppose R is a semi-local order in Q. Then the following are equivalent:

1. R is a Noetherian Asano order.
2. R is a principal ideal ring.

In particular, R is bounded if one of the conditions (1) or (2) is satisfied.

Proof. 2. $\Rightarrow$ 1. This is clear.

1. $\Rightarrow$ 2. Let P_i be the maximal ideals of R with $J(R) = P_1 \cap \cdots \cap P_n$ and I be a maximal left ideal of R. Then there exists a maximal ideal P_i, say $i = 1$, with $I \supset P_1, I \not\supset P_i$ and $I + P_i = R$ for any $i, 2 \leqslant i \leqslant n$. $P_i \supset I \cap P_i \supset P_i I$ imply $R \supset P_i^{-1}(I \cap P_i) \supset I$ and so either $R = P_i^{-1}(I \cap P_i)$ or $P_i^{-1}(I \cap P_i) = I$. Thus $I \cap P_i = P_i I$ since the first case implies $P_i = I \cap P_i$. It follows that the length of a composition series of R/P_i is the same as one of $I/P_i I$ for any i, $1 \leqslant i \leqslant n$ and thus $R/P_i \cong I/P_i I$ follows. Hence we have
$$R/J(R) \cong R/P_1 \oplus \cdots \oplus R/P_n \cong I/P_1 I \oplus \cdots \oplus I/P_n I \cong I/J(R)I,$$
since $J(R) = P_1 \cdots P_n$ and the P_i commute. Therefore I is left principal by Nakayama's lemma. Suppose R is not left principal and let I be a left ideal which is not left principal. Choose I maximal with this property. Then I is essential. Otherwise there exists a non-zero left ideal J such that $I \oplus J$ is essential, which is principal, a contradiction. We claim R/I is left Artinian. Let $R \supsetneqq I_1 \supsetneqq I_2 \supsetneqq \cdots \supsetneqq I$ be a strictly descending chain of left ideals. For a fixed regular element c in I, we have $R \subset I_1^{-1} \subset I_2^{-1} \subset \cdots \subsetneqq c^{-1}R$, where $I_i^{-1} = (R : I_i)_r$ and so $I_n^{-1} = I_{n+1}^{-1}$ for some n. But since I_i are left principal, we have $I_n = I_n^{-1-1} = I_{n+1}^{-1-1} = I_{n+1}$, a contradiction. Thus there is a left ideal $J \supsetneqq I$ such that J/I is a simple module and $J = Ra$ for some $a \in J$. Then Ia^{-1} is a maximal left ideal and is principal. Hence so is I which is a contradiction. Hence R is a left principal ideal ring and similarly R is a right principal ideal ring. To prove R is bounded, let I be an essential left ideal. If I is maximal, then it contains $J(R)$. Suppose that I does not contain non-zero ideals. Then we get the contradiction by the same method as the above. □

2.1.12 Remark

Let R be a Noetherian semi-local order in Q. Then R is hereditary if and only if $J(R)$ is invertible(cf. [29]).

2.1.13 Proposition

Let R be a Dubrovin valuation ring of Q.

1. R is a maximal order if and only if it is of rank one.
2. R is an Asano order if and only if it is a principal ideal ring.

Proof. 1. If rank $R \geq 2$, then there exists a Goldie prime P with $J(R) \supsetneqq P$ and $O_l(P) = R_P \supsetneqq R$ by Proposition 1.4.11 and Lemma 1.5.4. Hence R is not a maximal order. Conversely assume that R is not a maximal order. There exists an ideal A with $S = O_l(A) \supsetneqq R$ and $J(R) \supsetneqq J(S)$, a Goldie prime by Theorem 1.4.9. Hence rank $R \geq 2$

2. If $J(R) = J(R)^2$, then R is not Asano. If rank$R = 1$ and $J(R) \supsetneqq J(R)^2$, i.e. $J(R) = aR = Ra$ for some $a \in J(R)$, then R is a principal ideal ring, because R is bounded by Lemma 1.5.23. □

2.1.14 Remark

Suppose rank$R = 1$ and $J(R) = J(R)^2$. If R is Archimedean, then $D(R) \cong (\mathbb{R}, +)$ and so R is a maximal order but not satisfying the ascending chain conditions on divisorial ideals.

2.2 Krull Orders

In this section we define Krull orders in a simple Artinian ring and study the ideal theory and overrings of Krull orders. In order to define Krull orders we need some basic knowledge of torsion theories. Let R be a ring and E be an injective left R-module. Then a family $\mathcal{F}$ of left ideals I of R such that $\mathrm{Hom}_R(R/I, E) = 0$ satisfies the following:

(1) If $I \in \mathcal{F}$ and $r \in R$, then $I \cdot r^{-1} \in \mathcal{F}$, where $I \cdot r^{-1} = \{x \in R, xr \in I\}$.
(2) If I is a left ideal and there exists $J \in \mathcal{F}$ such that $I \cdot x^{-1} \in \mathcal{F}$ for every $x \in J$, then $I \in \mathcal{F}$.

Note that the properties of (1) and (2) entail the following:

(i) If $I \in \mathcal{F}$ and J is a left ideal with $I \subset J$, then $J \in \mathcal{F}$ and
(ii) If I and J belong to $\mathcal{F}$, then $I \cap J \in \mathcal{F}$.

A family $\mathcal{F}$ of left ideals satisfying (1) and (2) is said to be a **left Gabriel topology** on R. We also say that $\mathcal{F}$ is a left Gabriel topology corresponding to E and we refer the readers to [63, Chap. VI] for torsion theory and Gabriel topologies. We can define a right Gabriel topology on R in an obvious way. Let M be a left R-module. An element $m \in M$ is said to be **$\mathcal{F}$-torsion** if $Fm = 0$ for some $F \in \mathcal{F}$. Then the set $t_{\mathcal{F}}(M)$ of all $\mathcal{F}$-torsion elements in M is an R-submodule of M, which is called an **$\mathcal{F}$-torsion submodule** of M. If $t_{\mathcal{F}}(M) = M$ or $t_{\mathcal{F}}(M) = 0$, then M is called **$\mathcal{F}$-torsion** R-module or **$\mathcal{F}$-torsion-free** R-module, respectively. We say that $\mathcal{F}$ is **trivial** if either all modules are $\mathcal{F}$-torsion or $\mathcal{F}$-torsion-free, i.e. $\mathcal{F} \ni (0)$ or $\mathcal{F} = \{R\}$. In case of semi-prime Goldie rings, we have the following result:

2.2.1 Proposition

Let R be a semi-prime Goldie ring with its quotient ring Q. Every non-trivial left Gabriel topology on R consists of essential left ideals.

Proof. Let $\mathcal{F}$ be a non-trivial left Gabriel topology on R and assume that there exists $I \in \mathcal{F}$ such that I is not an essential left ideal of R. Then there is a uniform left ideal U of R with $I \cap U = 0$. By Zorn's Lemma, we can choose a left ideal J that is a maximal element in the $\{H,\ H$ is a left ideal with $H \supset I$ and $H \cap U = 0\}$. Then $J \in \mathcal{F}$ and J is a complement left ideal in the sense of Goldie with $\mathrm{u\,dim} R/J = \mathrm{u\,dim} R - \mathrm{u\,dim} J = 1$ (cf. [47, (2.2.10)]). Since R/J is torsion-free as a left R-module, we have a natural embedding $R/J \hookrightarrow Q \otimes R/J$. Now let $\mathrm{u\,dim} R = n$. Then $\sum^n \oplus R/J \hookrightarrow Q \otimes (\sum^n \oplus R/J) \cong Q$ as a left R-module and Q is an essential extension of $\sum^n \oplus R/J$. Hence there exists an $m \in \sum^n \oplus R/J$ with $l_R(m) = 0$. This implies $(0) \in \mathcal{F}$, because m is a $\mathcal{F}$-torsion element which is a contradiction. Hence $\mathcal{F}$ consists of essential left ideals of R. □

Let $\mathcal{F}$ be a non-trivial left Gabriel topology on a semi-prime Goldie ring R. Then $Q_{\mathcal{F}}R = \{q \in Q, Fq \subset R$ for some $F \in \mathcal{F}\}$ is an overring of R, which is called the **left quotients of R with respect to** $\mathcal{F}$. Similarly for any left R-submodule M of Q, we define $Q_{\mathcal{F}}M$ and it is a left $Q_{\mathcal{F}}R$-module. If I is a left ideal of R, then $Q_{\mathcal{F}}I$ is a left ideal of $Q_{\mathcal{F}}R$ and it is said to be **$\mathcal{F}$-closed** if $Q_{\mathcal{F}}I \cap R = I$.

In the remainder of this section, R is an order in a simple Artinian ring Q. To define a non-commutative Krull order, we introduce a specific Gabriel topology as follows: We denote by $\mathcal{F}_R(\mathcal{F}'_R)$ the left(right) Gabriel topology on R corresponding to $E(Q/R)(E'(Q/R))$, the injective hull of left(right) R-module Q/R. It turns out that $\mathcal{F}_R = \{F,\ F$ is a left ideal such that $(R : F \cdot r^{-1})_r = R$ for any $r \in R\}$ (cf. e.g. [63, p. 147, Proposition 5.5]). Similarly $\mathcal{F}'_R = \{F',\ F'$ is a right ideal such that $(R : r^{-1} \cdot F')_l = R$ for any $r \in R\}$, a right Gabriel topology on R. For any left R-ideal I, we define the left closure of I, $cl_\tau(I) = \{r \in R,\ Fr \subset I$ for some $F \in \mathcal{F}_R\}$, which is again a left R-ideal. I is said to be **τ-closed** if $I = cl_\tau(I)$. Similarly, for any right ideal J, we define the right closure $cl_\tau(J)$ of J and it is **τ-closed** if $J = cl_\tau(J)$ (we use the same notation $cl_\tau()$ for left and right ideals). An order in Q is said to be **τ-Noetherian** if it satisfies the ascending chain conditions on one-sided τ-closed ideals.

2.2.2 Definition

An order in a simple Artinian ring is **Krull** (in the sense of M. Chamarie [13, 14]) if it is a maximal order and τ-Noetherian.

Noetherian maximal orders are Krull. However there exists non-Noetherian Krull orders, e.g. $D = K[X_\lambda, \lambda \in \Lambda]$ is a commutative non-Noetherian Krull domain, where K is a field and $| \Lambda |= \infty$, an Ore extension $D[X; \sigma, \delta]$ over D is a non-Noetherian Krull orders it will be shown in Theorem 2.3.19.

2.2.3 Lemma

Let R be a Krull order in Q. Then:

1. Let I be a left R-ideal.
   (i) $cl_\tau(I) \subset {}^*I$. In particular, every divisorial left R-ideal is τ-closed.
   (ii) There is a finitely generated left R-ideal J such that $J \subset I$ and $J^{-1} = I^{-1}$.
   (iii) $II^{-1} \in \mathcal{F}_R$.
2. Let A be an R-ideal. Then
   (i) AA^{-1} and $A^{-1}A$ belong to $\mathcal{F}_R$.
   (ii) $cl_\tau(A) = {}^*A$.

Proof. 1. (i) Let $x \in cl_\tau(I)$. Then there is an $F \in \mathcal{F}_R$ with $Fx \subset I$ and $Fx(R : I)_r \subset I(R : I)_r \subset R$. Thus $x(R : I)_r \subset R$ and $x \in {}^*I$. Hence $cl_\tau(I) \subset {}^*I$. So if I is divisorial, then it is τ-closed.
(ii) Since R satisfies the ascending chain condition on divisorial one-sided ideals, there are a finite elements $a_i \in I$ such that ${}^*I = {}^*(Ra_1 + \cdots + Ra_n)$ (note every one-sided R-ideal is generated by regular elements ([47, (3.3.7)])). Put $J = Ra_1 + \cdots + Ra_n$. Then ${}^*I = {}^*J$ implies $I^{-1} = J^{-1}$.
(iii) Since II^{-1} is an ideal of R, it suffices to prove that $(R : II^{-1})_r = R$. If $II^{-1}q \subset R$, where $q \in Q$, then $II^{-1}qI \subset I$ and $I^{-1}q \subset I^{-1}$. Thus $q \in O_r(I^{-1}) = R$.

2. (i) $AA^{-1} \in \mathcal{F}_R$ follows from 1.(iii). If $A^{-1}Ax \subset R$, where x is a regular element in Q, then $x \in Rx = {}^*(A^{-1}A)x = {}^*(A^{-1}Ax) \subset R$. Hence $A^{-1}A \in \mathcal{F}_R$.
(ii) ${}^*A \subset cl_\tau(A)$ follows from $AA^{-1}\,{}^*A \subset A$ and $AA^{-1} \in \mathcal{F}_R$. □

2.2.4 Proposition

Let R be a Krull order in Q and let R' be an overring of R such that $Q_\mathcal{F}R = R' = RQ_{\mathcal{F}'}$ for some left(right) Gabriel topology $\mathcal{F}(\mathcal{F}')$ on R.

1. R' is a Krull order in Q.
2. For a left R-ideal I, $(I^{-1})Q_{\mathcal{F}'} = (R'I)^{-1} = (Q_{\mathcal{F}}I)^{-1}$.
3. The map $\varphi : I \longrightarrow Q_{\mathcal{F}}I$ is a bijection between the set of divisorial $\mathcal{F}$-closed left ideals of R and the set of divisorial left ideals of R'.

Proof. 1. Let A' be a non-zero ideal of R' and $q \in O_l(A')$. i.e. $qA' \subset A'$. There exists a finitely generated essential right ideal I such that $I \subset A' \cap R$ and $I^* = (A' \cap R)^*$. So there is an $F \in \mathcal{F}$ such that $FqI \subset R \cap A'$ and thus $Fq(A' \cap R)^* \subset (R \cap A')^*$. Hence $Fq \subset R$ and $q \in R'$, showing $O_l(A') = R'$ and similarly $O_r(A') = R'$. Hence R' is a maximal order. To prove that R' is τ-Noetherian, we claim that for every $F \in \mathcal{F}$ and $H \in \mathcal{F}_R$, $R'F$ and $R'H$ are both in $\mathcal{F}_{R'}$, the left Gabriel topology on R' corresponding to the injective left R'-module $E_{R'}(Q/R')$. Let q be any element in R'. Then $F_1q \subset R$ for some $F_1 \in \mathcal{F}$. For any $s \in (R'H \cdot q^{-1})^{-1}$ and $f \in F_1$, we have $(H \cdot (fq)^{-1})fs \subset (R'H \cdot q^{-1})s \subset R'$. By Lemma 2.2.3, there is a finitely generated left ideal H_1 such that $H_1 \subset H \cdot (fq)^{-1}$ and $H_1^{-1} = (H \cdot (fq)^{-1}))^{-1} = R$. Thus $H_1fsF' \subset R$ for some $F' \in \mathcal{F}'$ and so $fsF' \subset R$. This implies $F_1s \subset R'$ and $s \in Q_{\mathcal{F}}R' = R'$. Hence $R'H \in \mathcal{F}_{R'}$. Similarly we have $R'F \in \mathcal{F}_{R'}$ (the proof is a little easier).

Let I' be a τ-closed left ideals of R'. Then we prove that $I' = cl_\tau(R'(I' \cap R))$ and $I' \cap R$ is τ-closed. If $x \in I'$, then there is an $F \in \mathcal{F}$ such that $Fx \in I' \cap R$. Since $R'Fx \subset R'(I' \cap R)$ and $R'F \in \mathcal{F}_{R'}$, we have $x \in cl_\tau(R'(I' \cap R))$ and hence $I' = cl_\tau(R'(I' \cap R))$. That $I' \cap R$ is τ-closed follows easily since $R'H \in \mathcal{F}_{R'}$ for any $H \in \mathcal{F}_R$. Hence R' is left τ-Noetherian and similarly is right τ-Noetherian. Therefore R' is a Krull order in Q.

2. It is clear that $(I^{-1})Q_{\mathcal{F}'} \subset (Q_{\mathcal{F}}I)^{-1} \subset (R'I)^{-1}$. Conversely, let $b \in (R'I)^{-1}$. Then $Ib \subset R'Ib \subset R'$. Let J be a finitely generated left ideal such that $I \supset J$ and $I^{-1} = J^{-1}$. There exists an $F' \in \mathcal{F}'$ with $JbF' \subset R$ and so $bF' \subset J^{-1} = I^{-1}$. Hence $b \in (I^{-1})Q_{\mathcal{F}'}$.
3. Let I be an essential left ideal of R. Then it follows from (2) that $Q_{\mathcal{F}}(^*I) = {}^*(Q_{\mathcal{F}}I)$. So if I is divisorial $\mathcal{F}$-closed, then $Q_{\mathcal{F}}I$ is divisorial and $Q_{\mathcal{F}}I \cap R = I$. Conversely, let I' be an essential left ideal of R' and $I = I' \cap R$. It is clear that $I' \subset Q_{\mathcal{F}}I$. If I' is divisorial, then the converse inclusion is held since $R'F \in \mathcal{F}_{R'}$ for each $F \in \mathcal{F}$. □

2.2.5 Lemma

Let I be a τ-closed left ideal of a τ-Noetherian order R and let $\overline{R} = R/I$. Then $\operatorname{u\,dim}\overline{R}$ is finite.

Proof. Suppose there are left ideals I_i, $i \in \mathbb{N}$ properly containing I such that $\overline{I_1} \oplus \overline{I_2} \oplus \cdots \oplus \overline{I_n} \oplus \cdots$. Since $I_1 + \cdots + I_n \cap I_{n+1} = I$ for any $n \in \mathbb{N}$, we have

$cl_\tau(I_1 + \cdots + I_n) \cap cl_\tau(I_{n+1}) = cl_\tau(I) = I$. There exists a $m \in \mathbb{N}$ such that $cl_\tau(I_1 + \cdots + I_m) = cl_\tau(I_1 + \cdots + I_{m+1})$ since R is τ-Noetherian. It follows that $cl_\tau(I_{m+1}) = I$, a contradiction. Hence u dim$\overline{R}$ is finite. □

For any ideal A, we denoted by $\mathcal{C}(A) = \{c \in R,\ c \text{ is regular modulo } A\}$.

2.2.6 Lemma

Let A and B be divisorial ideals of a Krull order R. $\mathcal{C}(A \circ B) = \mathcal{C}(A) \cap \mathcal{C}(B)$.

Proof. Let $c \in \mathcal{C}(A \circ B)$ and $r \in R$. If $cr \in A$, then $crB \subset A \circ B$ and $rB \subset A \circ B$. Hence $r \in A$ which implies $c \in \mathcal{C}(A)$. Similarly $c \in \mathcal{C}(B)$ and $\mathcal{C}(A \circ B) \subset \mathcal{C}(A) \cap \mathcal{C}(B)$. Conversely, let $c \in \mathcal{C}(A) \cap \mathcal{C}(B)$ and $r \in R$. If $cr \in A \circ B$, then $crB^{-1} \subset A$ and $cr \in B$. Thus $r \in B$ entails $rB^{-1} \subset R$. Hence $rB^{-1} \subset A$ and $r \in A \circ B$. Therefore $c \in \mathcal{C}(A \circ B)$. □

Suppose S is a semi-prime Goldie ring with its quotient ring $Q(S)$, a semi-simple Artinian ring. For a left S-module M, the **reduced rank** of M is defined to be $\rho_S(M) = \text{u}\dim_{Q(S)}(Q(S) \otimes_S M)$. The definition is extended to non-semiprime rings as follows: Let S be a ring with a prime radical $N = N(S)$ such that S/N is Goldie and N is nilpotent. A **Loewy series** of left S-module M is a finite chain of submodules $M = M_0 \supset M_1 \supset \cdots \supset M_k = 0$ such that $NM_i \subset M_{i+1}$ for each i. The **reduced rank** of M is defined $\rho_S(M) = \sum_{i=0}^{k} \rho_{R/N}(M_i/M_{i+1})$. It is known that the reduced rank of a module is independent of the choice of Loewy series and that the reduced rank is additive on short exact sequences(cf. e.g. [47, (4.1.1) and (4.1.2)]).

We now use the reduced rank to prove the following lemma.

2.2.7 Lemma

Let A be a divisorial ideal of a Krull order R. Then $\overline{R} = R/A$ has an Artinian quotient ring.

Proof. Write $A = (P_1^{n_1})^* \circ \cdots \circ (P_k^{n_k})^*$ for some maximal divisorial ideals P_i, $1 \leq i \leq k$ and put $N = P_1 \cap \cdots \cap P_k$. Then N is divisorial and $\overline{N} = N/A$ is the nilpotent radical of $\overline{R}$. Since $\mathcal{C}(A) = \mathcal{C}(N)$ by Lemma 2.2.6, we have $\mathcal{C}_{\overline{R}}(\overline{0}) = \mathcal{C}_{\overline{R}}(\overline{N})$. To prove that $R' = R/N$ is left Goldie, let I' be a left annihilator, say, $I' = l_{R'}(X')$ for a subset X of R and $\varphi : R \longrightarrow R'$ be the canonical homomorphism. Then $I = \{r \in R,\ rX \subset N\}$ is a τ-closed left ideal with $\varphi(I) = I'$. Hence R' satisfies the a.c.c on left annihilators. By Lemma 2.2.5, u dimR' is finite and so R'

is left Goldie. Similarly R' is right Goldie. Consider the descending chain of ideals of $\overline{R}$:

$$\overline{R} \supset \overline{N} \supset \overline{(N^2)^*} \supset \cdots \supset \overline{(N^{m-1})^*} \supset \overline{(0)},$$

where $m = \max\{n_1, \cdots, n_k\}$. Note that $\overline{(N^i)^*}/\overline{(N^{i+1})^*} \cong (N^i)^*/(N^{i+1})^*$ and $R' \cong \bar{R}/\bar{N}$. Since $\mathrm{udim}_{R'}(N^i)^*/(N^{i+1})^*$ is finite we have $\rho_{\bar{R}}(\bar{R})$ is finite. Hence $\overline{R}$ has an Artinian quotient ring (cf. e.g. [47, (4.1.4)]).

2.2.8 *Lemma*

If A is a divisorial ideal of a Krull order R, then $\mathcal{C}(A) \cap F \neq \emptyset$ for any $F \in \mathcal{F}_R$.

Proof. We use the same notation as in Lemma 2.2.7. Since $\mathcal{C}(A) = \mathcal{C}(N)$, we may assume that A is semi-prime such that R/A is a Goldie ring. It suffices to prove that F is an essential left ideal modulo A by Goldie's Theorem. For a left ideal I, assume that $F \cap I \subset A$. Then for any $a \in I$, $(F \cdot a^{-1})a \subset F \cap I \subset A$ and so $a \in cl_\tau(A) = A$, i.e. $I \subset A$. Hence F is an essential left ideal modulo A and $\mathcal{C}(A) \cap F \neq \emptyset$ follows. □

2.2.9 *Lemma*

Let A be a divisorial ideal of a Krull order R and I be a left ideal of R. Then $Acl_\tau(I) \subset cl_\tau(AI)$.

Proof. Let $b \in cl_\tau(I)$. There is an $F \in \mathcal{F}_R$ with $Fb \subset I$. For any $a \in A$, we have $((AF) \cdot a^{-1})ab \subset AFb \subset AI$ and $ab \in cl_\tau(AI)$ if $(AF) \cdot a^{-1} \in \mathcal{F}_R$. For any $r \in R$ and $a \in A$, since $((AF) \cdot a^{-1}) \cdot r^{-1} = (AF) \cdot (ra)^{-1}$, in order to prove $(AF) \cdot a^{-1} \in \mathcal{F}_R$, it suffices to prove $((AF) \cdot a^{-1})^{-1} = R$ for any $a \in A$. Let $s \in ((AF) \cdot a^{-1})^{-1}$.

Then for any $q \in A^{-1}$, $qa \in R$ and $A(F \cdot (qa)^{-1})qs \subset A(F \cdot a^{-1})s \subset (AF) \cdot a^{-1}s \subset R$. Thus $A(F \cdot (qa)^{-1})qsA \subset A$ entails $F \cdot (qa)^{-1}qsA \subset O_r(A) = R$ and $qsA \subset R$. Thus $A^{-1}sA \subset R$ and $A^{-1}s \subset A^{-1}$. Hence $s \in R$ as desired. □

2.2.10 *Lemma*

With the same notation as in Lemma 2.2.9, $^*(A^n) \cap I \subset cl_\tau(AI)$ for some $n \in \mathbb{N}$.

Proof. There exists an $n \in \mathbb{N}$ such that $cl_\tau(\sum_{i \in \mathbb{N}} A^{-i}(I \cap {}^*(A^i))) = cl_\tau(\sum_{i=1}^{n-1} A^{-i}(I \cap {}^*(A^i)))$. Thus $A^{-n}(I \cap {}^*(A^n)) \subset cl_\tau(\sum_{i=1}^{n-1} A^{-i}(I \cap {}^*(A^i)))$ and

$A^nA^{-n}(I \cap {}^*(A^n)) \subset A^n cl_\tau(\sum_{i=1}^{n-1} A^{-i}(I \cap {}^*(A^i))) \subset cl_\tau(A^n(\sum_{i=1}^{n-1} A^{-i}(I \cap {}^*(A^i)))) \subset cl_\tau(AI)$. Hence $I \cap {}^*(A^n) \subset cl_\tau(AI)$ since $A^nA^{-n} \in \mathcal{F}_R$. □

2.2.11 Lemma

Let A be a divisorial ideal of a Krull order R. Then

1. $\cap_{n\in\mathbb{N}} {}^*(A^n) = (0)$.
2. $\mathcal{C}(A) \subset \mathcal{C}(0)$.

Proof. 1. Suppose $B = \cap {}^*(A^n) \neq (0)$ and let c be a regular element in B. Then $Rc = Rc \cap {}^*(A^n) \subset cl_\tau(Ac) = cl_\tau(A)c = Ac$ for some $n \in \mathbb{N}$ and so $R = A$, a contradiction.
2. This easily follows from (1), since $\mathcal{C}(A) = \mathcal{C}({}^*(A^n))$ for all $n \in \mathbb{N}$. □

2.2.12 Proposition

Let $A = {}^*(P_1^{e_1} \cdots P_n^{e_n})$ be a divisorial ideal of a Krull order R, where P_i are maximal divisorial ideals. Then A is localizable and R_A is a bounded principal ideal ring such that $R_A = R_{P_1} \cap \cdots \cap R_{P_n}$ and $J(R_A) = P_1R_A \cap \cdots \cap P_nR_A$.

Proof. To prove that A is localizable, let $r \in R$ and $c \in \mathcal{C}(A)$. Then for any $i \in \mathbb{N}$ there exists $r_i \in R$ and $c_i \in \mathcal{C}(A)$ such that $s_i = r_ic - c_ir \in {}^*(A^i)$ by Lemma 2.2.7. Put $I = \sum_{i\in\mathbb{N}} Rs_i$. By Lemma 2.2.10, $cl_\tau(AI) \supset I \cap {}^*(A^n)$ for some $n \in \mathbb{N}$. Since $s_n \in I \cap {}^*(A^n)$, there is an $F \in \mathcal{F}_R$, $Fs_n \subseteq AI$. Thus for any $d \in F \cap \mathcal{C}(A)$ we have $ds_n = \sum_{j=1}^m a_js_j$ for some $a_j \in A$. It follows that $(dr_n - \sum_{j=1}^m a_jr_j)c = dr_nc - \sum_{j=1}^m a_jr_jc = d(s_n + c_nr) - \sum_{j=1}^m a_j(s_j + c_jr) = dc_nr - \sum_{j=1}^m a_jc_jr = (dc_n - \sum_{j=1}^m a_jc_j)r$ and $dc_n - \sum_{j=1}^m a_jc_j \in \mathcal{C}(A)$. Hence $\mathcal{C}(A)$ is left Ore and by symmetry it is right Ore. To prove that R_A is left Noetherian, let I' be a left ideal of R_A and let $I = I' \cap R$. It is clear that $I' = R_AI$ and so I is τ-closed by Lemma 2.2.8. Thus R_A is left Noetherian and it is right Noetherian by symmetry. Note that for any ideal B, R_AB is an ideal of R_A (cf. e.g. [16]). Put $J = P_1 \cap \cdots \cap P_n$ and $\overline{R} = R/A$. Since $\overline{\mathcal{C}(A)} = \mathcal{C}_{\overline{R}}(\overline{0})$, R_A/R_AA is the Artinian quotient ring of $\overline{R}$ by Lemma 2.2.7. It is clear that $J' = R_AJ = R_AP_1 \cap \cdots \cap R_AP_n$ and R_AP_i are prime ideals of R_A. To prove that R_A is semi-local with $J(R_A) = R_AP_1 \cap \cdots \cap R_AP_n$, let I' be a maximal left ideal of R_A and suppose that $I' \not\supset J'$. Then we have $1 = a' + b'$ for some $a' \in I'$ and $b' \in J'$. There exists a $c \in \mathcal{C}(A)$ with $ca', cb' \in R$ and so $ca' = c(1 - b')$. This entails $ca' \in \mathcal{C}(A)$, since $\mathcal{C}(A) = \mathcal{C}(J)$ and $\mathcal{C}(J) \subset \mathcal{C}_{R_A}(J')$ and so $R_A = R_Aca' = R_Aa' \subset I'$, a contradiction. Thus $I' \supset J'$, and hence $J(R_A) = R_AP_1 \cap \cdots \cap R_AP_n$ and R_A is semi-local. To prove that R_A is an Asano order, let B' be any non-zero ideal of R_A and $B = B' \cap R$. Then we have $R_A = R_A BB^{-1} = B'B^{-1}$ by Lemmas 2.2.3 and 2.2.8. Hence B' is invertible and so R_A is an Asano order. It follows from Proposition 2.1.11 that R_A is a principal ideal ring. It is clear that $R_A = R_{P_1} \cap \cdots \cap R_{P_n}$ since $\mathcal{C}(A) = \mathcal{C}(P_1) \cap \cdots \cap \mathcal{C}(P_n)$. □

Let $\mathcal{A}$ be a family of non-zero ideals closed under the multiplication and let $R(\mathcal{A}) = \cup\{A^{-1}, \in \mathcal{A}\})$. It is easy to see that $R(\mathcal{A})$ is an overring of R. In particular, $S(R) = \cup\{A^{-1}, A$ is non-zero ideal $\}$ is an overring, which is called an **Asano overring**.

2.2.13 Lemma

Let $\mathcal{A}$ be a set of non-zero ideals of a Krull order R which is closed under the multiplication and $R(\mathcal{A}) = \cup\{A^{-1}, A \in \mathcal{A}\}$. Then

1. $R(\mathcal{A}) = Q_{\mathcal{F}}R = RQ_{\mathcal{F}'}$, where $\mathcal{F}(\mathcal{F}')$ is a left(right) Gabriel topology on R corresponding to $E(Q/R(\mathcal{A}))(E'(Q/R(\mathcal{A})))$, the injective hull of left(right) R-module $Q/R(\mathcal{A})$.
2. The map: $B \longrightarrow Q_{\mathcal{F}}B$, where B is a divisorial $\mathcal{F}$-closed ideal, is a bijection between the set of divisorial $\mathcal{F}$-closed ideals and the set of divisorial ideals of $R(\mathcal{A})$. In particular, $Q_{\mathcal{F}}B = \cup\{A^{-1} \circ B, A \in \mathcal{A}\}$.

Proof. 1. We will show that an essential left ideal I of R is an element in $\mathcal{F}$ if and only if, for any $r \in R$, $^*(I \cdot r^{-1}) \supset A$ for some $A \in \mathcal{A}$. Suppose $I \in \mathcal{F}$, i.e. $\mathrm{Hom}(R/I, E(Q/R(\mathcal{A}))) = 0$. If $(I \cdot r^{-1})^{-1} \not\subset R(\mathcal{A})$, then for any $q \in (I \cdot r^{-1})^{-1}$ but not in $R(\mathcal{A})$, we have a non-zero homomorphism $\varphi : R/I \cdot r^{-1} \longrightarrow Q/R(\mathcal{A})$ given by $\varphi(x + I \cdot r^{-1}) = xq + R(\mathcal{A})$, where $x \in R$. This is a contradiction. Hence $(I \cdot r^{-1})^{-1} \subset R(\mathcal{A})$. Now let J be a finitely generated right R-ideal being contained in $(I \cdot r^{-1})^{-1}$ such that $J^{-1} = (I \cdot r^{-1})^{-1-1} = {}^*(I \cdot r^{-1})$. Then there is an $A \in \mathcal{A}$ with $AJ \subset R$, i.e. $A \subset J^{-1} = {}^*(I \cdot r^{-1})$. Conversely suppose that for any $r \in R$ there is an $A \in \mathcal{A}$ with $^*(I \cdot r^{-1}) \supset A$. To prove that $I \in \mathcal{F}$, it suffices to prove that $\mathrm{Hom}(R/I, Q/R(\mathcal{A})) = 0$. Let $f \in \mathrm{Hom}(R/I, Q/R(\mathcal{A}))$. Then there is a $q \in Q$ such that $f(r + I) = rq + R(\mathcal{A})$ and $Iq \subset R(\mathcal{A})$. For a finitely generated left ideal $J \subset I$ with $J^{-1} = I^{-1}$ i.e. $^*J = {}^*I$, there is a $B \in \mathcal{A}$ with $JqB \subset R$ and $^*IqB \subset R$. Since $^*I \supset A$ for some $A \in \mathcal{A}$, we have $AqB \subset R$ and so $q \in A^{-1} \circ B^{-1} = (BA)^{-1} \subset R(\mathcal{A})$. Hence $f = 0$. It is now clear that $Q_{\mathcal{F}}R = R(\mathcal{A})$, because any element in $\mathcal{A}$ belongs to $\mathcal{F}$. Similarly $\mathcal{F}' = \{I, I$ is an essential right ideal such that for any $r \in R$, $(r^{-1} \cdot I)^* \supset A$ for some $A \in \mathcal{A}\}$ and $RQ_{\mathcal{F}'} = R(\mathcal{A})$.

2. For any divisorial left R-ideal I and divisorial R-ideal A, we define a multiplication "∘": $A \circ I = {}^*(AI)$. Then it is easy to see that $Q_{\mathcal{F}}I = \cup\{A^{-1} \circ I, A \in \mathcal{A}\}$. If I is a divisorial ideal, then $Q_{\mathcal{F}}I = \cup\{A^{-1} \circ I, A \in \mathcal{A}\} = \cup\{I \circ A^{-1}, A \in \mathcal{A}\} = IQ_{\mathcal{F}'}$. Hence the statement is clear from Proposition 2.2.4. □

2.2.14 Lemma

Let I be a divisorial and bounded left ideal of a Krull order R and let P be a maximal divisorial ideal. Then the following are equivalent:

1. $I \cap \mathcal{C}(P) = \emptyset$.
2. $A \subset P$, where $A = \{a \in R, aR \subset I\}$.

Proof. 1. $\Rightarrow$ 2. Suppose that $A \not\subset P$. Then since R/P is a prime Goldie ring, $0 \neq A \cap \mathcal{C}(P) \subset I \cap \mathcal{C}(P)$.
2. $\Rightarrow$ 1. Suppose that there exists a $c \in I \cap \mathcal{C}(P)$. Then since $A^{-1} \supset I^{-1}$ and $c^{-1}R \supset I^{-1}$, we have $I^{-1}AP^{-1}P \subset R \cap c^{-1}P = P$ and $I^{-1}A \subset P$. Thus $I^{-1} \subset P \circ A^{-1} = A^{-1} \circ P$, $P^{-1}AI^{-1} \subset R$ and $P^{-1}A \subset I^{-1-1} = {}^*I = I$. Thus $P^{-1}A \subset A$ and $P^{-1} \subset O_l(A) = R$. This is a contradiction. □

If an overring R' is a ring of left and right quotients of R with respect to left and right Gabriel topologies, then it is a Krull order by Proposition 2.2.4. In the case $R' \subset S(R)$, we will have a more detailed structure of R':

2.2.15 Proposition

Let R' be an overring of a Krull order R. Suppose that R' is contained in $S(R)$. Then the following are equivalent:

1. $R' = Q_{\mathcal{F}}R$ for a left Gabriel topology $\mathcal{F}$ on R.
2. $R' = R(\mathcal{A})$ for some $\mathcal{A}$, a family of non-zero ideals closed under multiplication.
3. $cl_\tau(R') = R'$, i.e. R' is τ-closed as a left R-module.
4. For any R-ideal A with $A \subset R'$, ${}^*A \subset R'$.
5. There exists a family $\mathcal{P}$ of maximal divisorial ideals of R such that $R' = (\cap R_P) \cap S(R)$, where $P \in \mathcal{P}$. □

Proof. 1. $\Rightarrow$ 2. Put $\mathcal{A} = \{A, A$ is an ideal with $A \in \mathcal{F}\}$ which is closed under multiplication and $R(\mathcal{A}) \subset R'$. To prove the converse inclusion, let $q \in R'$. Then RqR is an R-ideal since $q \in S(R)$. For a finitely generated right R-ideal I such that $I \subset RqR$ and $I^* = (RqR)^*$, there is an $F \in \mathcal{F}$ with $FI \subset R$ and so $FRq \subset FRqR \subset R$. Thus $q \in (FR)^{-1}$ and $FR \in \mathcal{A}$. Hence $q \in R(\mathcal{A})$.
2. $\Rightarrow$ 3. Let $q \in cl_\tau(R')$. Then $Fq \subset R'$ for some $F \in \mathcal{F}_R$. Since R is Krull, we may assume that F is finitely generated and so $AFq \subset R$ for some $A \in \mathcal{A}$. Thus $A^{-1}AFq \subset A^{-1}$ and hence $q \in cl_\tau(A^{-1}) = A^{-1} \subset R'$ by Lemma 2.2.3. Therefore R' is τ-closed as a left R-module.
3. $\Rightarrow$ 4. Let A be an R-ideal with $A \subset R'$. Then $AA^{-1}A^* \subset A$ and $AA^{-1} \in \mathcal{F}_R$. Hence $A^* \subset cl_\tau(A) \subset cl_\tau(R') = R'$.
4. $\Rightarrow$ 5. Put $\mathcal{P} = \{P, P$ is a maximal divisorial ideal such that $R' \subset R_P\}$. It is clear that $R' \subset (\cap_{P\in\mathcal{P}} R_P) \cap S(R)$. Conversely let $q \in (\cap_{P\in\mathcal{P}} R_P) \cap S(R)$ and I be the greatest left ideal with $Iq \subset R$. We may assume that ${}^*I = I$

and is bounded since $q \in S(R)$. If A is the greatest ideal with $I \supset A$, then $A \not\subset P$ for all $P \in \mathcal{P}$ by Lemma 2.2.14, because $I \cap C(P) \neq \emptyset$. Write $A = {}^*(P_1^{n_1} \cdots P_k^{n_k})$, where P_i are maximal divisorial ideals with $P_i \notin \mathcal{P}$, i.e. $R' \not\subset R_{P_i}$. Let $q_i \in R' - R_{P_i}$ and $C_i = (R + Rq_i R)$. Then $C_i^{-1} \subset P_i$, because if $C_i^{-1} \not\subset P_i$, then $q_i \in C_i R_{P_i} = C_i C_i^{-1} R_{P_i} = R_{P_i}$, a contradiction. Thus, since $C_i \subset R'$, we have $R' \supset {}^*C_i \supset P_i^{-1}$ for each i, $1 \leqq i \leqq k$. Hence $q \in A^{-1} = {}^*(P_1^{-n_1} \cdots P_k^{-n_k}) \subset R'$.

5. $\Rightarrow$ 1. By Lemma 2.2.13, $S(R)$ is a left quotient of R with respect to a left Gabriel topology. Hence R' is a left quotient of R with respect to a left Gabriel topology. This follows from the following more general result; if $\mathcal{F}_\alpha$ are left Gabriel topologies on R, then so is $\mathcal{F} = \cap \mathcal{F}_\alpha$ and $Q_{\mathcal{F}} R = \cap_\alpha Q_{\mathcal{F}_\alpha} R$. □

An order S in Q is said to be **divisorially simple** if it does not have non-trivial divisorial ideals. Now we describe the structure of Krull orders in terms of localizations.

2.2.16 Theorem

Let R be a Krull order in a simple Artinian ring Q. Then

1. $R = \cap R_P \cap S(R)$, where P ranges over all maximal divisorial ideals.
2. R_P is a local principal ideal ring, i.e. a rank one Dubrovin valuation ring whose Jacobson radical is not idempotent, and $S(R)$ is a divisorially simple Krull order in Q.
3. Any regular element c in R is a non-unit in only finitely many of R_P, i.e. R has a finite character property.
4. For any essential left ideal I, ${}^*I = \cap R_P I \cap {}^*(S(R)I)$ and ${}^*(S(R)I) = \cup\{A^{-1} \circ {}^*I$, A ranges over all non-zero ideals of $R\}$.

Proof. 1. It is clear that $R \subset \cap R_P \cap S(R)$. Let $x \in \cap R_P \cap S(R)$. Then there is a divisorial ideal A with $Ax \subset R$. We may assume that A is maximal with respect to this property. If $A \neq R$, there exists a maximal divisorial ideal P with $P \supset A$ and $R \supset P^{-1}A \supsetneqq A$. Let $c \in C(P)$ with $xc \in R$. Then we have $PP^{-1}Ax \subset R \cap Px \subset R \cap Pc^{-1} \subset P$ and so $P^{-1}Ax \subset R$. Hence $(P^{-1} \circ A)x \subset R$, which is a contradiction. Hence $A = R$ and $x \in R$.

2. This follows from Propositions 2.2.4, 2.2.12 and Lemma 2.2.13.
3. Look at the left ideal $\sum_P (P^{-1}c \cap R)$. There are a finite number of maximal divisorial ideals P_i, $1 \leqq i \leqq n$ such that: ${}^*(\sum_P (P^{-1}c \cap R)) = {}^*(\sum_{i=1}^n (P_i^{-1}c \cap R))$. So for any maximal divisorial ideal P, we have $P^{-1}c \cap R \subset (\cap_{i=1}^n P_i)^{-1}c$. If $rc \in P$, where $r \in R$, then $P^{-1}rc \subset (\cap_{i=1}^n P_i)^{-1}c$ and $P^{-1}r \subset (\cap_{i=1}^n P_i)^{-1}$. Thus $r(\cap_{i=1}^n P_i) \subset P$ and hence $r \in P$ if $P \neq P_i$. i.e. $c \in C(P)$. Hence c is a unit in R_P for all $P \neq P_i$, $1 \leqq i \leqq n$.

4. Since R is τ-Noetherian and R_P is hereditary, it easily follows that $R_P I = {}^*(R_P I) = R_P{}^*I$. Let $\mathcal{F}$ be the left Gabriel topology corresponding to the injective left R-module $E(Q/S(R))$. Then, by Proposition 2.2.4 and its right version, ${}^*(Q_{\mathcal{F}}I) = Q_{\mathcal{F}}({}^*I)$. $Q_{\mathcal{F}}I \supset S(R)I$ entails ${}^*(Q_{\mathcal{F}}I) \supset {}^*(S(R)I)$. To prove the converse inclusion, let $x \in Q_{\mathcal{F}}I$. Then $Ax \subset I$ for some non-zero ideal A and $S(R)A^{-1}Ax \subset S(R)I$. Hence $x \in {}^*(S(R)I)$, because $S(R)A^{-1}A \in \mathcal{F}_{S(R)}$, the left Gabriel topology on $S(R)$ which corresponds to $E_{S(R)}(Q/S(R))$, by the proof of Proposition 2.2.4 and Lemma 2.2.3. Thus we obtained $Q_{\mathcal{F}}({}^*I) = {}^*(S(R)I)$ and so ${}^*I \subset \cap R_P I \cap {}^*(S(R)I)$. The converse inclusion follows; $(\cap R_P I \cap {}^*(S(R)I))I^{-1} \subset \cap R_P \cap S(R)$. □

2.2.17 *Proposition*

With the same notation and assumptions as in Theorem 2.2.16: $T = \cap R_P$ is a bounded Krull order and the map: $I \longrightarrow \cap R_P I = {}^*(TI)$ is a bijection between the set of bounded divisorial left ideals and the set of divisorial left ideals of T. Furthermore, $D(R) \cong D(T)$.

Proof. Since $C(P)$ is an Ore set, $\mathcal{F} = \{F,\ F$ is an essential left ideal such that $F \cap C(P) \neq \emptyset$ for all $P\}$ is a left Gabriel topology on R and $Q_{\mathcal{F}}R = \cap R_P$, and by right version, T is a right quotient of R with respect to a right Gabriel topology on R. Hence T is a Krull order. Let c be a regular element in T. To prove that T is bounded, we may assume that $c \in R$ since there is a $d \in R$ with $dc \in R$. There are only finite number of P_i such that $R_{P_i} \neq R_{P_i}c$. Since R_{P_i} is bounded, $R_{P_i}c \supset A_i$ for some non-zero ideal A_i of R_{P_i}. Hence $Tc \supset \cap A_i \cap T$, a non-zero ideal of R, i.e. T is left bounded and similarly it is right bounded. For an essential left ideal I of R, we have $\cap R_P I = Q_{\mathcal{F}}I$ and $Q_{\mathcal{F}}I \cap R$ is bounded since T is bounded. Thus if I is $\mathcal{F}$-closed then it is bounded. Conversely if I is bounded and divisorial then $I = \cap R_P I \cap S(R) = \cap R_P I \cap R = Q_{\mathcal{F}}I \cap R$ which is $\mathcal{F}$-closed. Hence the map is a bijection by Proposition 2.2.4. Let I be a divisorial left ideal of R. It is clear that $Q_{\mathcal{F}}I \supset {}^*(TI)$. The converse inclusion follows from the property; $(\cap R_P I)(TI)^{-1} \subset T$. Let A be a divisorial ideal of R. Then $A = \cap R_P A \cap {}^*(S(R)A) = \cap R_P A \cap S(R) = \cap R_P A \cap R$. Hence we have $D(R) \cong D(T)$. □

2.2.18 *Corollary*

For a Krull order R, the following are equivalent:

1. R is bounded.
2. $S(R) = Q$.
3. $R = \cap R_P$, where P ranges over all maximal divisorial ideals.

The corollary shows that bounded Krull orders may be considered as a global theory of rank one Dubrovin valuation rings whose Jacobson radicals are not idempotent.

Each maximal divisorial ideal of a Krull order is a minimal prime ideal (see Theorem 2.1.2). The converse is also true in case of bounded Krull orders.

2.2.19 Corollary

A non-zero ideal of a bounded Krull order is a maximal divisorial ideal if and only if it is a minimal prime ideal.

Proof. Let P be a minimal prime ideal and c be a regular element in P. Since R is bounded Rc contains a divisorial ideal. Hence P is a maximal divisorial ideal by Theorem 2.1.2. □

Let R be an order in Q and C_i be a regular Ore set. Suppose $R = \cap R_{C_i}$. For a left R-ideal I we define a w-operation: ${}_wI = \cap R_{C_i} I$ and I is called w**-closed** if $I = {}_wI$. Similarly w-closed right ideals are defined and R is said to be w**-Noetherian** if R satisfies the ascending chain conditions on w-closed left ideals as well as w-closed right ideals. If I is an essential left ideal, then ${}_wI \subset {}^*I$ since ${}^*I = \cap\{Rc,\ Rc \supset I$ for some $c \in U(Q)\}$, which is clear from the proof of Proposition 1.5.8.

2.2.20 Proposition

Let R be an order in Q satisfying the following conditions:

(i) $R = \cap R_{C_i}$, where C_i are regular Ore sets
(ii) R_{C_i} is a local principal ideal ring for each i
(iii) R has a finite character property

Then R is a bounded Krull order.

Proof. It is clear from the conditions that R is a maximal order and is w-Noetherian. We first prove that ${}_wI = {}^*I$ for every left R-ideal I. Note that $(R : I)_r = (R : {}^*I)_r \subset (R : {}_wI)_r \subset (R : I)_r$, i.e. $(R : I)_r = (R : {}_wI)_r$. Since ${}_wI$ is finitely generated as w-operations we have $R_{C_i}{}^*I = {}^*(R_{C_i} I)$ which is equal to $R_{C_i} I$, because $R_{C_i} I$ is projective. Thus ${}^*I \subset \cap R_{C_i}{}^*I = \cap R_{C_i} I = {}_wI$ and hence ${}^*I = {}_wI$.

Suppose $I \subset R$ and to prove $cl_\tau(I) = {}^*I$, let $x \in cl_\tau(I)$. Then $Fx \subset I$ for some $F \in \mathcal{F}_R$ and $Fx(R : I)_r \subset I(R : I)_r \subset R$. Thus $x(R : I)_r \subset R$ and hence $x \in {}^*I$, showing $cl_\tau(I) \subset {}^*I$.

To prove the converse inclusion, let $x \in {}_wI$, i.e. $x \in R_{C_i} I$ and $c_i x \in I$ for some $c_i \in C_i$. Put $F = \sum Rc_i$. Then $Fx \subset I$ and $R = {}_wF = {}^*F$ and so

$(R : F)_r = R$. Hence $F \in \mathcal{F}_R$ since each $\mathcal{C}_i$ is an Ore set. Therefore $x \in cl_\tau(I)$, showing $cl_\tau(I) = {}^*I$. Hence R is τ-Noetherian since it is w-Noetherian and so it is a Krull order. □

The center of maximal orders are completely integrally closed (see Proposition 2.1.4). In case of Krull orders we have the following.

2.2.21 *Proposition*

The center of a Krull order in Q is a Krull domain.

Proof. Let D be the center of a Krull order R and K be the quotient field of D. For any divisorial prime ideal P of R, R_P is a local principal ideal ring with the unique maximal ideal $P' = PR_P$. For any $x \in K$, we have $xR_P = P'^{v_P(x)}$, where $v_P(x) \in \mathbb{Z}$. Then it is easy to see that the map $v_P : K \longrightarrow \mathbb{Z}$ is a valuation on K and so $O_{v_P} = \{x, x \in K, v_P(x) \geq 0\}$ is a discrete rank one valuation ring containing D. Thus $D \subset \cap O_{v_P}$, where P runs over all divisorial prime ideals of R. If $x \in O_{v_P}$ for all P, then $x \in R_P$ and $x \in S(R)$, because xR is invertible. Hence $x \in R$ and so $x \in D$ follows. Thus $D = \cap O_{v_P}$. For any non-zero $a \in D$, $aR_P = R_P$ for almost all P, i.e. $v_P(a) = 0$ which entails $aO_{v_P} = O_{v_P}$ for almost all P, i.e. D has a finite character property. Hence D is a Krull domain (cf. e.g. [25, (44.9) Theorem]). □

2.2.22 *Remark*

In the sequel of this section we restrict attention to the consideration of maximal orders in central simple algebras. Throughout the remainder of this section let Q be a central simple algebra, i.e. a simple Artinian ring with $[Q : K] < \infty$, where $K = Z(Q)$. A subring R of Q with $D = Z(R)$ is called a **D-order** in Q if the following are satisfied:

(i) $K = Q(D)$, the quotient field of D and $KR = Q$.
(ii) Every element of R is integral over D.

We say that a D-order is **maximal** if it is maximal in all D-orders by inclusion and a D-order is a **classical** D-order if it is a finitely generated D-module. The theory of classical maximal D-orders is well known, e.g. [57]. In case D is a Krull domain, the theory of maximal D-orders as well as tame D-orders is well-documented in [77]. In this section, we introduce basic properties of maximal D-orders in a self-contained way as an application of foregoing sections.

2.2.23 Proposition

Let R be a D-order in Q.

1. R is a bounded order in Q and $Q = \{c^{-1}r, r \in R \text{ and } 0 \neq c \in D\}$.
2. There exists a maximal D-order in Q containing R.

Proof. 1. Let $q \in Q$. Then $q = \sum k_i r_i$ for some $k_i \in K$ and $r_i \in R$. There is a non-zero $c \in D$ with $ck_i = d_i \in D$. Thus $q = c^{-1}r$, where $r = \sum d_i r_i$. Hence R is an order in Q. Let I be an essential left ideal of R. Then I contains a regular element by Goldie's Theorem, say, c and $c^{-1} = d^{-1}r$ for some $d \in D$ and $r \in R$. Hence $d = rc \in I$, showing R is a left bounded. Similarly R is right bounded.
2. This follows from Zorn's Lemma. □

The reduced trace $tr : Q \longrightarrow K$ induces a K-vector space isomorphism $t : Q \longrightarrow \text{Hom}_K(Q, K)$ by $t(q)(s) = tr(qs)$ for $q, s \in Q$. Note that if $q \in Q$ is integral over D, then $tr(q) \in D$. For any K-basis $\{q_1, \cdots, q_n\}$ of Q we may find $q_1^*, \cdots, q_n^*$ such that $tr(q_i^* q_j) = \delta_{ij}$. We write $L = Dq_1 \oplus \cdots \oplus Dq_n$ and $L^c = Dq_1^* \oplus \cdots \oplus Dq_n^*$. Note that t restricted to L^c induces an isomorphism of L^c with $L^* = (D : L)_l = \{x \in Q, xL \subset D\} \cong \text{Hom}_D(L, D)$. Recall that a D-submodule M of Q is said to be a **D-lattice** in Q if M contains an K-basis for Q and M is contained in a finitely generated D-submodule of Q.

2.2.24 Proposition

Let R be a D-order in Q and S be a D-order with $S \supset R$. Then

1. There exists a finitely generated D-submodule of Q containing S. In particular S is equivalent to R.
2. R is a D-lattice in Q.

Proof. 1. Let $\{q_i\}$ be a K-basis contained in R and L be as before. $LS \subset S$ implies $tr(LS) \subset tr(S) \subset D$ since S is integral over D. Hence $S \subset L^c$, because $L^c = \{x \in Q, tr(Lx) \subset D\}$. The last statement is now clear.
2. This is clear from (1).

We know from Proposition 2.2.21 that D is a Krull Domain if R is a Krull order in Q. In the remainder of this section, we assume that D is a Krull domain.

2.2.25 Proposition

Let R be a D-order in Q and suppose D is a Krull domain.

1. If S is an order in Q equivalent to R and $S \supset D$, then S is also a D-order.
2. R is a maximal D-order in Q if and only if it is a maximal order in Q.

Proof. 1. There are regular elements $x, y \in R$ such that $xSy \subset R$. Let $0 \neq d \in Rx \cap K$ and $0 \neq d_1 \in yR \cap K$. Then $cS \subset R$, where $c = dd_1$. Thus for any $s \in S$, we have $cD[s] \subset cS \subset R \subset L^c$, where L^c is a finitely generated D-module in Q. For any minimal prime ideal p of D, we have $cD_p[s] \subset c(L^c)_p$ which is a finitely generated D_p-module. Since $D_p[s]$ is also finitely generated and thus s is integral over D_p. Hence it is integral over D because $D = \cap D_p$. Therefore S is integral over D. There are regular elements x_1, y_1 in S such that $x_1Ry_1 \subset S$ and $Q = Kx_1Ry_1 \subset KS$. Thus $Q = KS$ and it is now clear that $Z(S) = D$ and hence S is a D-order in Q.
2. This is clear. □

2.2.26 *Lemma*

Suppose R is a maximal D-order and D is a discrete rank one valuation ring. Then R is a local principal ideal ring.

Proof. It follows that R is a classical order in Q. Hence R is a local principal ideal ring by the classical theory, e.g. [57]. □

2.2.27 *Lemma*

Let R be a D-order and D be a Krull domain. Then R_p is a Noetherian D_p-order for every minimal prime ideal p of D.

Proof. It is clear that R_p is a D_p-order. Put $\overline{D_p} = D_p/pD_p$ and $\overline{R_p} = R_p/pR_p$. Then, $[\bar{R}_p : \bar{D}_p] < \infty$ since R_p is a finitely generated D_p-module and so $\overline{R_p}$ is Artinian. Let I be an essential one-sided ideal of R_p. Then $I \cap D = p^n D_p$ for some $n \geqq 1$. Since $R_p/p^n R_p$ is Artinian and $p^n R_p$ is principal, it follows that I is finitely generated. Hence R_p is Noetherian, because R_p is a prime Goldie ring. □

2.2.28 *Proposition*

Let R be a D-order and D be a Krull domain. Then R is a maximal D-order if and only if

(i) $R = \cap R_p$, where p runs over all minimal prime ideals p of D.
(ii) R_p is a maximal D_p-order in Q for every minimal prime ideal p of D.

Proof. If R satisfies (i) and (ii), then it is clear that R is a maximal D-order in Q. Conversely suppose R is a maximal D-order in Q. Then (i) is clear since $\cap R_p$ is also a D-order in Q. To prove (ii) let S be a D_p-order in Q with $S \supset R_p$. Then $(R_p : S)_l = \{q \in Q, qS \subset R_p\}$ is a left R_p and right S-module in Q and we know from Proposition 2.2.23 that $(R_p : S)_l \cap D \neq 0$. Now let $A = (R_p : S)_l \cap R$ is a non-zero ideal of R. For every $s \in S$, we have $As \subset R_p$ and ${}_wAs \subset R_p$ follows, where ${}_wA = \cap R_pA$. Since R_p is Noetherian and R has a finite character property, R is w-Noetherian. There is a finitely generated left ideal I such that $A \supset I$ and ${}_wI = {}_wA$. So there is a $c \in D - p$ with $Isc \subset R$. Furthermore $Isc \subset (R_p : S)_l sc \subset (R_p : S)_l$ entails $Isc \subset A$. Thus ${}_wAsc = {}_wIsc = {}_w(Isc) \subset {}_wA$ and so $sc \in R$ since ${}_wA$ is an ideal of R. Hence $s \in R_p$ and $S = R_p$. □

2.2.29 *Proposition*

Suppose R is a D-order in Q. Then R if a Krull order if and only if

(i) D is a Krull domain.
(ii) R is a maximal D-order.

Proof. Suppose R is a Krull order. Then D is a Krull domain by Proposition 2.2.21 and R is a maximal D-order. Conversely suppose R satisfies the conditions (i) and (ii). Then R is a maximal order and R_p is a local principal ideal ring for every minimal prime ideal p of D by Lemma 2.2.26 and Proposition 2.2.28. Since $R = \cap R_p$, it follows that R is a bounded Krull order by Proposition 2.2.20. □

2.2.30 *Corollary*

Suppose R is a maximal D-order and D is a Krull domain. Then

1. The map: $P \longrightarrow p = P \cap D$ is a bijection between the set of all minimal prime ideals of R and the set of all minimal prime ideals of D. The converse map is given by: $p \longrightarrow P = J(R_p) \cap R$.
2. For every minimal prime ideal P, $R_P = R_p$, where $p = P \cap D$.

Proof. 1. Let P be a minimal prime ideal of R. Then P is a divisorial ideal by Corollary 2.2.19. If $p = P \cap D$ is not a minimal prime ideal of D, then $R = {}_wP \subset {}^*P = P$ which is a contradiction by Proposition 2.2.28. Hence p is a minimal prime ideal of D. Note $J(R_p) = P_p$ since $J(R_p)$ is a minimal prime ideal of R_p. hence the map is one-to-one. For every minimal prime ideal p of D, it is easy to see that $P = J(R_p) \cap R$ is a minimal prime ideal of R.
2. It is clear that $R_P \supset R_p$. However since there are no proper overrings of R_p, we have $R_P = R_p$. □

From Theorem 2.1.2 and Corollary 2.2.30 we have:

2.2.31 Corollary

If R is a maximal D-order and D is a Krull domain, then $D(R) \cong D(D)$.

2.2.32 Lemma

Let S be a ring and D be a subring of $Z(S)$. Suppose S is integral over D. Then

1. (Lying over) For any prime ideal p of D, there exists a prime ideal P of S with $P \cap D = p$.
2. (Going up) Given prime ideals $p_0 \subsetneqq p$ of D and a prime ideal P_0 of R with $P_0 \cap D = p_0$, there exists a prime ideal P of R such that $P_0 \subsetneqq P$ and $P \cap D = p$.

Proof. 1. The set of ideals A with $A \cap (D - p) = \emptyset$ has a maximal one by Zorn's Lemma, say, P, which is obviously a prime ideal and $P \cap D \subset p$. It suffices to prove $P \cap D = p$. We assume, on the contrary, that $P \cap D \subsetneqq p$. Take $a \in p$ and $a \notin P \cap D$. Then $(P + aR) \cap (D - p) \neq \emptyset$ and take $b = c + ar \in D - p$, where $c \in P$ and $r \in R$. Since r is integral over D, we have $r^n + d_{n-1}r^{n-1} + \cdots + d_0 = 0$ for some $d_i \in D$ and $n > 0$. Multiplying the equation by a^n we obtain $(ar)^n + d_{n-1}a(ar)^{n-1} + \cdots + d_0 a^n = 0$. Now $ar = b - c$ implies; $b^n + d_{n-1}ab^{n-1} + \cdots + d_0 a^n \in P \cap D \subset p$. Since $a \in p$, $b^n \in p$ and so $b \in p$, a contradiction. Hence $D \cap P = p$ follows.
2. $P_0 \cap D = p_0$ implies $P_0 \cap (D - p) = \emptyset$. Thus the set of ideals $\{A,\ A \cap (D-p) = \emptyset$ and $A \supset P_0\}$ is non-empty and the maximal one in the set, say, P is a prime ideal with $P \cap D = p$ by (1).

2.2.33 Proposition

Let R be a D-order in Q. Then the following are equivalent:

1. R is a maximal D-order and D is a Dedekind domain.
2. R is an Asano order in Q.
3. R is a Dedekind order in Q.

Proof. Let R be a maximal order in Q. Then it is easy to check that R is a Asano if and only if any maximal ideal is divisorial.

1. $\Longrightarrow$ 2.: Let P be a maximal ideal. Then $p = P \cap D$ is a minimal prime ideal of D. Since R_p is local, we have $J(P_p) = P_p$ and $P = J(R_p) \cap R$ which is a minimal prime ideal and so it is divisorial. Hence R is an Asano order.
2. $\Longrightarrow$ 3.: Let P be a maximal ideal. Then R/P is Artinian by Posner's theorem for PI prime rings, e.g. [47, (13.6.5)]. Thus each non-zero factor ring of R is also

Artinian. Hence R is Noetherian since it is bounded. By Proposition 2.1.10 R is a Dedekind order.

3. $\Longrightarrow$ 1.: By Proposition 2.2.29, R is a maximal D-order and D is a Krull domain. It follows from Lemma 2.2.32 that non-zero prime ideals of D are maximal and hence D is a Dedekind domain. □

2.3 Ore Extensions Over Krull Orders

Throughout this section R denotes an order in a simple Artinian ring Q. Let σ be an automorphism of R. Then σ is extended to an automorphism of Q by $\sigma(c^{-1}a) = \sigma(c)^{-1}\sigma(a)$, where $a, c \in R$ and c is regular, which is denoted by the same symbol σ. An additive map δ from R to R is said to be a **left σ-derivation** of R if $\delta(ab) = \sigma(a)\delta(b) + \delta(a)b$ for $a, b \in R$. δ is also extended to a left σ-derivation of Q by $\delta(c^{-1}) = -\sigma(c^{-1})\delta(c)c^{-1}$ for any regular $c \in R$.

An **Ore extension** $S = R[X;\sigma,\delta]$ is the ring of polynomials in an indeterminate X over R with multiplication: $Xa = \sigma(a)X + \delta(a)$ for every $a \in R$. For a natural number n and $a \in R$, $X^n a = \sigma^n(a)X^n + a_{n-1}X^{n-1} + \cdots + a_0$ for some $a_i \in R$. If $\delta = 0$, then $R[X;\sigma,0]$ is denoted by $R[X;\sigma]$ which is called a **skew polynomial ring**. If $\sigma = 1$, then we write $R[X;\delta]$ for $R[X;1,\delta]$ and it is said to be a **differential polynomial ring**. Put $\sigma' = \sigma^{-1}$ and $\delta' = -\delta\sigma^{-1}$. Then δ' is a right σ'-derivation of R and $S = R[X;\sigma',\delta'] = \{f(X) = X^n a_n + \cdots + a_0,\ a_i \in R\}$. We sometimes use the polynomials with coefficients on the right in order to obtain right hand properties. We first show that $T = Q[X;\sigma,\delta]$ is a principal ideal ring and we explicitly describe the generator of any ideal of T, which are used to study the structure of S: if R is a Krull order then so is S. In case either $R[X;\sigma]$ or $R[X;\delta]$ we obtain detailed results on the group of divisorial fractional ideals and class groups.

2.3.1 *Lemma*

Let $Q = M_n(D)$ be a simple Artinian ring, where D is a skewfield. If $\{e_{ij}\}$ and $\{e'_{ij}\}$ are complete sets of matrix units for Q, then there is an inner automorphism i_u such that $e'_{ij} = i_u(e_{ij}) = ue_{ij}u^{-1}$ for some $u \in U(Q)$.

Proof. We may suppose Q is the endomorphism ring of the left D-module $M = D^n$. Put $M_i = e_{ii}(M)$ for any i. Since $e_{11} + \cdots + e_{nn} = 1$, the identity and e_{ii} are orthogonal idempotents, we have $M = M_1 \oplus \cdots \oplus M_n$ such that M_i are a free D-module of rank one. For $i > 1$, $e_{i1} : M_1 \longrightarrow M_i$ and $e_{1i} : M_i \longrightarrow M_1$ are both isomorphisms with $e_{i1}e_{1i} = 1_{M_i}$ and $e_{1i}e_{i1} = 1_{M_1}$. If v_1 is a base of M_1, then $v_i = e_{i1}(v_1)$ is a base of M_i and $\{v_1, \cdots, v_n\}$ is a base of M such that $e_{ij}(v_k) = e_{ij}e_{k1}(v_1) = \delta_{jk}v_i$, where δ_{jk} is the Kronecker's delta. Similarly, there is a base $\{v'_1, \cdots, v'_n\}$ of M such that $e'_{ij}(v'_k) = \delta_{jk}v'_i$. Thus there exists an automorphism u of M with $u(v_i) = v'_i$ for all i. It is clear that $e'_{ij} = ue_{ij}u^{-1} = i_u(e_{ij})$. □

Any map $f : D \longrightarrow D$ is naturally extended to a map $\widehat{f} : M_n(D) \longrightarrow M_n(D)$ by $\widehat{f}((a_{ij})) = (f(a_{ij}))$, where $(a_{ij}) \in M_n(D)$. Let σ be an automorphism of D and δ be a left σ-derivation. Then it is easy to see that $\widehat{\sigma}$ is an automorphism of $M_n(D)$ and $\widehat{\delta}$ is a left $\widehat{\sigma}$-derivation.

2.3.2 Lemma

Let $\{e_{ij}\}$ be a complete set of matrix units for $M_n(D)$ and σ be an automorphism of $M_n(D)$. Then the following are equivalent:

1. $\sigma(e_{ij}) = e_{ij}$ for all i, j.
2. There exists an automorphism σ_0 of D such that $\sigma = \widehat{\sigma_0}$.

Proof. The proof is easy.

2.3.3 Lemma

If σ is an automorphism of $M_n(D)$, then there exist a unit u in $M_n(D)$ and an automorphism σ_0 of D such that $\sigma = i_u\widehat{\sigma_0}$.

Proof. Let $\{e_{ij}\}$ be a complete set of matrix units for $M_n(D)$ and put $e'_{ij} = \sigma(e_{ij})$. Then $\{e'_{ij}\}$ is a complete set of matrix units for $M_n(D)$. By Lemma 2.3.1, we have $e'_{ij} = i_u(e_{ij})$ for some unit u in $M_n(D)$ and so $i_u^{-1}\sigma(e_{ij}) = e_{ij}$ for all i, j. Hence $i_u^{-1}\sigma = \widehat{\sigma_0}$ for an automorphism σ_0 of D by Lemma 2.3.2 and $\sigma = i_u\widehat{\sigma_0}$ follows. □

The following lemma is also easy to prove.

2.3.4 Lemma

Let σ_0 be an automorphism of D and δ be a left $\widehat{\sigma_0}$-derivation of $M_n(D)$. Then the following are equivalent:

1. $\delta(e_{ij}) = 0$ for all i, j.
2. There exists a left σ_0-derivation δ_0 of D such that $\delta = \widehat{\delta_0}$.

Let σ be an automorphism of a ring R and $d \in R$. Define $\delta_d(a) = da - \sigma(a)d$ for all $a \in R$. It is easy to see that δ_d is a left σ-derivation, which is called an **inner σ-derivation** induced by d. Note that if δ_1 and δ_2 are left σ-derivation of R, then so is $\delta_1 + \delta_2$.

2.3.5 Lemma

Let σ_0 be an automorphism of D and δ be a left $\widehat{\sigma_0}$-derivation of $M_n(D)$. Then there exists a $v \in M_n(D)$ and a left σ_0-derivation of D such that $\delta = \widehat{\delta_0} + \delta_v$, where δ_v is an inner $\widehat{\sigma_0}$-derivation induced by v.

Proof. As before $\{e_{ij}\}$ is a complete set of matrix units for $M_n(D)$. Put $v = \Sigma_{i=1}^n \delta(e_{i1})e_{1i}$. Then for any e_{hk}, $\delta_v(e_{hk}) = ve_{hk} - e_{hk}v = \delta(e_{h1})e_{1k} - e_{hk}v$. Operating δ to $1 = \Sigma_{i=1}^n e_{i1}e_{1i}$, we have $v = -\Sigma_{i=1}^n e_{i1}\delta(e_{1i})$. Thus $\delta_v(e_{hk}) = \delta(e_{h1})e_{1k} + e_{h1}\delta(e_{1k}) = \delta(e_{hk})$ and so $\delta - \delta_v$ is a left $\widehat{\sigma_0}$-derivation such that $(\delta - \delta_v)(e_{hk}) = 0$ for all h, k. Hence $\delta - \delta_v = \widehat{\delta_0}$ for a left $\widehat{\sigma_0}$-derivation δ_0 of D by Lemma 2.3.4 and $\delta = \widehat{\delta_0} + \delta_v$ follows. □

2.3.6 Proposition

Let $Q = M_n(D)$ be a simple Artinian ring, where D is a skewfield, σ be an automorphism of Q, δ be a left σ-derivation and $T = Q[X;\sigma,\delta]$. Then there exist an automorphism σ_0 of D and a left σ_0-derivation δ_0 of D such that

$$T = Q[Y;\widehat{\sigma_0},\widehat{\delta_0}] = M_n(D[Y;\sigma_0,\delta_0]),$$

where $Y = uX + v$ for some $u \in U(Q)$ and $v \in Q$.

Proof. By Lemma 2.3.3, $\sigma = i_u^{-1}\widehat{\sigma_0}$, where $u \in U(Q)$ and σ_0 is an automorphism of D. Put $X_1 = uX$ and $\delta_1 = u\delta$. It is easy to see that δ_1 is a left $\widehat{\sigma_0}$-derivation and that $T = Q[X;\sigma,\delta] = Q[X_1;\widehat{\sigma_0},\delta_1]$. It follows from Lemma 2.3.5 that $\delta_1 = \delta_v + \widehat{\delta_0}$ for some $v \in Q$ and a left σ_0-derivation δ_0 of D. Hence if we put $Y = X_1 - v = uX - v$, then we have $T = Q[Y;\widehat{\sigma_0},\widehat{\delta_0}]$ which is naturally isomorphic to $M_n(D[Y;\sigma_0,\delta_0])$. □

By the division algorithm, $D[Y;\sigma_0,\delta_0]$ is a principal ideal ring and hence so is T.

2.3.7 Corollary

With the same notation and assumptions, $T = Q[X;\sigma,\delta]$ is a principal ideal ring.

Since T is a prime and principal ideal ring, it is a prime Goldie ring and so T has the classical quotient ring $Q(T)$ which is a simple Artinian ring. We will first study the structure of ideals of T in order to study the structure of $S = R[X;\sigma,\delta]$.

2.3.8 Lemma

Let $p(X)$ be a polynomial of T. Suppose there exists an automorphism σ' of Q such that $p(X)a = \sigma'(a)p(X)$ for all $a \in Q$. Then the leading coefficient of $p(X)$ is a unit in Q. If $\delta = 0$, then every non-zero coefficients of $p(X)$ is a unit in Q.

Proof. Write $p(X) = a_n X^n + \cdots + a_0$. Then for every $a \in Q$, we have $a_n\sigma^n(a) = \sigma'(a)a_n$. Since σ^n and σ' are both automorphism of Q, we have $a_n Q = Qa_n = Q$ and so a_n is a unit in Q. The last statement is shown in a similar way. □

2.3.9 Lemma

Every non-zero ideal of T is principal and generated by a monic polynomial.

Proof. Let $b(X) = b_n X^n + \cdots + b_0$ be a polynomial of minimal degree of an ideal A of T. Since $L_n(A) = \{c_n, c(X) = c_n X^n + \cdots + c_0 \in A\} \cup \{0\}$ is an ideal of Q, there exists a monic polynomial $a(X)$ of degree n in A. By division algorithm $A = Ta(X)$ and similarly $A = a(X)T$. □

Let $a(X)$ be a monic polynomial with $n = \deg a(X)$. It is easy to see that $Ta(X)$ is an ideal if and only if $a(X)b = \sigma^n(b)a(X)$ for all $b \in Q$ and $a(X)X = (X + c)a(X)$ for some $c \in Q$.

2.3.10 Theorem

Let Q be a simple Artinian ring with an automorphism σ and a left σ-derivation δ, and let $T = Q[X; \sigma, \delta]$. If $p(X)$ is a monic, non-constant, of minimal degree such that $P = Tp(X)$ is an ideal of T, then every ideal of T is of the form: $Tw(X)p(X)^m$, where $m \geq 0$ and $w(X) \in Z(T)$, the center of T. Note if there are no such $p(X)$, then T is a simple ring.

Proof. Let A be a non-zero ideal of T and $a(X)$ be a monic polynomial in A such that $A = Ta(X)$. It follows that $a(X) = b(X)p(X)^m$ for some $m \geq 0$ and $b(X)$ with $b(X) \notin Tp(X)$. Note that $B = Tb(X)$ is also an ideal of T with $B \not\subset P$. Thus it suffices to prove that $B = Tw(X)$ for some $w(X) \in Z(T)$.

(i) In case $\delta = 0$, TX is an ideal and $p(X) = X + u$ for some $u \in Q$. Suppose $P = TX$ and write $b(X) = X^l + b_{l-1}X^{l-1} + \cdots + b_0$ with $b_0 \neq 0$. By Lemma 2.3.8, $b_0 \in U(Q)$ and then it is easy to see that $w(X) = b_0^{-1}b(X) \in Z(T)$ such that $B = Tw(X)$. Suppose $p(X) = X + u$ with $u \neq 0$. As before, $Y = u^{-1}p(X) = u^{-1}X + 1$ is central and $T = Q[Y]$. Then it is clear that any ideal of T is of the form: $Tw(Y)$ with $w(Y) \in Z(T)$.

(ii) In case $\delta \neq 0$. By division algorithm, $b(X) = q(X)p(X) + r(X)$ with $r(X) \neq 0$ and $\deg r(X) < \deg p(X)$. Let $n = \deg p(X)$ and $l = \deg b(X)$. For any $a \in Q$, $p(X)a = \sigma^n(a)p(X)$, $b(X)a = \sigma^l(a)b(X)$, $p(X)X = (X + c)p(X)$ and $b(X)X = (X + d)b(X)$ for some $c, d \in Q$. For any $a \in Q$, $b(X)a = \sigma^l(a)b(X) = \sigma^l(a)q(X)p(X) + \sigma^l(a)r(X)$ and also $b(X)a = (q(X)p(X) + r(X))a = q(X)\sigma^n(a)p(X) + r(X)a$. Therefore $(\sigma^l(a)q(X) - q(X)\sigma^n(a))p(X) = r(X)a - \sigma^l(a)r(X)$. A comparison of degrees on both sides of last equation entails that $r(X)a = \sigma^l(a)r(X)$ for all $a \in Q$, and so the leading coefficient of $r(X)$ is a unit in Q by Lemma 2.3.8.

Suppose $\deg r(X) < \deg p(X) - 1$. From the equations $p(X)X = (X + c)p(X)$ and $b(X)X = (X + d)b(X)$, we have $((q(X)(X + c) - (X + d)q(X))p(X) = (X + d)r(X) - r(X)X$ and so $r(X)X = (X + d)r(X)$, i.e. $Tr(X)$ is an ideal of T. By the choice of $p(X)$, $r(X) = r \in Q$ which is a unit in Q. Let $w(X) = r^{-1}b(X)$. Since $Tw(X) = Tb(X)$, it suffices to prove $w(X) \in Z(T)$. $rX = (X + d)r$ entails $r = \sigma(r)$ and $\delta(r) = -dr$. Hence $w(X)a = r^{-1}b(X)a = r^{-1}\sigma^l(a)b(X) = aw(X)$ since $ra = \sigma^l(a)r$. Furthermore $w(X)X = r^{-1}(X + d)b(X) = r^{-1}(X + d)rw(X) = (Xr^{-1} - \delta(r^{-1}))rw(X) + r^{-1}drw(X) = Xw(X) + (-\delta(r^{-1})r + r^{-1}dr)w(X) = Xw(X)$ since $\delta(r^{-1}) = -r^{-1}\delta(r)r^{-1}$. Hence $w(X) \in Z(T)$ as desired. Suppose $\deg r(X) = \deg p(X) - 1$. Then $p(X) = (uX + v)r(X) + s(X)$ with $s(X) = 0$ or $\deg s(X) < \deg r(X)$, where $u \in U(Q)$ and $v \in Q$. Put $Y = uX + v$. Then $\sigma^n(a)p(X) = \sigma^n(a)(Yr(X) + s(X))$ and $p(X)a = Y\sigma^l(a)r(X) + s(X)a$. Thus $(\sigma^n(a)Y - Y\sigma^l(a))r(X) = s(X)a - \sigma^n(a)s(X)$ yields $Y\sigma^l(a) = \sigma^n(a)Y$ for all $a \in Q$, because the leading coefficient of $r(X)$ is a unit in Q. Hence $Yb = \sigma^{n-l}(b)Y$ for all $b \in Q$. Therefore $T = [Y; \sigma^{n-l}]$ and we are back to case (i). □

In case $\sigma = 1$ or $\delta = 0$, we have the following

2.3.11 Corollary

1. Every ideal of $Q[X;\sigma]$ is of the form $Q[X;\sigma]w(X)X^m$, where $w(X)$ is a central element and $m \geq 0$.
2. Every ideal of $Q[X;\delta]$ is generated by central element if $Q[X;\delta]$ is not a simple ring.

Proof. 1. Clearly $p(X) = X$ is a monic, non-constant, of minimal degree such that $Q[X;\sigma]X$ is an ideal. Hence the statement follows from Theorem 2.3.10.

2. Let $p(X)$ be a monic, non-constant, of minimal degree such that $Q[X;\delta]p(X)$ is an ideal and $\deg p(X) = l$. Then $p(X)X = (X + c)p(X)$ for some $c \in Q$ and a comparison of coefficient of X^l entails $c = 0$. Hence $p(X)X = Xp(X)$ and $p(X)a = ap(X)$ for all $a \in Q$. Hence $p(X)$ is a central element. □

2.3.12 Remark

Let Δ be a simple ring with its quotient ring $Q(\Delta)$. We do not assume that Δ is Artinian, e.g. nth Weyl algebra with char$(\Delta) = 0$. Theorem 2.3.10 holds for Ore extensions $\Delta[X;\sigma,\delta]$ (cf.[12]).

We will next study Ore extensions over R: Let $S = R[X;\sigma,\delta]$ and $\mathcal{C} = \mathcal{C}_R(0)$. Then $\mathcal{C}$ is a regular Ore set of S such that $S_{\mathcal{C}} = T = Q[X;\sigma,\delta]$ and so S has a classical quotient ring $Q(S) = Q(T)$ which is a simple Artinian ring, i.e. S is an order in $Q(S)$. For a right R-ideal $\mathfrak{a}$, $\mathfrak{a}S = \mathfrak{a}[X;\sigma,\delta] = \{a_nX^n + \cdots + a_0, a_i \in \mathfrak{a}\}$ is a right S-ideal. Similarly for a left R-ideal $\mathfrak{b}$, $S\mathfrak{b} = \mathfrak{b}[X;\sigma',\delta'] = \{X^nb_n + \cdots + b_0, b_i \in \mathfrak{b}\}$ is a left S-ideal.

2.3.13 Lemma

Let $S = R[X;\sigma,\delta]$ and $T = Q[X;\sigma,\delta]$. If $\mathfrak{a}$ is a right R-ideal and $\mathfrak{b}$ be a left R-ideal, then $(S : \mathfrak{a}S)_l = S(R : \mathfrak{a})_l$ and $(S : S\mathfrak{b})_r = (R : \mathfrak{b})_r S$. In particular $(\mathfrak{a}S)^* = \mathfrak{a}^*S$ and $^*(S\mathfrak{b}) = S^*\mathfrak{b}$.

Proof. It is clear that $(S : \mathfrak{a}S)_l \supset S(R : \mathfrak{a})_l$. Conversely let $q \in (S : \mathfrak{a}S)_l$, i.e. $q\mathfrak{a}S \subset S$. Then $q \in qT = q\mathfrak{a}T \subset T$. Write $q = X^nq_n + \cdots + q_0$ with $b_i \in Q$. Then $q_i\mathfrak{a} \subset R$ and $q_i \in (R : \mathfrak{a})_l$ for each i, $1 \leqq i \leqq n$. Hence $q \in S(R : \mathfrak{a})_l$. The other statements are now clear. □

$\mathcal{F}_S$ and $\mathcal{F}'_S$ denote the left and right Gabriel topologies on S corresponding to injective hull $E_S(Q(S)/S)$ of the left S-module $Q(S)/S$ and $E'_S(Q(S)/S)$ of the right S-module $Q(S)/S$, respectively.

2.3.14 Lemma

If $F \in \mathcal{F}_R$ then $SF \in \mathcal{F}_S$.

Proof. Because of Lemma 2.3.13, it suffices to prove that $SF\cdot f^{-1} \cap R \in \mathcal{F}_R$ for any $f \in S$. We prove this by induction on $n = \deg f$. If $\deg f = 0$ then $SF \cdot f^{-1} \cap R = F \cdot f^{-1} \in \mathcal{F}_R$. Write $f = X^na + g$ with $\deg g < n$. By induction hypothesis, $G = SF \cdot (X^{n-1}a)^{-1} \cap R \in \mathcal{F}_R$. Thus for any $r \in R$, $G \cdot (\sigma^{-1}(r))^{-1} \in \mathcal{F}_R$ and let $\sigma^{-1}(s) \in G \cdot (\sigma^{-1}(r))^{-1}$, i.e. $\sigma^{-1}(sr) \in G$. Then $srX = X\sigma^{-1}(sr) - \delta(\sigma^{-1}(sr))$. For any $\alpha \in (G\cdot(\delta(\sigma^{-1}(sr)))^{-1}$ we have $\alpha srX = \alpha X\sigma^{-1}(sr) - \alpha\delta(\sigma^{-1}(sr)) \in SG$ and so $\alpha s \in (SG\cdot X^{-1} \cap R)\cdot r^{-1}$. i.e. $(G\cdot(\delta(\sigma^{-1}(sr)))^{-1})s \subset (SG\cdot X^{-1} \cap R)\cdot r^{-1}$. Thus for any $t \in (R : (SG\cdot X^{-1} \cap R)\cdot r^{-1})_r$, we have $G\cdot(\delta(\sigma^{-1}(sr)))^{-1}st \subset R$ and $st \in R$. Since s is any element in $\sigma(G \cdot (\sigma^{-1}(r))^{-1}) \in \mathcal{F}_R$, $t \in R$ follows and hence $SG \cdot X^{-1} \cap R \in \mathcal{F}_R$. Furthermore, by the induction hypothesis, $SF \cdot g^{-1} \cap R \in \mathcal{F}_R$.

It is easy to see that $SF \cdot g^{-1} \cap SG \cdot X^{-1} \subset SF \cdot f^{-1}$ and hence $SF \cdot f^{-1} \cap R \in \mathcal{F}_R$ as desired. □

2.3.15 Proposition

R is τ-Noetherian if and only if $S = R[X;\sigma,\delta]$ is τ-Noetherian.

Proof. Suppose R is τ-Noetherian. Consider an ascending chain $I_1 \subset I_2 \subset \cdots \subset I_n \subset \cdots$ of τ-closed left ideals of S. For any left ideal I of S and any $m \in \mathbb{N}$ allowing $m = 0$, $L_m(I) = \{a_m, \exists f = a_m X^m + \cdots + a_0 \in I\} \cup \{0\}$ is a left ideal of R. There is a $p \in \mathbb{N}$ such that $cl_\tau(L_p(I_p)) = cl_\tau(L_m(I_m))$ for any $m \geq p$. Then for any n, m with $n \geq p$, $m \geq p$, we have $cl_\tau(L_p(I_p)) = cl_\tau(L_m(I_n))$. Furthermore, for any $j < p$, there is an $r(j) \geq 0$ such that $cl_\tau(L_j(I_r)) = cl_\tau(L_j(I_{r(j)}))$ for any $r \geq r(j)$. Put $s = max\{r(0), r(1), \cdots, r(p-1), p\}$. then we have $cl_\tau(L_m(I_t)) = cl_\tau(L_m(I_s))$ for any $m \geq 0$ and $t \geq s$. Thus to prove the ascending chain is stationary, it suffices to prove if $I \subset J$ are τ-closed left ideals with $cl_\tau(L_m(I)) = cl_\tau(L_m(J))$ for all $m \geq 0$, then $I = J$. Let $f \in J - I$ and $f = a_n X^n + \cdots + a_0$ is of minimal degree with $f \notin I$. There is an $F \in \mathcal{F}_R$ such that $Fa_n \in L_n(I)$ and so for any $r \in F$ there exists $g = ra_n X^n +$(the lower degree part) in I. Then $\deg(rf - g) < n$ and $rf - g \in J$. Hence $rf - g \in I$ and $rf \in I$ for any $r \in F$, which implies $SFf \subset I$. Hence $f \in I$ by Lemma 2.3.14, which is a contradiction.

Suppose S is τ-Noetherian. We prove $E_S(Q(S)/S)$ is injective as a left R-module; let $\mathfrak{a}$ be a left ideal of R and φ be any R-homomorphism from $\mathfrak{a}$ to $E_S(Q(S)/S)$. The map $\varphi' : S\mathfrak{a} \longrightarrow E_S(Q(S)/S)$ given by $\varphi'(X^n a_n + \cdots + a_0) = X^n \varphi(a_n) + \cdots + \varphi(a_0)$, where $a_i \in \mathfrak{a}$ is additive. To prove that φ' is an S-homomorphism, it suffices to prove $\varphi'(X^i r X^j a) = X^i r \varphi'(X^j a)$ for $r \in R$, $a \in \mathfrak{a}$ and $i, j \geq 0$. Write $rX^j = X^j r_j + \cdots + r_0$ for some $r_j \in R$. Then $\varphi'(X^i r X^j a) = \varphi'(X^{i+j} r_j a + \cdots + X^i r_0 a) = X^{i+j}\varphi(r_j a) + \cdots + X^i \varphi(r_0 a) = X^{i+j} r_j \varphi(a) + \cdots + X^i r_0 \varphi(a) = X^i(X^j r_j + \cdots + r_0)\varphi(a) = X^i r X^j \varphi(a) = X^i r \varphi'(X^j a)$. Hence φ' is an S-homomorphism. Thus there exists a $t \in E_S(Q(S)/S)$ such that $\varphi'(g(X)) = g(X)t$ for any $g(X) \in S\mathfrak{a}$. In particular $\varphi'(a) = at$ for any $a \in \mathfrak{a}$ and hence $E_S(Q(S)/S)$ is injective as a left R-module. Since $Q/R \subset Q(S)/S \subset E_S(Q(S)/S)$ it follows $E(Q/R) \subset E_S(Q(S)/S)$. Therefore R is τ-Noetherian by [63, Proposition 2.4, p. 264]. □

2.3.16 Lemma

Suppose R is τ-Noetherian. Let $S = R[X;\sigma,\delta]$ and $T = Q[X;\sigma,\delta]$. If I is a left S-ideal and J is a right S-ideal, then $(T : TI)_r = (S : I)_r T$ and $(T : JT)_l = T(S : J)_l$. Hence $^*(TI) = T{}^*I$ and $(JT)^* = J^* T$.

Proof. It is clear that $(S : I)_r T \subset (T : TI)_r$. To prove the converse inclusion, let $q \in (T : TI)_r$, i.e. $TIq \subset T$. Since S is τ-Noetherian, there exists a finitely generated left S-ideal I_0 such that $I_0 \subset I$ and $cl_\tau(I) = cl_\tau(I_0)$. Thus $I_0 qc \subset S$ for some $c \in \mathcal{C}_R(0)$ and $Iqc \subset S$. Hence $qc \in (S : I)_r$ and so $q \in (S : I)_r T$. Similarly $(T : JT)_l = T(S : J)_l$. Therefore $^*(TI) = T\,^*I$ and $(JT)^* = J^* T$. □

Spec(T) denotes the set of all prime ideals of T and $\text{Spec}_0^*(S) = \{P, P$ is a divisorial prime ideal with $P \cap R = (0)\}$.

2.3.17 Proposition

Suppose R is τ-Noetherian. The map: $P' \longrightarrow P = P' \cap S$ is a bijection between Spec(T) and $\text{Spec}_0^*(S)$, where $0 \neq P' \in \text{Spec}(T)$.

Proof. Let $P' \in \text{Spec}(T)$ with $P' \neq 0$ and $P = P' \cap S$. It is clear that $P' = PT = TP$ and that P is a prime ideal with $P \cap R = (0)$. Since T is a principal ideal ring, $P' = {}^*P'$ and so $P = {}^*P$ by Lemma 2.3.16. Similarly $P = P^*$ and hence $P \in \text{Spec}_0^*(S)$. Conversely let $P \in \text{Spec}_0^*(S)$ and $P_1 = TP \cap S$. Then $TP_1 = TP = PT$ and as before there exists a finitely generated left ideal P_0 such that $P_1 \supset P_0$ and ${}^*P_1 = {}^*P_0$. Thus $P_0 c \subset P$ for some $c \in \mathcal{C}_R(0)$ and so ${}^*P_1 c = {}^*P_0 c \subset P$. Hence ${}^*P_1 \subset P$ since $P \in \text{Spec}_0^*(S)$ and so $P_1 = P$. Therefore $TP \in \text{Spec}(T)$ and $TP \cap S = P$. □

2.3.18 Proposition

Suppose R is τ-Noetherian, and let $S = R[X;\sigma,\delta]$ and $T = Q[X;\sigma,\delta]$. If $P \in \text{Spec}_0^*(S)$, then P is localizable and $S_P = T_{P'}$ which is a local and principal ideal ring, where $P' = TP$.

Proof. The following properties are easily checked:

(i) $\mathcal{C}_R(0) \subset \mathcal{C}(P)$
(ii) $\mathcal{C}_S(P) \subset \mathcal{C}_T(P')$
(iii) For any $t \in \mathcal{C}_T(P')$ with $ct \in S$ for some $c \in \mathcal{C}_R(0)$, it follows that $ct \in \mathcal{C}(P)$

By Proposition 2.2.12, P' is localizable and $T_{P'}$ is a local and principal ideal ring. For any $s(X) \in S$ and $c(X) \in \mathcal{C}(P)$, $d(X)s(X) = t(X)c(X)$ for some $d(X) \in \mathcal{C}_T(P')$ and $t(X) \in T$. There exists a $c \in \mathcal{C}_R(0)$ such that $cd(X), ct(X) \in S$. Since $cd(X) \in \mathcal{C}(P)$, $\mathcal{C}(P)$ is a left Ore set and similarly is a right Ore set. By (ii) $S_P \subset T_{P'}$. Let $q = t(X)c(X)^{-1} \in T_{P'}$, where $t(X) \in T$ and $c(X) \in \mathcal{C}_T(P')$. Then there is a $d \in \mathcal{C}_R(0)$ with $t(X)d \in S$, $c(X)d \in \mathcal{C}(P)$ and hence $q = (t(X)d)(c(X)d)^{-1} \in S_P$. □

An R-ideal $\mathfrak{a}$ is called **σ-stable** if $\sigma(\mathfrak{a}) \subset \mathfrak{a}$ and it is **σ-invariant** if $\sigma(\mathfrak{a}) = \mathfrak{a}$. An order R is said to be a **σ-maximal order** if $O_l(\mathfrak{a}) = R = O_r(\mathfrak{a})$ for any σ-invariant ideal $\mathfrak{a}$, and R is a **σ-Krull order** if it is σ-maximal and τ-Noetherian. Let $\mathfrak{a}$ be a σ-stable ideal of a τ-Noetherian order R. Note that $\mathfrak{a}$ is a σ-invariant if it is τ-closed as a one sided ideal.

2.3.19 Theorem

Let R be an order in Q and consider the following conditions:

1. R is a σ-Krull order.
2. $R[X;\sigma]$ is a Krull order.
3. $R[X;\sigma,\delta]$ is a Krull order.

Then (1) and (2) are equivalent and (1) implies (3).

Proof. 1. $\Longrightarrow$ 3.: Put $S = R[X;\sigma,\delta]$ and $T = Q[X;\sigma,\delta]$. Let A be a non-zero ideal of S. Then $TA = AT$ is an ideal of T and so $AT = Tg$ for some $g \in T$. Let $q \in O_l(A)$, i.e. $qA \subset A$ and $qg \in qAT \subset AT = Tg$. Thus $q \in T$. $L(A) = \{a,\, aX^n + \cdots + a_0 \in A\} \cup \{(0)\}$ is a σ-stable ideal. Write $q = s_mX^m + \cdots + s_0$ with $s_i \in Q$. Then for any $a \in L(A)$, there is an $h(X) = aX^n + a_{n-1}X^{n-1} + \cdots + a_0$ in A and $qh(X) = s_m\sigma^m(a)X^{n+m} +$ (the lower degree part). Since $s_m\sigma^m(a) \in L(A)$ it follows that $s_m\sigma^m(L(A)) \subset L(A)$ and, by taking τ-closure, $s_m cl_\tau(\sigma^m(L(A))) \subset cl_\tau(L(A))$. Hence $s_m \in O_l(cl_\tau(L(A)))$ since $cl_\tau(\sigma^m(L(A))) = \sigma^m(cl_\tau(L(A))) = cl_\tau(L(A))$. Thus $(q - s_mX^m)A \subset A$ and by induction on $m = \deg q$, we have $q \in S$, i.e. $O_l(A) = S$. Similarly $O_r(A) = S$ and hence S is a Krull order in $Q(S)$.
1. $\Longrightarrow$ 2. This is a special case.
2. $\Longrightarrow$ 1. Let $\mathfrak{a}$ be a σ-invariant. Then $A = \mathfrak{a}S = \mathfrak{a}[X;\sigma]$ is an ideal of $R[X;\sigma]$ and $R[X;\sigma] = O_l(A) = O_l(\mathfrak{a})[X;\sigma]$. Hence $O_l(\mathfrak{a}) = R$ and $O_r(\mathfrak{a}) = R$ similarly. Therefore R is a σ-Krull order. □

If R is a Krull order then it is a σ-Krull order. Thus we have the following.

2.3.20 Corollary

If R is a Krull order, then so is $R[X;\sigma,\delta]$.

In case either $\sigma = 1$ or $\delta = 0$, we describe all divisorial ideals of $R[X;\sigma]$ and $R[X;\delta]$ if they are Krull orders. A σ-invariant ideal $\mathfrak{p}$ is called **σ-prime** if $\mathfrak{a}\mathfrak{b} \subset \mathfrak{p}$, where $\mathfrak{a}$ and $\mathfrak{b}$ are σ-stable ideals, then either $\mathfrak{a} \subset \mathfrak{p}$ or $\mathfrak{b} \subset \mathfrak{p}$. We can define **$\delta$-stable R-ideals** and **δ-prime ideals** in an obvious way. R is said to be a **δ-maximal order** if $O_l(\mathfrak{a}) = R = O_r(\mathfrak{a})$ for all δ-stable ideals $\mathfrak{a}$. A δ-maximal order which is τ-Noetherian is called a **δ-Krull order**.

2.3.21 Lemma

1. Suppose R is a σ-Krull order in Q, let $S = R[X;\sigma]$ and $T = Q[X;\sigma]$. If A is a divisorial S-ideal, then $\sigma(A) = A$. Furthermore if $A \subset T$ and $\mathfrak{a} = A \cap Q \neq (0)$, then $\mathfrak{a}$ is a divisorial R-ideal and σ-invariant.
2. Suppose R is a δ-Krull order in Q, let $S = R[X;\delta]$ and $T = Q[X;\delta]$. If A is an S-ideal, then $\mathfrak{a} = A \cap Q$ is δ-stable.

Proof. 1. Note that σ is extended to an automorphism of $S = R[X;\delta]$ by $\sigma(s) = XsX^{-1}$ for any $s \in S$. Since S is a Krull order and XS is invertible, we have $^*(\sigma(A)) = {}^*(XAX^{-1}) = XS \circ A \circ X^{-1}S = A$ by Theorem 2.1.2. Thus $\sigma(A) \subset A$ and so $\sigma(A) = A$ since S is τ-Noetherian. If $A \subset T$ and $\mathfrak{a} = A \cap Q \neq (0)$, then $\mathfrak{a}$ is σ-invariant. Furthermore since $A = {}^*A \supset {}^*(S\mathfrak{a}) = S{}^*\mathfrak{a}$ by Lemma 2.3.13, it follows that $\mathfrak{a} = {}^*\mathfrak{a}$ and similarly $\mathfrak{a} = \mathfrak{a}^*$, i.e. $\mathfrak{a}$ is divisorial.
2. For any $a \in \mathfrak{a}$, we have $A \ni Xa - aX = \delta(a)$ and so $\mathfrak{a}$ is δ-stable. □

We denote by $D_\sigma(R)$ the set of divisorial σ-invariant R-ideals. Similarly $D_\delta(R)$ is the set of divisorial δ-stable R-ideals. Then following proposition is proved using the same method as Theorem 2.1.2 with respect to properties of σ and δ, respectively:

2.3.22 Proposition

1. If R is a σ-Krull order, then $D_\sigma(R)$ is an Abelian group generated by maximal divisorial σ-prime ideals.
2. If R is a δ-Krull order, then $D_\delta(R)$ is an Abelian group generated by maximal divisorial δ-prime ideals.

2.3.23 Lemma

1. Suppose R is a σ-Krull order, let $S = R[X;\sigma]$ and $\mathfrak{p}$ be a divisorial ideal which is σ-invariant. Then $P = \mathfrak{p}S$ is a prime ideal if and only if $\mathfrak{p}$ is σ-prime.
2. Suppose R is a δ-Krull order, let $S = R[X;\delta]$ and $\mathfrak{p}$ be a divisorial ideal which is δ-stable. Then $P = \mathfrak{p}S$ is a prime ideal if and only if $\mathfrak{p}$ is δ-prime ideal. In particular P is divisorial if and only if $\mathfrak{p}$ is.

Proof. 1. It is clear that P is an ideal of S. Assume that P is prime. Then clearly $\mathfrak{p}$ is a σ-prime ideal. Conversely assume that $\mathfrak{p}$ is a σ-prime ideal. It follows from Lemma 2.3.13 that P is divisorial. To prove that P is prime, assume that $AB \subset P$, where A and B are ideals of S. We may assume that A and B are both divisorial. If $A \supsetneq P$ and take $a(X) = a_nX^n + \cdots + a_0 \in A - P$ with $a_n \notin \mathfrak{p}$, then for any $b(X) \in B$, $a(X)b(X) \in P$ entails $a_n\sigma^n(L(B)) \subset \mathfrak{p}$. Note that $L(B)$

is σ-invariant since $\sigma(B) = B$ by Lemma 2.3.21. So $\sigma^{-n}(a_n)L(B) \subset \mathfrak{p}$ and $L(B) \subset \mathfrak{p}$, i.e. $b_l X^l \in P$, where $b(X) = b_l X^l + \cdots + b_0$. Then $a(X)(b(X) - b_l X^l) \in P$ and inductively we have $b(X) \in P$. Hence P is a prime ideal.
2. This is proved in a similar way as in (1). □

The proof of the following proposition is similar to the proof of Theorem 2.3.19

2.3.24 *Proposition*

An order R is a δ-Krull order if and only if $R[X;\delta]$ is a Krull order.

A divisorial ideal of a Krull order R is maximal if and only if it is a prime ideal, because $D(R)$ is generated by maximal divisorial ideals of R. Similarly a divisorial σ-invariant ideal of a σ-Krull order is maximal if and only if it is a σ-prime ideal. In case of δ-Krull orders, a divisorial δ-stable ideal is maximal if and only if it is a δ-prime ideal.

2.3.25 *Theorem*

1. Suppose R is a σ-Krull order in Q and $S = R[X;\sigma]$, $T = Q[X;\sigma]$. Then $D(S) \cong D_\sigma(R) \oplus D(T)$.
2. Suppose R is a δ-Krull order in Q and let $S = R[X;\delta]$, $T = Q[X;\delta]$. Then $D(S) \cong D_\delta(R) \oplus D(T)$.

Proof. 1. Let P be an ideal of S with $\mathfrak{p} = P \cap R$. Because of Proposition 2.3.17 and Lemma 2.3.23, it suffices to prove that P is a divisorial prime ideal if and only if either $P \in \mathrm{Spec}_0^*(S)$ or $\mathfrak{p}$ is a divisorial σ-prime ideal and $P = \mathfrak{p}S$. Suppose P is a divisorial prime ideal. If $\mathfrak{p} = (0)$, then $P \in \mathrm{Spec}_0^*(S)$ by Proposition 2.3.17. If $\mathfrak{p} \neq (0)$, then it is divisorial and σ-invariant by Lemma 2.3.21. It is easily checked that $\mathfrak{p}$ is a σ-prime ideal. Hence $\mathfrak{p}S$ is a divisorial prime ideal and so $P = \mathfrak{p}S$ by Theorem 2.1.2. Conversely suppose $\mathfrak{p}$ is a divisorial σ-prime ideal with $P = \mathfrak{p}S$. Then P is a divisorial prime ideal by Lemmas 2.3.13 and 2.3.23.
2. This is proved in a similar way to the proof of (1). □

2.3.26 *Lemma*

1. Let $S = R[X;\sigma]$ be a Krull order and $T = Q[X;\sigma]$. If A is a divisorial S-ideal such that $T \supset A$ and $\mathfrak{a} = A \cap Q \neq (0)$, then $A = S\mathfrak{a} = \mathfrak{a}S$.
2. Let $S = R[X;\delta]$ be a Krull order and $T = Q[X;\delta]$. If A is a divisorial S-ideal such that $T \supset A$ and $\mathfrak{a} = A \cap Q \neq (0)$, then $A = S\mathfrak{a} = \mathfrak{a}S$.

Proof. 1. By Theorem 2.3.25, $A = {}^*(\mathfrak{b}[X;\sigma]P_1^{e_1}\cdots P_k^{e_k})$, where $\mathfrak{b} \in D_\sigma(R)$ and $P_i \in \mathrm{Spec}_0^*(S)$. Since $\mathfrak{a} \neq (0)$, it follows from Lemma 2.3.16 that $T = TA = TP_1^{e_1}\cdots P_k^{e_k}$ and so $e_1 = \cdots = e_k = 0$. Thus $A = \mathfrak{b}[X;\sigma] = S\mathfrak{b} = \mathfrak{b}S$ and hence $\mathfrak{b} = \mathfrak{a}$.
2. This is proved in the same way as in the proof of (1). □

Let R be a Krull order in Q. The set of principal R-ideals forms a subgroup $P(R)$ of $D(R)$. The factor group $D(R)/P(R)$ is called the **class group** of R and is denoted by $\mathcal{C}(R)$. If R is a σ-Krull order, then we define $\mathcal{C}_\sigma(R) = D_\sigma(R)/P_\sigma(R)$ which is called the **σ-class group** of R, where $P_\sigma(R)$ is the subgroup of σ-invariant principal R-ideals. If R is a δ-Krull order, then $\mathcal{C}_\delta(R) = D_\delta(R)/P_\delta(R)$ is said to be the **δ-class group** of R, where $P_\delta(R)$ is the set of δ-stable principal R-ideals.

2.3.27 Theorem

1. Suppose R is a σ-Krull order in Q and let $S = R[X;\sigma]$. Then the map $\varphi : D_\sigma(R) \longrightarrow D(S)$ defined by $\varphi(\mathfrak{a}) = \mathfrak{a}S$, where $\mathfrak{a} \in D_\sigma(R)$ induces an isomorphism: $\mathcal{C}_\sigma(R) \cong \mathcal{C}(S)$.
2. Suppose R is a δ-Krull order in Q and let $S = R[X;\delta]$. Then the map $\varphi : D_\delta(R) \longrightarrow D(S)$ defined by $\varphi(\mathfrak{a}) = \mathfrak{a}S$, where $\mathfrak{a} \in D_\delta(R)$ induces a surjective map: $\mathcal{C}_\delta(R) \longrightarrow \mathcal{C}(S)$. If R is a domain, then $\mathcal{C}_\delta(R) \cong \mathcal{C}(S)$.

Proof. 1. The map $\varphi : D_\sigma(R) \longrightarrow D(S)$ induces the map $\overline{\varphi} : \mathcal{C}_\sigma(R) \longrightarrow \mathcal{C}(S)$. If $\mathfrak{a} \in D_\sigma(R)$ such that $S\mathfrak{a} = Sf = fS$ for some $f \in Q(S)$, then $T = Tf = fT$ implies $f \in T$ and $\mathfrak{a} = Rf_0 = f_0R$, where f_0 is a constant term of f, showing $\overline{\varphi}$ is injective. To prove that $\overline{\varphi}$ is surjective, let $A \in D(S)$. Then $TA = Tw(X)X^m$ by Corollary 2.3.11, where $w(X)$ is a central element of $T = Q[X;\sigma]$. $TAX^{-m}w(X)^{-1} = T$ entails $\mathfrak{a} = AX^{-m}w(X)^{-1} \cap Q \neq (0)$ and is an element in $D_\sigma(R)$ by Lemma 2.3.21. It follows from Lemma 2.3.26 that $AX^{-m}w(X)^{-1} = S\mathfrak{a}$ and hence $A = S\mathfrak{a} \circ Sw(X)X^m$, i.e. A and $S\mathfrak{a}$ are the same class in $\mathcal{C}(A)$. Hence $\overline{\varphi}$ is surjective.
2. As in (1), we define the map $\varphi : D_\delta(R) \longrightarrow D(S)$ by $\varphi(\mathfrak{a}) = S\mathfrak{a}$ for all $\mathfrak{a} \in D_\delta(R)$ and $\overline{\varphi} : \mathcal{C}_\delta(R) \longrightarrow \mathcal{C}(S)$. Every ideal of T is principal generated by a central element by Corollary 2.3.11 and so $\overline{\varphi}$ is surjective as in (1). Suppose R is a domain and let $\mathfrak{a} \in D_\delta(R)$ such that $S\mathfrak{a} = Sf = fS$ for some $f \in T$. Then the leading coefficient of f is regular and so $\deg f = 0$. Hence $\mathfrak{a} = Rf = fR$ which shows $\overline{\varphi}$ is injective. □

2.3.28 Remark

A τ-Noetherian prime Goldie ring R is called a **unique factorization ring** (a UFR) if every prime ideal P with ${}^*P = P$ or $P^* = P$ is principal. It turns out that R is a UFR if and only if it is a Krull order and every divisorial ideal is principal.

We define σ-UFRs and δ-UFRs in an obvious way (cf. [1]). Theorem 2.3.27 shows that R is a σ-UFR if and only if $R[X;\sigma]$ is a UFR. Furthermore R is a δ-UFR, then $R[X;\delta]$ is a UFR and the converse also holds if R is a domain.

2.3.29 *Remark*

The following are interesting open questions:

1. Let $S = R[X;\sigma,\delta]$ and $T = Q[X;\sigma,\delta]$. What is a necessary and sufficient condition for S to be Krull (cf. Theorem 2.3.19). In case S is a Krull order, describe explicitly the group $D(S)$ of all divisorial S-ideals and the class groups $\mathcal{C}(S)$ in terms of the properties of R and $T = Q[X;\sigma,\delta]$.
2. If $R[X;\delta]$ is a UFR, then is R a δ-UFR in case R is not necessarily a domain?

2.4 Non-commutative Valuation Rings in $K(X;\sigma,\delta)$

In Chap. 1, we defined three different non-commutative valuation rings, i.e. valuation rings, total valuation rings and Dubrovin valuation rings, and studied elementary properties of them. Let V be a non-commutative valuation ring of a simple Artinian ring K, σ be an automorphism of K and δ be a left σ-derivation of K. We denote by $K(X;\sigma,\delta)$ the quotient ring of Ore extension $K[X;\sigma,\delta]$.

The aim of this section is to give a partial answer to the following question:

"Find out all non-commutative valuation rings S of $K(X;\sigma,\delta)$ with $S \cap K = V$ and study the structure of them".

In the end of the section we propose a program for the question. Let us begin with the following case; V is a total valuation ring of a skewfield K, $\sigma(V) = V$, $\delta(V) \subset V$ and $\delta(J(V)) \subset J(V)$. In this case (σ,δ) is said to be **compatible** with V.

2.4.1 *Proposition*

With the assumption above, let $R = V[X;\sigma,\delta]$ and $\mathcal{C} = R - J(V)[X;\sigma,\delta]$. Then $\mathcal{C}$ is an Ore set of R and $R^{(1)} = R_{\mathcal{C}}$ is a total valuation ring of $K(X;\sigma,\delta)$ with $R^{(1)} \cap K = V$.

Proof. We let $\overline{V} = V/J(V)$. Then (σ,δ) naturally induces an automorphism σ of $\overline{V}$, a left σ-derivation δ of $\overline{V}$ and $V/J(V)[X;\sigma,\delta] \cong \overline{V}[X;\sigma,\delta]$ is a domain. Thus $\mathcal{C}$ is multiplicatively closed. To prove that $\mathcal{C}$ is a left Ore set, let $c(X) \in \mathcal{C}$ and $0 \neq f(X) \in R$. Since $K(X;\sigma,\delta)$ is the quotient ring of R, we have $f(X)c(X)^{-1} = d(X)^{-1}g(X)$ for some $d(X), g(X) \in R$, i.e.

$d(X)f(X) = g(X)c(X)$. The elements $d(X)$ and $g(X)$ can be factored as follows: $d(X) = d_1 d_1(X)$ and $g(X) = g_1 g_1(X)$ with $d_1, g_1 \in V$ and $d_1(X), g_1(X) \in \mathcal{C}$. Furthermore we have either $g_1 = d_1 r$ or $d_1 = g_1 s$ for some $r, s \in V$. In the first case it follows that $d_1(X)f(X) = rg_1(X)c(X)$ with $d_1(X) \in \mathcal{C}$ and $rg_1(X) \in R$.

In the second case we have $sd_1(X)f(X) = g_1(X)c(X)$ and so $sd_1(X) \in \mathcal{C}$. Hence $\mathcal{C}$ is a left Ore set and similarly it is a right Ore set.

To prove that $R^{(1)}$ is a total valuation ring, let $f(X)^{-1}g(X) \in K(X;\sigma,\delta)$, where $f(X)$, $g(X) \in R$. As before $f(X) = f_1 f_1(X)$ and $g(X) = g_1 g_1(X)$ with $f_1, g_1 \in V$ and $f_1(X), g_1(X) \in \mathcal{C}$. Either $f_1^{-1}g_1 \in V$ or $g_1^{-1}f_1 \in V$ entails that either $f(X)^{-1}g(X) \in R^{(1)}$ or $g(X)^{-1}f(X) \in R^{(1)}$. Hence $R^{(1)}$ is a total valuation ring of $K(X;\sigma,\delta)$. Let $q \in R^{(1)} \cap K$. If $q \notin V$, then $q^{-1} \in J(V) \subset J(V)[X;\sigma,\delta]R^{(1)} = J(R^{(1)})$ which is a contradiction. Hence $R^{(1)} \cap K = V$.

2.4.2 Remark

Proposition 2.4.1 is obtained in case σ is an endomorphism (cf. [10, Theorem 1]).

An ideal $\mathfrak{a}$ of V is called **(σ,δ)-stable** if $\sigma(\mathfrak{a}) \subset \mathfrak{a}$ and $\delta(\mathfrak{a}) \subset \mathfrak{a}$. A (σ,δ)-stable ideal $\mathfrak{a}$ is called **strongly** if $\sigma(\mathfrak{a}) = \mathfrak{a}$.

2.4.3 Proposition

The correspondence $\mathfrak{a} \longrightarrow A = R^{(1)}\mathfrak{a}$ is a one-to-one correspondence between strongly (σ,δ)-stable ideals of V and ideals of $R^{(1)}$.

Proof. Let A be an ideal of $R^{(1)}$ and $\mathfrak{a} = A \cap V$. For any $a \in \mathfrak{a}$, $Xa = \sigma(a)X + \delta(a)$ and we have either $\delta(a) = \sigma(a)r$ or $\sigma(a) = \delta(a)s$ for some $r, s \in V$. In the former case $Xa = \sigma(a)(X + r)$ with $X + r \in \mathcal{C}$. Hence $\sigma(a) \in \mathfrak{a}$ and $\delta(a) \in \mathfrak{a}$. In the latter case $Xa = \delta(a)(sX + 1)$ with $(sX + 1) \in \mathcal{C}$. So $\delta(a) \in \mathfrak{a}$ and $\sigma(a) \in \mathfrak{a}$ since $A \ni (Xa - \delta(a))X^{-1} = \sigma(a)$. Hence $\mathfrak{a}$ is (σ,δ)-stable and similarly $\mathfrak{a}$ is (σ',δ')-stable, where $\sigma' = \sigma^{-1}$ and $\delta' = -\delta\sigma^{-1}$ as in Sect. 2.3. Hence $\mathfrak{a}$ is strongly (σ,δ)-stable. It is clear that $R^{(1)}\mathfrak{a} \subset A$. To prove the converse inclusion, let $\alpha = c(X)^{-1}f(X) \in A$, where $c(X) \in \mathcal{C}$ and $f(X) = f_1(X)f_1 \in R$ with $f_1(X) \in \mathcal{C}$ and $f_1 \in V$. Then $R^{(1)}\alpha = R^{(1)}f_1$ and $f_1 \in \mathfrak{a}$. Thus $\alpha \in R^{(1)}\mathfrak{a}$ which shows $A = R^{(1)}\mathfrak{a}$.

Conversely let $\mathfrak{a}$ be a strongly (σ,δ)-stable ideal of V. Since $R^{(1)}\mathfrak{a} = R^{(1)}\mathfrak{a}[X;\sigma,\delta]$, each element in $R^{(1)}\mathfrak{a}$ is of the form; $c(X)^{-1}a$, where $c(X) \in \mathcal{C}$ and $a \in \mathfrak{a}$. Suppose $b = c(X)^{-1}a \in R^{(1)}\mathfrak{a} \cap V$. Then $c(X)b = a$ and so $\deg c(X) = 0$, i.e. $c(X) = c_0 \in U(V)$. Thus $b \in \mathfrak{a}$ and $R^{(1)}\mathfrak{a} \cap V = \mathfrak{a}$ follows. To prove that $R^{(1)}\mathfrak{a}$ is an ideal of $R^{(1)}$, it is enough to show that $\mathfrak{a}c(X)^{-1} \subset R^{(1)}\mathfrak{a}$ for any $c(X) \in \mathcal{C}$. For any $a \in \mathfrak{a}$ and $c(X) \in \mathcal{C}$, we have $ac(X)^{-1} = d(X)^{-1}b(X)$ for some $d(X) \in \mathcal{C}$ and

$b(X) \in R$, i.e. $d(X)a = b(X)c(X)$. As before $b(X) = b_1(X)b_1$ with $b_1(X) \in C$ and $b_1 \in V$ and so $d(X)a = b(X)c(X) = b_1(X)b_1c(X) = b_1(X)c_1(X)b_2$ for some $c_1(X) \in C$ and $b_2 \in V$, where $b_1c(X) = c_1(X)b_2$. Thus $R^{(1)}a = R^{(1)}b_2$ and $b_2 \in \mathfrak{a}$. Hence $b_1c(X) \in \mathfrak{a}[X;\sigma,\delta]$ and so $b_1 \in \mathfrak{a}$ since $c(X) \in C$, i.e. $b(X) \in \mathfrak{a}[X;\sigma,\delta]$. Therefore $ac(X)^{-1} = d(X)^{-1}b(X) \in R^{(1)}\mathfrak{a}[X;\sigma,\delta] = R^{(1)}\mathfrak{a}$. □

A strongly (σ,δ)-stable ideal $\mathfrak{p}$ of V is called **(σ,δ)-prime** if $\mathfrak{a}\mathfrak{b} \subset \mathfrak{p}$, where $\mathfrak{a}$ and $\mathfrak{b}$ are (σ,δ)-stable ideals of V, implies either $\mathfrak{a} \subset \mathfrak{p}$ or $\mathfrak{b} \subset \mathfrak{p}$.

2.4.4 Proposition

Let $\mathfrak{p}$ be a strongly (σ,δ)-stable ideal of V. Then $\mathfrak{p}$ is a (σ,δ)-prime ideal of V if and only if $R^{(1)}\mathfrak{p}$ is a prime ideal of $R^{(1)}$. In particular, $\mathfrak{p}$ is completely prime if and only if so is $R^{(1)}\mathfrak{p}$.

Proof. Let $\mathfrak{a}$ be a (σ,δ)-stable ideal of V. We prove that $R^{(1)}\mathfrak{a}[X;\sigma,\delta]$ is an ideal of $R^{(1)}$. It suffices to prove that $a(X)c(X)^{-1} \in R^{(1)}\mathfrak{a}[X;\sigma,\delta]$ for any $a(X) \in \mathfrak{a}[X;\sigma,\delta]$ and $c(X) \in C$. There are $a \in V$ and $d(X), e(X) \in C$ such that $a(X)c(X)^{-1} = d(X)^{-1}ae(X)$. So $\mathfrak{a}[X;\sigma,\delta] \ni d(X)a(X) = ae(X)c(X)$. Thus $a \in \mathfrak{a}$, because $e(X)c(X) \in C$ and hence $a(X)c(X)^{-1} \in R^{(1)}\mathfrak{a}[X;\sigma,\delta]$. Therefore the first part of proposition easily follows from Proposition 2.4.3. If $R^{(1)}\mathfrak{p}$ is completely prime, then so is $\mathfrak{p}$ since $R^{(1)}\mathfrak{p} \cap V = \mathfrak{p}$. Conversely suppose $\mathfrak{p}$ is completely prime. We show that $\mathfrak{p}[X;\sigma,\delta]$ is localizable. Let $f(X) \in R$ and $c(X) \in R-\mathfrak{p}[X;\sigma,\delta]$. Then $d(X)f(X) = g(X)c(X)$ for some $d(X), g(X) \in R$. As before $d(X) = d_1d_1(X)$ and $g(X) = g_1g_1(X)$ where $d_1, g_1 \in V$ and $d_1(X), g_1(X) \in C$. We have either $d_1 = g_1r$ or $g_1 = d_1s$ for some $r, s \in V$. If $d_1 = g_1r$, then $rd_1(X)f(X) = g_1(X)c(X) \notin \mathfrak{p}[X;\sigma,\delta]$ and so $rd_1(X) \notin \mathfrak{p}[X;\sigma,\delta]$. If $g_1 = d_1s$, then $d_1(X)f(X) = sg_1(X)c(X)$ and $d_1(X) \in C \subset R-\mathfrak{p}[X;\sigma,\delta]$. Hence $\mathfrak{p}[X;\sigma,\delta]$ is left localizable and similarly it is right localizable. Put $S = V[X;\sigma,\delta]_{\mathfrak{p}[X;\sigma,\delta]}$ which contains $R^{(1)}$ so that S is a total valuation ring with $J(S) = S\mathfrak{p}[X;\sigma,\delta]$. It follows from the proof of Theorem 1.4.9 that $R^{(1)} \supset J(S)$. So each element $\alpha \in J(S)$ is of the form; $\alpha = d(X)^{-1}p(X) = f(X)c(X)^{-1}$, where $d(X) \in R - \mathfrak{p}[X;\sigma,\delta]$, $p(X) \in \mathfrak{p}[X;\sigma,\delta]$, $f(X) \in R$ and $c(X) \in C$, i.e. $\mathfrak{p}[X;\sigma,\delta] \ni p(X)c(X) = d(X)f(X)$. Thus $\alpha \in \mathfrak{p}[X;\sigma,\delta]R^{(1)} = R^{(1)}\mathfrak{p}[X;\sigma,\delta] = R^{(1)}\mathfrak{p}$. This entails $J(S) = R^{(1)}\mathfrak{p}$ and it is localizable by Theorem 1.4.9. Hence $R^{(1)}\mathfrak{p}$ is completely prime. □

2.4.5 Proposition

$R^{(1)}$ is a valuation ring of $K(X;\sigma,\delta)$ if and only if

(i) V is a valuation ring
(ii) $\delta(a) \in Va$ and $Va = V\sigma(a)$ for all $a \in K$

Proof. For any $a \in K$ it is easily seen that $R^{(1)}a \cap K = Va$. Suppose $R^{(1)}$ is a valuation ring. Then V is a valuation ring. For any $a \in K$ we have either $\sigma(a)V \supset \delta(a)V$ or $\sigma(a)V \subset \delta(a)V$. If $\sigma(a)V \supset \delta(a)V$, then $\sigma(a^{-1})\delta(a) \in V$ and $X + \sigma(a^{-1})\delta(a) \in U(R^{(1)})$. So we have $R^{(1)}a = R^{(1)}Xa = R^{(1)}\sigma(a)(X + \sigma(a^{-1})\delta(a)) = \sigma(a)R^{(1)}(X + \sigma(a^{-1})\delta(a)) = \sigma(a)R^{(1)}$ and hence $Va = V\sigma(a) \ni \delta(a)$ follows. If $\sigma(a)V \subset \delta(a)V$, i.e. $\sigma(a) = \delta(a)r$ for some $r \in V$, then $R^{(1)}a = R^{(1)}Xa = R^{(1)}\delta(a)(rX+1) = R^{(1)}\delta(a)$ and so $Va = V\delta(a)$. Since $\delta(a) = sa$ for some $s \in U(R^{(1)})$, we have $Xa = \sigma(a)X + \delta(a) = \sigma(a)X + sa$ and so $(X-s)a = \sigma(a)X$. Thus $R^{(1)}a = R^{(1)}(X-s)a = R^{(1)}\sigma(a)X = R^{(1)}\sigma(a)$ and hence $Va = V\sigma(a)$.

Suppose that (i) and (ii) hold. Then for any $a \in V$, $\mathfrak{a} = aV = Va$ is a strongly (σ,δ)-ideal and so $R^{(1)}\mathfrak{a} = \mathfrak{a}R^{(1)}$ is an ideal of $R^{(1)}$ by Proposition 2.4.3. Hence $R^{(1)}a = aR^{(1)}$ follows. Let $f(X)$ be any element in R and $f(X) = f_1 f_1(X)$ for some $f_1 \in V$ and $f_1(X) \in \mathcal{C}$. Then we have $R^{(1)}f(X) = R^{(1)}f_1 f_1(X) = f_1 R^{(1)} f_1(X) = f_1 R^{(1)}$ and $f(X)R^{(1)} = f_1 R^{(1)}$. Thus $R^{(1)}f(X) = f(X)R^{(1)}$ and hence $R^{(1)}$ is a valuation ring. □

A total valuation ring S of $K(X;\sigma,\delta)$ with $S \cap K = V$ is called a **Gauss extension** of V in $K(X;\sigma,\delta)$ if for any $f(X) = X^n a_n + \cdots + a_0 \in K[X;\sigma,\delta]$ $Sf(X) = SX^i a_i$ for some $X^i a_i$ with $SX^i a_i \supset SX^j a_j$ for all j, $0 \leqq j \leqq n$.

2.4.6 Remark

$R^{(1)}$ is a Gauss extension of V in $K(X;\sigma,\delta)$. We will give an example of a total valuation ring of $K(X;\sigma)$ which is not a Gauss extension of V at the end of this section.

We will study the structure of $R = V[X;\sigma,\delta]$.

2.4.7 Lemma

Let $R = V[X;\sigma,\delta]$ and $S = K[X;\sigma,\delta]$. Then

1. $R = R^{(1)} \cap S$.
2. For any $f(X), g(X) \in R$, put $I = Rf(X) + Rg(X)$. Then $R^{(1)}I \cap SI = Rc(X)$ for some $c(X) \in R$.

Proof. 1. Let $g(X) = d(X)^{-1}f(X)$, where $d(X) \in \mathcal{C}$, $f(X) \in R$ and $g(X) \in S$. As before we have $g(X) = ag_1(X)$ for some $a \in K$ and $g_1(X) \in \mathcal{C}$. Thus $a = d(X)^{-1}f(X)g_1(X)^{-1} \in R^{(1)} \cap K = V$ and hence $g(X) \in R$.

2. Since $R^{(1)}$ is a total valuation ring we may suppose that $R^{(1)}f(X) \supset R^{(1)}g(X)$. As in (1), $f(X) = f_1(X)a$ for some $f_1(X) \in \mathcal{C}$ and $a \in V$. Thus $R^{(1)}I = R^{(1)}a$ and also $SI = Sb(X)$ for some $b(X) \in S$. Furthermore we have $b(X)a^{-1} =$

$bb_1(X)$ for some $b \in K$ and $b_1(X) \in \mathcal{C}$. It follows that $R^{(1)} \cap Sb(X)a^{-1} = R^{(1)} \cap Sbb_1(X) = R^{(1)} \cap Sb_1(X) = R^{(1)}b_1(X) \cap Sb_1(X) = Rb_1(X)$. Hence $R^{(1)}I \cap SI = R^{(1)}a \cap Sb(X) = (R^{(1)} \cap Sb(X)a^{-1})a = Rb_1(X)a = Rc(X)$, where $c(X) = b_1(X)a$. □

In Chap. 1, we defined a Bezout order in a simple Artinian ring, i.e. any finitely generated one-sided ideal is principal. We generalize this concept from divisorial ideal's point of view: An order in a simple Artinian ring is said to be a **generalized left Bezout order** if every finitely generated essential and divisorial left ideal is principal. We can similarly define a generalized right Bezout order and an order is **generalized Bezout** if it is generalized left and right Bezout.

2.4.8 *Proposition*

$R = V[X;\sigma,\delta]$ is a generalized Bezout order in $K(X;\sigma,\delta)$.

Proof. For any $f(X), g(X) \in R$, we put $I = Rf(X) + Rg(X)$. Then $(R^{(1)}I \cap SI)(R : I)_r \subseteq R^{(1)}I(R : I)_r \cap SI(R : I)_r \subset R^{(1)} \cap S = R$. This implies $R^{(1)}I \cap SI \subseteq {}^*I$ and $R^{(1)}I \cap SI = Rc(X)$ for some $c(X) \in R$ by Lemma 2.4.7. It follows ${}^*I = {}^*({}^*I) \supset {}^*(R^{(1)}I \cap SI) = {}^*(Rc(X)) \supset {}^*I$ and hence ${}^*I = Rc(X)$. Then, by induction on generators, we can prove that any finitely generated and divisorial left ideal is principal. It follows from the right version that any finitely generated and divisorial right ideal is principal. Hence R is a generalized Bezout order. □

Let R be an order in a simple Artinian ring Q. Let a, b, c and d be regular elements in R. We say that d is a left greatest common divisor of a and b (a left GCD for short) if

(i) $Rd \supset Ra$ and $Rd \supset Rb$ and
(ii) If $Rc \supset Ra$ and $Rc \supset Rb$ then $Rc \supset Rd$

which is written $d = l\text{-GCD}\{a,b\}$. Note that $c = l\text{-GCD}\{a,b\}$ if and only if $Rc = Rd$. A right GCD is defined similarly. An order in a simple Artinian ring is said to be a **GCD-order** if any two regular elements have a left GCD as well as a right GCD. As a dual concept of GCDs, we define a left least common multiple m of a and b (a l-LCM for short and written; $m = l\text{-LCM}\{a,b\}$) and similarly we defined a right least common multiple. An order is called a **LCM-order** if any two regular elements have a left LCM and a right LCM.

2.4.9 *Lemma*

Let R be an order in a simple Artinian ring Q. Let a, b, c and d be regular elements in R such that $d = l\text{-GCD}\{ac, bc\}$. Then $dc^{-1} = l\text{-GCD}\{a,b\}$.

Proof. First note that $dc^{-1} \in R$, because $Rc \supset Rac$ and $Rc \supset Rbc$. It is clear that $Rdc^{-1} \supset Ra$ and $Rdc^{-1} \supset Rb$. Suppose that $Re \supset Ra$ and $Re \supset Rb$ for some $e \in R$. Then $Rec \supset Rac$ and $Rec \supset Rbc$. Thus $Rec \supset Rd$, i.e. $Re \supset Rdc^{-1}$. Hence $dc^{-1} = l\text{-GCD}\{a,b\}$. □

2.4.10 Proposition

Let R be an order in a simple Artinian ring Q. Then the following are equivalent:

1. R is a generalized Bezout order.
2. R is a GCD order.
3. R is a LCM order.

Proof. 1. ⇒ 2. For any regular elements a, b in R, let $I = Ra + Rb$. Then ${}^*I = Rd$ for some $d \in R$. Suppose that $Rc \supset Ra$ and $Rc \supset Rb$. Then $Rc \supset I$ and so $Rc \supset {}^*I$. Hence $d = l\text{-GCD}\{a,b\}$. Similarly we can prove that a and b has a right GCD.

2. ⇒ 1. For any regular elements a and b in R, put $I = Ra + Rb$ and $d = l\text{-GCD}\{a,b\}$. Then $Rd \supset {}^*I$. Suppose $Rq \supset I$ for some $q \in Q$. If $q \in R$, then $Rq \supset Rd$. If $q \notin R$, then $q = ec^{-1}$ for some c, e in R. $Rec^{-1} \supset I$ implies $Re \supset Rac$ and $Re \supset Rbc$. Set $g = l\text{-GCD}\{ac, bc\}$. Then $gc^{-1} = l\text{-GCD}\{a,b\} = d$ by Lemma 2.4.9. So $Rq = Rec^{-1} \supset Rgc^{-1} = Rd$ and thus $Rd = {}^*I$ as in Proposition 1.5.8. Hence R is a generalized left Bezout order by induction on the generators of essential left ideals, because any essential one-sided ideal is generated by regular elements ([16, Theorem 1.21]). Similarly R is a right generalized Bezout order.

1. ⇒ 3. : For any regular elements a and b in R, put $I = a^{-1}R + b^{-1}R$. There is a regular element c in R with ca^{-1}, cb^{-1} in R. It follows that $cI^* = (cI)^* = (ca^{-1}R + cb^{-1}R)^* = dR$ for some $d \in R$, i.e. $I^* = c^{-1}dR$. Thus $Ra \cap Rb = (R : I)_l = (R : I^*)_l = Rd^{-1}c$ and so $d^{-1}c = l - \text{LCM}\{a,b\}$. Hence R is a left LCM order and similarly R is a right LCM order.

3. ⇒ 1. : For any regular elements a and b in R, put $I = Ra + Rb$. There is a regular element $c \in R$ with $ca^{-1}, cb^{-1} \in R$. Then $c(R : I)_r = c(a^{-1}R \cap b^{-1}R) = (ca^{-1}R \cap cb^{-1}R) = dR$ for some $d \in R$, i.e. $(R : I)_r = c^{-1}dR$. Thus ${}^*I = Rd^{-1}c$. Hence R is a left generalized Bezout order and similarly R is a right generalized Bezout order. □

Let R be an order in a simple Artinian ring Q and σ be an automorphism of Q. Furthermore we let $P = XQ[X;\sigma]$ and $T = Q[X;\sigma]_P$. It is easy to see that $c(X) = c_nX^n + \cdots + c_0 \in C(P)$ if and only if $c_0 \in U(Q)$. Thus we have a map $\varphi : T \longrightarrow Q$ defined by $\varphi(c(X)^{-1}f(X)) = c_0^{-1}f_0$, where $c(X) \in C(P)$ and $f(X) = f_mX^m + \cdots + f_0 \in Q[X;\sigma]$.

2.4.11 Lemma

φ is a ring epimorphism with $\ker \varphi = XT = J(T)$.

Proof. By using Ore condition for $\mathcal{C}(P)$ we have φ is well defined and a ring homomorphism. It is also clear that φ is an epimorphism with Ker $\varphi = XT$. □

We denote by $R^{(2)}$ the pre-image of R by φ, i.e. $R^{(2)} = R + J(T)$.

2.4.12 Lemma

$T = R^{(2)}_{\mathcal{C}}$, where $\mathcal{C} = \mathcal{C}_R(0)$.

Proof. For any $t \in T$, there exists $c \in \mathcal{C}$ with $\varphi(tc) = \varphi(t)c \in R$ and so $tc \in R^{(2)}$, i.e. $t = \alpha c^{-1}$ for some $\alpha \in R^{(2)}$. For any $\beta \in R^{(2)}$ and $d \in \mathcal{C}$, we have $d^{-1}\beta \in T$ and so there are $c \in \mathcal{C}$ and $\alpha \in R^{(2)}$ such that $d^{-1}\beta = \alpha c^{-1}$. Hence $\mathcal{C}$ is a right Ore set of $R^{(2)}$ and similarly it is left Ore. Now it is clear that $T = R^{(2)}_{\mathcal{C}}$. □

2.4.13 Lemma

$J(R^{(2)}) = J(R)$+J(T) and $R^{(2)}/J(R^{(2)}) \cong R/J(R)$.

Proof. Since $R^{(2)}/J(T) \cong R$, it suffices to prove that $I \supset J(T)$ for any maximal left ideal of $R^{(2)}$. Suppose on the contrary that I is a maximal left ideal of $R^{(2)}$ with $I \not\supset J(T)$. Then $I + J(T) = R^{(2)}$ and $TI = T$ by Nakayama Lemma. It follows that $I = R^{(2)}I \supset J(T)I = J(T)T = J(T)$ which is a contradiction. Hence $J(R^{(2)}) = J(R) + J(T)$ with $R^{(2)}/J(R^{(2)}) \cong R/J(R)$.

2.4.14 Lemma

The following are equivalent:

1. $\sigma(R^{(2)}) = R^{(2)}$.
2. $X \in \mathrm{st}(R^{(2)}) = \{q \in Q : qR^{(2)} = R^{(2)}q\}$.
3. $\sigma(R) = R$.

Proof. 1. ⇔ 2. This is obvious, because $\sigma(R^{(2)}) = XR^{(2)}X^{-1}$.
1. ⇒ 3. $\sigma(R) = \sigma(R^{(2)} \cap Q) = \sigma(R^{(2)}) \cap \sigma(Q) = R^{(2)} \cap Q = R$.
3. ⇒ 1. Since $XT = TX$, σ induces an automorphism of T and so $\sigma(J(T)) = J(T)$. Hence $\sigma(R^{(2)}) = \sigma(R + J(T)) = R + J(T) = R^{(2)}$. □

2.4.15 Lemma

1. $\mathrm{st}(R) = \mathrm{st}(R^{(2)}) \cap Q$.
2. Let $\alpha = c(X)^{-1}f(X) \in Q(X;\sigma)$ with $f(X), c(X) \in Q[X;\sigma]$ and $f_0, c_0 \in U(Q)$, where $f_0 = f(0)$ and $c_0 = c(0)$. Then $R^{(2)}\alpha = R^{(2)}c_0^{-1}f_0$ and $\alpha R^{(2)} = c_0^{-1}f_0R^{(2)}$. In particular, $\alpha \in \mathrm{st}(R^{(2)})$ if and only if $c_0^{-1}f_0 \in \mathrm{st}(R)$.

Proof. 1. This is obvious.
2. Since $\varphi(\alpha) = c_0^{-1}f_0$, we have $\alpha - c_0^{-1}f_0 \in J(T)$ by Lemma 2.4.11 and so $\alpha f_0^{-1}c_0 - 1 \in J(T) \subset J(R^{(2)})$ by Lemma 2.4.12. Hence $\alpha f_0^{-1}c_0 \in U(R^{(2)})$, i.e. $R^{(2)}\alpha = R^{(2)}c_0^{-1}f_0$. Similarly $\alpha R^{(2)} = c_0^{-1}f_0R^{(2)}$ follows. Now the last statement is clear.

2.4.16 Proposition

Let R be an order in a simple Artinian ring Q, σ be an automorphism of Q and $T = Q[X;\sigma]_P$, where $P = Q[X;\sigma]X$. Then

1. $R^{(2)}$ is a Dubrovin valuation ring of $Q(X;\sigma)$ if and only if so is R.
2. If R is a Dubrovin valuation ring of $Q(X;\sigma)$, then $\{R^{(2)}\mathfrak{p}$ and $J(T)$, $\mathfrak{p}$ is a prime ideal of $R\}$ is the set of prime ideals of $R^{(2)}$.
3. Assume that Q is a skewfield.
   (i) $R^{(2)}$ is a total valuation ring if and only if so is R.
   (ii) $R^{(2)}$ is a valuation ring if and only if R is a valuation ring and $\sigma(R) = R$.

Proof. 1. Since T is a Dubrovin valuation ring, the statement follows from Theorem 1.4.9 and Proposition 1.4.17 since $R^{(2)}/J(T) \cong R$.
2. It is clear that any prime ideal of $R^{(2)}$ containing $J(T)$ is of the form $\mathfrak{p} + J(T)$, where $\mathfrak{p}$ is a prime ideal of R and that $\mathfrak{p}R^{(2)} = \mathfrak{p} + J(T) = R^{(2)}\mathfrak{p}$. $J(T)$ is a minimal prime ideal of $R^{(2)}$ since $\cap X^nT = 0$.
3. (i) If Q is a skewfield, then T is a total valuation ring and so the statement is clear.
   (ii) Any element of $Q(X;\sigma)$ is of the form $X^nc(X)^{-1}f(X)$, where $n \in \mathbb{Z}$ and $f(X), c(X) \in Q[X;\sigma]$ such that $f(0) \neq 0$, $c(0) \neq 0$. Hence the statement follows from Lemmas 2.4.14, 2.4.15 and 3.(i).

2.4.17 Remark

Let R be a Dubrovin valuation ring of a simple Artinian ring Q with finite dimension over its center. Then we have the following ideal-theoretic properties:

1. $\Gamma_R = \mathrm{st}(R)/U(R)$ is Abelian.

2. If $Ra \supset aR$ for some $a \in Q$, then $Ra = aR$ and $Sa = aS$ for any overring S of R.
3. For any R-ideal I of Q, $O_l(I) = O_r(I)$.

Proposition 2.4.16 is used to obtain an example of Dubrovin valuation ring of Q in which these properties do not necessarily hold in case Q is infinite dimensional over its center [33].

2.4.18 Remark

Let V ba a commutative valuation ring of a field K and $T = K[X]_{(X+1)}$ be a localization of $K[X]$ at prime ideal $(X+1)$. Consider the natural epimorphism φ from T to $\overline{T} = T/(X+1)T$, which contains K and $[\overline{T} : K] = 2$. Let W be any valuation ring of $\overline{T}$ with $W \cap K = V$ and $R^{(2)} = \varphi^{-1}(W)$, the preimage of W by φ. Then $R^{(2)}$ is a valuation ring of $K(X)$ with $R^{(2)} \cap K = V$. But it is not a Gauss extension of V in $K(X)$.

Let R be a Dubrovin valuation ring of a simple Artinian ring Q, σ be an automorphism of Q and δ be a left σ-derivation. Suppose that (σ, δ) is compatible with R. In the beginning of this section we proposed an open question on non-commutative valuation rings of $Q(X;\sigma,\delta)$ lying over R. The observation in this section suggests us a program for the question:

2.4.19 Remark

1. $J(R)[X;\sigma,\delta]$ is localizable or not. It it is localizable, then $R^{(1)} = R[X;\sigma,\delta]_{J(R)[X;\sigma,\delta]}$ is a Dubrovin valuation ring or not.
2. Let P be a prime ideal of $Q[X;\sigma,\delta]$. Find out all Dubrovin valuation rings of $Q(X;\sigma,\delta)$ which are subrings of $Q[X;\sigma,\delta]_P$ and lying over R.
3. Find out Dubrovin valuation rings of $Q(X;\sigma,\delta)$ which are different types from ones in (1) and (2).

The following is also an open question:

2.4.20 Remark

If R is a generalized Bezout order in Q, then so is $R[X]$. Moreover so is $R[X;\sigma,\delta]$.

2.5 Arithmetical Pseudovaluations and Divisors

Let C be a commutative domain with field of fractions K and let A be a c.s.a. over K with $n = \dim_K A$. We assume $R \subset A$ is a C-order such that fractional ideals of R commute and R is projective as a C-module. In foregoing section we have seen that a maximal order Λ over a Dedekind domain satisfies the latter condition. We say that R is an **arithmetical order** if the set of fractional ideals of R, $F(R)$ say, commutes.

In [69] more general arithmetical rings are considered but we shall restrict attention to arithmetical orders in c.s.a.

Let Γ be a totally ordered semigroup written additively. A **pseudovaluation** v on $F(R)$ is a function $v : F(R) \to \Gamma \cup \{\infty\}$, satisfying:

- *PV.1.* For $I, J \in F(R), v(IJ) \geq v(I) + v(J)$
- *PV.2.* For $I, J \in F(R), v(I + J) \geq \min\{v(I), v(J)\}$
- *PV.3.* $v((R)) = 0$ and $v(0) = \infty$
- *PV.4.* If $I \subset J$ in $F(R)$ then $v(I) \geq v(J)$

From PV.4 it follows that PV.2 may be changed to $v(I + J) = \min\{v(I), v(J)\}$ for $I, J \in F(R)$.

A pseudovaluation on $F(R)$ is said to be **arithmetical** (a.p.v.) if $v(IJ) = v(I) + v(J)$ for $I, J \in F(R)$. Two a.p.v.'s are said to be equivalent, say $v_1 \sim v_2$, if $v_1(RxR) \geq o$ if and only if $v_2(RxR) \geq 0$ for all $x \in A$. We may define a pseudovaluation v^* on A as a function $v^* : A \to \Gamma \cup \{\infty\}$, for some totally ordered semigroup Γ, satisfying:

PV^*_1. For $a, b \in A$, $v^*(ab) \geq v^*(a) + v^*(b)$
PV^*_2. For $a, b \in A$, $v^*(a + b) = \min\{v^*(a), v^*(b)\}$
PV^*_3. We have $v^*(1) = 0$ and $v(0) = \infty$

2.5.1 Lemma

If v is a pseudovaluation $F(R) \to \Gamma \cup \{\infty\}$ then we define $v^* : A \to \Gamma \cup \{\infty\}$ by putting $v^*(a) = v(RaR)$. Then v^* is a pseudovaluation on A such that $v^*(r) \geq 0$ for all $r \in R$.

Proof. Follows directly from: $RabR \subset RaRbR, R(a + b)R \subset RaR + RbR$. □

2.5.2 Proposition

If v is an a.p.v. on A defined over R, then $P = \{a \in A, v(RaR) > 0\}$ defines a prime (P, A^P) in A. Conversely if (P, A^P) is a prime in A such that $R \subset A^P$, then there exists an a.p.v., v say, such that $P = \{a \in A, v(RaR) > 0\}$ and $\{b \in A, v^*(b) \geq 0\}$ is contained in A^P.

Proof. Let v be an a.p.v. defined on $F(R)$, look at $P = \{a \in A, v(RaR) > 0\}$. If $a, b \in P$ then $v(R(a-b)R) \geq v(RaR + RbR) \geq \{v(RaR), v(RbR)\} > 0$, hence $a - b \in P$. So we only have to establish that $A - P$ is an m-system for A^P. Suppose $a, b \in A$ are such that $aA^Pb \subset P$ then $aRb \subset P$. Thus $v(RaRbR) > 0$ and the a.p.v. property then yields: $0 < v(RaRbR) = v(RaR) + v(RbR)$, then either $v(RaR) > 0$ or $v(RbR) > 0$. The latter entails that either a or b is in P.

Conversely if (P, A^P) is a prime with $R \subset A^P$ we define for $I \in F(R)$, $v(I) = \{a \in A, aI \subset P\}$ and put $\Gamma = \{v(I), I \in F(R)\}$. We define a partial order on Γ by $v(I) \leq v(J)$ if and only if $v(I) \subset v(J)$. Then Γ is totally ordered. Indeed, suppose $I, J \in F(R)$ are such that $v(I) \not\subset v(J)$ and $v(J) \not\subset v(I)$. Then there is an $a \in A$ such that $aI \subset P$ and $aJ \not\subset P$, and a $b \in A$ such that $bJ \subset P$ but $bI \not\subset P$. Thus $RaRJA^P RbRI \not\subset P$ as P is a prime in A. Thus for some $z \in A^P$ we have: $RaRJRzRbRI \not\subset P$ but by the commutativity of fractional R-ideals in A we then have: $RaRJRzRbRI = RbRJRzRaRI \subset P$, a contradiction. Thus Γ is a totally ordered semigroup. Define: $v(I) + v(J) = v(IJ)$. This is an addition with unit element $v(R)$ turning Γ into a totally ordered semigroup. Indeed if $v(I) \supset v(J)$ then we look at $v(IH)$ and $v(JH)$ for $H \in F(R)$; if $q \in v(JH)$ then $qJH \subset P$, hence $RqRJH = RqRHJ \subset P$ or $qH \subset v(J) \subset v(I)$, thus $qHI \subset P$ or $q \in v(HI)$, hence $v(IH) \subset v(JH)$, similarly $v(HI) \supset v(HJ)$. For the well definedness look at $v(I) = v(I')$ and $v(J) = v(J')$ and $x \in v(IJ)$. Hence, $RaRIJ \subset P$, $RaRI \subset v(J) = v(J')$ then yields: $RaRIJ' = RaRJ \subset P$. Again, since $v(I') = v(I)$, then $RaRJ = RaRI'J' \subset P$ and $a \in v(I'J')$ follows. Consequently $v(IJ) = v(I'J')$ follows by repeating the foregoing argument interchanging the role of I, J and I', J' resp. The properties PV.1...PV.4 follow directly from the definition of v, $P = \{a, v(RaR) > 0\}$. If $b \in A$ is such that $v(RbR) = 0$ and $a \in P$ then $v^*(ba) \geq v^*(b) + v^*(a) = v(RbR) + v(RaR) > 0$ yields $ba \in P$: similarly $ab \in P$, hence $b \in A^P$. This also establishes $\{b \in A, v^*(b) \geq 0\} \subset A^P$. □

2.5.3 Remarks

1. If (P, A^P) is a prime of A such that $R \subset A^P$ then $P \cap R$ is a prime ideal of R. This is clear from the proof above.
2. If an $a.p.v.$ v corresponds to a prime (P_v, A^{P_v}) and v_p is the $a.p.v.$ associated to that prime, then v_p is equivalent to v as is easily checked.
3. Lat v_P be the $a.p.v.$ of the prime (P, A^P), $R \subset A^P$. For every $I \in F(R)$, $v_P(I) = \inf\{v_P(RaR, a \in I\}$. Indeed since it is clear that $v_P(RaR) \geq v_P(I)$ for every $a \in I$ and if $J \in F(R)$ is such that $v_P(J) \geq v_P(I)$ but $v_P(J) \leq v_P(RaR)$ for all $a \in I$, then $y \in v_P(J)$ yields $yRaR \subset P$ for all $a \in I$, or $yI \subset P$ and $y \in v_P(I)$. Then $v_P(J) \leq v_P(I)$ and $v_P(J) = v_P(I)$, thus $v_P(I)$ is $\inf\{v_P(RaR), a \in I)$.

4. The restriction of v^* to K defines a valuation of K. If $\pi = (p, K^p)$ is the restriction of (P, A^P) to K then every prime (P, A^P) with $R \subset A^P$ is π-fractional.
5. If $Ra = aR$ then for an $a.p.v$ v and all $b \in A$ we have $v(ba) = v(b) + v(a)$, $v(ab) = v(a) + v(b)$ (since $RbaR = RbRaR$).

2.5.4 *Lemma*

Let v be an $a.p.v.$ in A, Γ_v its value semigroup and (P, A^P) its associated prime. The following are equivalent:

1. Γ_v is a totally ordered group.
2. For every fractional ideal $I \in F(R)$ there is an $a \in A$ such that $aI \subset A^P$ and $aI \not\subset P$.

Proof. 1. $\Rightarrow$ 2. If Γ is a group every $v(I)$ has an inverse, thus for any $I \in F(R)$ there is a $J \in F(R)$ such that $v(IJ) = v(R) = 0$. Since $P \subset v(R)$, $P \subset v(IJ)$ hence $IJP \subset P$ or $IJ \subset A^P$ follows. Since $aIJ \subset P$ yields $aR \subset P$ hence $a \in P$ it follows that $IJ \not\subset P$. Then there is a $y \in J$ such that $RyR\,I = IRyR \not\subset P$ but $RyRI \subset A^P$.

2. $\Rightarrow$ 1. Let a be as in (2) and take $y \in v(RaRI)$. Then we have that $RyR\,RaRI \subset P$ and for every $z \in A^P$ we obtain:
$RzR\,RyR\,RaRI = RyRRzRaRRI \subset P$. Since $RaRI \not\subset P$ by the choice of a we must have $RyR \subset P$ (since P is prime), hence we arrive at $yR \subset P$ or $y \in v(R)$. From $RaRI \subset A^P$ the inclusion $v(RaRI) \supset v(R)$ follows (in fact $v(R) = P$ is 0 of Γ_v). Consequently $0 = v(R) = v(RaRI) = v(RaR) + v(I)$, so $v(I)$ has an inverse in Γ_v. □

2.5.5 *Proposition*

Let v be an $a.p.v.$ in A with corresponding Γ_v and (P, A^P) as before. If Γ_v is a group then $A^P = \{a \in A, v(RaR) \geq 0\}$.

Proof. That $\{a \in A, v(RaR) \geq 0\} \subset A^P$ was established in Proposition 2.5.2. For $a \in A^P$, (RaR) is invertible in Γ_v hence $v(J) + v(RaR) = 0$ in Γ_v for some $J \in F(R)$. From the proof of Lemma 2.5.4 we may assume that $J = RbR$. Hence $v(RaRRbR) = 0$ so if $v(RaR) < 0$ then $v(RbR) > 0$ i.e. $b \in P$. Now $a \in A^P$ yields $RaRRbR \subset P$ or $v(RaRRbR) > 0$, contradiction. Thus $v(RaR) \geq 0$ or $a \in \{a \in A, v(RaR) \geq 0)$. □

2.5.6 *Corollary*

If (P, A^P), $R \subset A^P$ is a prime of A with associated $a.p.v.$, v then (P, A^P) is a dominating prime if Γ_v is a group.

Proof. Consider $I \in F(R), I \not\subset A^P$. From Lemma 2.5.4.(2) it follows that there is an $a \in A$ such that $aI \subset A^P, aI \not\subset P$. Then $Ia \in A^P, Ia \not\subset P$. From the fact $I \not\subset A^P$ it follows that either $IP \not\subset P$ or $PI \not\subset P$, say $IP \not\subset P$. Then $IPA^P \not\subset P$. Consider IPA^Pa. For all $p \in PA^P$ we have: $Ipa \subset IpRaR \subset IRaRp \subset IaRp \subset P$, thus $IPA^Pa \subset P$. Since (P, A^P) is a prime such that $IP \not\subset P$ we must have $a \in P$.

If (Q, A^Q) dominates (A, A^P) take $b \in A^Q - A^P$ and $a \in P$ such that $aRbR \subset A^P$ but $aRbR \not\subset P$. From $a \in P \subset Q$ and $RbR \subset A^P$ it follows that $aRbR \subset Q \cap A^P = P$, contradiction. Hence $(P, A^P) = (Q, A^Q)$ or (P, A^P) is a dominating prime of A. □

The section so far depends heavily on R having a commutative fractional ideal theory. From the foregoing section we retain that if R is a maximal order over a Dedekind domain D in K this situation is realized. Also if R is a Dubrovin valuation of A it holds.

As before A is assumed to be finite dimensional over K.

2.5.7 *Proposition*

Let v be an $a.p.v.$ in A such that Γ_v is a totally ordered group, then (P, A^P) the associated prime of v is a Dubrovin valuation ring of A.

Proof. By Remark 2.5.3.(4) the restriction of v^* to K is a valuation of K. From Proposition 2.5.5 it follows that $A^P \cap K = 0_{v^*}$ is the valuation ring of $v^*|K$. Let us check that A^P/P is Artinian, in fact finite dimensional over k, the residue field of v^* on K. If not then $B = A^P/m_{v^*}A^P$ is also not finite dimensional over k as $m_{v^*}A^P \subset P$, where m_{v^*} is the maximal ideal $P \cap K$ of O_{v^*}. Let $\{\bar{x}_\alpha, \alpha \in \mathcal{A}\}$ be an infinite number of k-independent elements in B and $x_\alpha, \alpha \in \mathcal{A}$ be chosen representatives in A^P. Since A is finite dimensional over K we have a relation:

$$\sum_{i=1}^{n} \lambda_i x_{\alpha_i} = 0 \text{ with } \lambda_i \in K$$

We may multiply $\lambda_i, i = 1, \ldots, n$, by some λ such that $\lambda\lambda_i = \lambda_i^* \in O_v$ for all $i = 1, \ldots, n$, but not **all** in m_{v^*}. Then $\Sigma\lambda_i^* x_{\alpha_i} = 0$ yields $\Sigma\bar{\lambda_i^*}\, \bar{x_{\alpha_i}} = 0$ with $\bar{\lambda_i^*} \in k = O_{v^*}/m_{v^*}$ not all zero reducing modulo $m_{v^*}A^P$. The latter is a contradiction, hence A^P/P is finite dimensional over k, thus Artinian. Since P is prime A^P/P is prime Artinian and therefore simple Artinian. The second condition for a Dubrovin valuation ring follows from Lemma 2.5.4.(2). Indeed for any $q \in A$ there is an a such that $aRq \subset A^P$ and $aRq \not\subset P$, so for some $r \in R$ we have $arq \in A^P$. Conversely, since $RqR\ RaR = RaR\ RqR$ we have $qRa \subset A^P$ and $qRa \not\subset P$ hence $qr'a \in A^P$ for some $r' \in R$. Thus A^P is a Dubrovin valuation ring. □

2.5.8 Corollary

In the situation of the foregoing proposition v^* is a value function associated to A^P if $\Gamma_v = v^*(stA^P)$.

Proof. It is clear that the fourth axiom for a value function is exactly given by $\Gamma_v = v^*(\mathrm{st}(A^P))$, i.e. for every $q \in A$ there is an $a \in A$ such that $RqR = Ra = aR$. □

2.5.9 Definition

A prime (P, A^P) with $R \subset A^P$ is **discrete** if A^P has the ascending chain condition on ideals and P is the unique nonzero prime ideal of A^P and $P = \pi A^P = A^P\pi$ for some invertible element π of A.

2.5.10 Lemma

If (P, A^P) with $R \subset A^P$ is a discrete prime then for every integral ideal I of R, IA^P is an ideal of A^P, moreover every ideal of A^P is $P^n = \pi^n A^P$ for some $n \in \mathbb{N}$.

Proof. If $I \in F(R)$ then IA^P is a two-sided A^P module because for all $a \in A^P$, $aI \subset RaRI \subset IRaR \subset IA^P$, consequently for an integral ideal I of R, IA^P is an ideal of A^P.

If J is an ideal of A^P then $J = A^P a_1 A^P + \ldots + A^P a_d A^P$ for some $a_1, \ldots, a_d \in J$ because of the $a.c.c.$ on ideals in A^P. Now each $a_i = \pi^{n_i} b_i$ with $b_i \in A^P - P$, $i = 1, \ldots, 1$, for some $n \in \mathbb{N}$, otherwise $a_i \in \cap_n P^n$. Thus either $J \subset \cap_n P^n$ or $J = P^{n_i}$ where $n_i = \min\{n_1, \ldots, n_d\}$. We claim that every nonzero ideal of A^P contains some P^m for a certain $m \in \mathbb{N}$. Indeed, suppose the set of ideals of A^P not containing some P^m is nonempty, then by the $a.c.c.$ on ideals for A^P this set is inductively ordered hence it has maximal elements. Let H be such a maximal element, then $H \neq P$ and hence there are ideals I_1 and I_2 of A^P such that $I_1, I_2 \supsetneq H$ and $I_1 I_2 \subset H$. Now $P^{m_1} \subset I_1$, $P^{m_2} \subset I_2$ for some $m_1, m_2 \in \mathbb{N}$, thus $P^{m_1+m_2} \in H$ contradiction. Thus if $\cap_n P^n$ is a nonzero ideal then $\cap_n P^n = P^m$ for some $m \in \mathbb{N}$ or $P^m = P^{m+1}$ yielding: $A^P\pi^m = A^P\pi^{m+1}$ or $A^P = A^P\pi = P$, a contradiction.

2.5.11 Proposition

If (P, A^P) is a discrete prime then its associated Γ is a group and $\Gamma \cong \mathbb{Z}$ with generator $v(A^P)$.

Proof. If I is an integral ideal of R such that $IA^P = \pi^n A^P$, put $J = R\pi^{-n}R$ and calculate $v(IJ) = v(A^P) = P = v(R) = 0$ in Γ. If $I \in F(R)$ let $c \in Z(R)$ be such that $cI \subset R$ and let $v(J)$ be the inverse of $v(cI)$ in Γ, then $v(I) + v(cR) + v(J) = 0$ in Γ, then $v(I) + v(cR) + v(J) = 0$ in Γ, hence $v(I)$ has an inverse in Γ.

If $J \in F(R)$ is such that $v(J) \geq 0$ then $J \subset A^P$ hence $JA^P = \pi^n A^P$ for some $n \in \mathbb{N}$. However, if $v(J) < 0$ then $v(J)^{-1} = v(I) \geq 0$ hence $IA^P = \pi^m A^P$ for some $m \in \mathbb{N}$, thus $JIA^P = A^P$ (JI is not in P). Consequently: $JA^P = \pi^{-m}A^P$. Associating $v(J)$ to n if $JA^P = \pi^n A^P$ for $n \in \mathbb{Z}$ defines an isomorphism $\Gamma \cong \mathbb{Z}$.

□

2.5.12 Remark

If (P, A^P) with $A^P \supset R$ is a discrete prime then A^P has centre $Z(A^P) = K \cap A^P$ a discrete valuation ring.

Proof. Remark 2.5.3.(4) yields that the restriction of v^* to K yields a valuation with value group Γ and valuation ring $A^P \cap K = Z(A^P) = \{\lambda \in K, v^*(\Lambda) \geq 0\}$. Since $\Gamma \cong \mathbb{Z}$ the valuation induced on K is also discrete. □

2.5.13 Proposition

As before assume that R is a C-order in the $K-c.s.a.$ A (we assume $ch(K) = 0$) which is projective as a C-module. The following are equivalent:

1. Every ideal of R is finitely generated as a two-sided ideal.
2. The set of (integral) ideals of R satisfies the $a.c.c.$
3. $Z(R)$ is a Noetherian domain.
4. R is a finitely generated module over the Noetherian $Z(R)$.
5. R is (left and right) Noetherian.

Proof. 1. $\Rightarrow$ 2. If $I_0 \subset I_1 \subset \ldots \subset I_n \subset \ldots \subset R$ is a chain of ideals then $J = \cup_i I_i$ is an ideal of R hence $J = Ra_1R + \ldots + Ra_dR$ for a finite set $\{a_1, \ldots, a_d\} \subset J$. Take n such that $a_1, \ldots, a_d \in I_n$, then $I_n = J$ and the chain terminates.

2. $\Rightarrow$ 3. Since R is projective as a C-module we have for any ideal I of C that $RI \cap C = I$, hence C satisfies the ascending chain condition because a chain $I_0 \subset I_1 \subset \ldots \subset C$ yields $RI_0 \subset RI_1 \subset \ldots \subset R$ and the latter terminates.

3. ⇒ 4. R is a prime $P.I.$-ring as an order of a $c.s.a$. A result of G. Cauchon yields that if $Z(R) = V$ is Noetherian ($ch(K) = 0$ holds too); then R is a finitely generated C-module.
4. ⇒ 5. Well-known
5. ⇒ 1. Obvious. □

2.5.14 Corollary

If $F(R)$ is a group then R is Noetherian and a finitely generated $Z(R)$-module over a Noetherian centre.

Proof. If I is an ideal of R and I^{-1} its inverse in $F(R)$ then there is a finitely generated $Ra_1R+\ldots+Ra_dR \subset I$ such that we have: $(Ra_1R+\ldots+Ra_dR)I^{-1} \ni 1$ hence $(Ra_1R+\ldots+Ra_dR)I^{-1} = R$. Then $Ra_1R+\ldots+Ra_dR = I$ follows from multiplying the foregoing by I. So every ideal of R is finitely generated (two-sided) so we may apply the foregoing proposition. □

Now we look at sets of discrete primes in a finite dimensional K-algebra A. First we observe that any discrete prime is a Dubrovin valuation ring with an associated value function v^* inducing a discrete valuation on K; indeed its fractional ideals form a group (abelian) and every fractional ideal is a principle module on the left (using π^{-1} and multiplying by suitable powers of it). If (P, A^P) is a discrete prime then A^P is a finitely generated (left) module over $O_{v^*} = A^P \cap K$ and hence a free O_{v^*}-module of finite rank, as a Dubrovin valuation it is then also a maximal order over the discrete valuation ring O_{v^*}. So in case A is finite dimensional over K: the set of **discrete primes of A is identical to the set of maximal orders over discrete valuation rings of K!**

A set D of discrete primes of A is said to be **proper** if they yield inequivalent a.p.v. (recall that v_1 is an equivalent a.p.v. to v_2 exactly if: $v_1(I) \leq v_1(J)$ if and only if $v_2(I) \leq v_2(J)$ for fractional R-ideals I and J) restricting to inequivalent valuations of K. A proper set of discrete primes is **divisorial** if for every $q \in A$, $v^*(q) = 0$ for almost all $v \in \mathcal{D}$ (we may view $\mathcal{D}$ as a set of discrete a.p.v.). If $\mathcal{D}$ is divisorial then its elements are called **prime divisors**. A **divisor** δ of A with divisorial set $\mathcal{D}$ is a formal sum $\delta = \sum_{v\in\mathcal{D}} \gamma_v v$ where $\gamma_v \in \mathbb{Z}$ is 0 for almost all $v \in \mathcal{D}$.

We call γ_v the **order of δ in** v, written $\gamma_v = \mathrm{ord}_v\delta$. In case $\gamma_v \geq 0$ for all $v \in \mathcal{D}$ we call δ an **integral divisor**. A divisor δ_1 divides a divisor δ_2, written $\delta_1|\delta_2$ if for all $v \in \mathcal{D}$, $\mathrm{ord}_v\delta_1 \leq \mathrm{ord}_v\delta_2$.

Any divisor is determined by the function $\mathrm{ord} : \mathcal{D} \to \mathbb{Z}, v \mapsto \gamma_v$. We define $\delta_1 + \delta_v$ by $\mathrm{ord}_v(\delta_1+\delta_2) = \min\{\mathrm{ord}_v(\delta_1), \mathrm{ord}_v(\delta_2)\}$; $\delta_1.\delta_2$ is defined by $\mathrm{ord}_v(\delta_1.\delta_2) = \mathrm{ord}_v(\delta_1) + \mathrm{ord}_v(\delta_2)$. The set of divisors for $\mathcal{D}$, $\mathrm{div}(\mathcal{D})$, equipped with the multiplication introduced above is an abelian group.

Consider a proper set $\mathcal{D}$ of discrete primes such that for almost all $(P, A^P) \in \mathcal{D}$, A^P contains a fixed arithmetical ring R (e.g. a maximal order over a Dedekind domain D in K). Let $D(K)$ be the set of discrete valuation of K induced by the $v \in \mathcal{D}$.

2.5.15 *Lemma*

If $\mathcal{D}(K)$ is divisorial then $\mathcal{D}$ is divisorial.

Proof. Take $q \neq 0$ in A, $q = c^{-1}r$ for some $c \in K$, $r \in R$. Since $D(K)$ is divisorial $v^*(c^{-1}) = 0$ for almost all $v^*|K$, consequently $q \in 0_{v^*}R \subset A^{P_v}$ for almost all (P_v, A^{P_v}) in $\mathcal{D}$. For every invertible $q \in A$ it then follows that $v^*(q) = 0$ for almost all $v \in \mathcal{D}$ since $0 = v^*(qq^{-1}) \geq v^*(q) + v^*(q^{-1}) \geq 0$ for almost all $v \in \mathcal{D}$ (each v^* is a value function!). Observe it also follows that for every fractional ideal I of R, $v(I) = 0$ for almost all $v \in \mathcal{D}$! □

For a fractional ideal I of R define the divisor: $\partial_I = \sum_{v \in \mathcal{D}} v(I)v$, the **ideal divisor** of I.

In case $I = RqR$ for some $q \in A$, then ∂_I is said to be **principal**. *A* divisor as before is principal if $\partial_q = \sum_{v \in \mathcal{D}} v(q)v$. For a principal divisor we define the **divisor of zeros** as: $\partial_{z(q)} = \sum_{v(q)>0} v(q)v$, the **divisor of poles** is defined as: $\partial_{p(q)} = \sum_{v(q)<0} -v(q)v$. Note that, in case R is contained in all A^P appearing in $\mathcal{D}$, then principal divisors are exactly all principal ideal divisors.

2.5.16 *Corollary*

With R and $\mathcal{D}$ as before, the map:

$$\partial : \mathcal{F}(R) \rightarrow \mathrm{Div}(\mathcal{D}), I \mapsto \partial_I$$

is compatible with sum and product (similar result for principal divisors fails if $R \not\subset A^P$ for some A^P appearing in $\mathcal{D}$!).

If R is a maximal order in A over the Dedekind domain D in K let $\mathcal{D}(R)$ consist of all discrete primes (P, A^P) of A such that $R \subset A^P$. Then $Z(A^P) = O_{v^*}$ (A is assumed to be a $c.s.a.$ over K), and $m_{v^*} \cap D = p = P \cap D$ is a maximal ideal of D. Then O_{v^*} is the localization of D at p and $Q_{D-p}(R)$ is a maximal O_{v^*}-order in R. Since $Q_{D-p}(R) \subset A^P$ it follows that $A^P = Q_{D-p}(R)$ and we have a bijective correspondence $\mathcal{D}(R) \leftrightarrow \mathrm{Spec}\, D - \{0\}$. Moreover $R = \cap_{v \in \mathcal{D}} A^{P_v}$ and the approximation property holds for $\mathcal{D}$:

2.5.17 Definition. Approximation Property(A)

Given a finite set $v_1, \ldots, v_n$ of $a.p.v.$ in $\mathcal{D}(R), \lambda_1, \ldots, \lambda_n \in \mathbb{Z}$ and $q_1, \ldots, q_n \in A$, there exists an $x \in A$ such that: $v_i(x - q_i) \geq \lambda_i, i = 1, \ldots, n$ and $v(x) \geq 0$ for all $v \in \mathcal{D}(R)$-$\{v_1, \ldots, v_n\}$.

The property (A) in the situation described above can be established completely similar to the commutative case.

A modification of the proofs in the commutative case (see e.g. [69]) leads to the following.

2.5.18 Lemma

In the situation of Definition 2.5.17 there exists an $x \in A$ such that $v_i(x - q_i) = \lambda_i, i = 1, \ldots, n$ and $v(x) \geq 0$ for all $v \in \mathcal{D}(R) - \{v_1, \ldots, v_n\}$, ($A_1$-property). Also, for v_i, Λ_i as before there exists an invertible $x \in A$ that is right invariant for $A^{P_{v_1}}$: such that $v_1(x) = \lambda_1$, $v_j(x) = \lambda_j$ for $j = 2, \ldots, n$ and $v(x) \geq 0$ for all $v \in \mathcal{D}(R) - \{v_1, \ldots, v_n\}$, ($A_2$-property).

This leads to the following, in the situation of the lemma.

2.5.19 Proposition

The map $\partial : \mathcal{F}(R) \to \mathrm{div}(\mathcal{D}(R))$ is bijective, i.e. it is an isomorphism between the abelian groups $\mathrm{div}(\mathcal{D}(R))$ with multiplication of divisors and $\mathcal{F}(R)$ with multiplication of fractional ideals. Integral ideals of R correspond to integral divisors, prime ideals correspond to prime divisors i.e. $\mathcal{D}(R)$.

For a $c.s.a.$ A over K look at $\mathcal{D}_k(K)$, the set of discrete valuations of K containing some subfield k. It is well-known that there exists a finite subset $V_K \subset \mathcal{D}_k(K)$ such that $\mathcal{D} = \mathcal{D}_k(K) - V_K$ is divisorial in K and satisfies the (A)-property. For each $v \in \mathcal{D}$ choose a maximal O_v-order Λ_v in A, then this is a discrete prime and so we obtain a set $\mathcal{D}(A)$ which is divisorial in A. There is a finite subset $V_A \subset \mathcal{D}(A)$, consisting of those Λ_v for $v \in V_K$, such that $D(A) - V_A$ satisfies condition (A). We define the divisors of A as $\mathrm{div}\mathcal{D}(A)$. The subset $I\,\mathrm{div}\mathcal{D}(A)$ consists of ideal divisors i.e. divisors defined with respect to the divisorial set $\mathcal{D}(A) - V_A$; the elements of $I\,\mathrm{div}\mathcal{D}(A)$ may be viewed as ideals of some maximal order over a Dedekind domain in A, (cf. Proposition 2.5.19). When K is a function field in one variable over k, the foregoing is used in a Riemann–Roch theorem for $c.s.a.$ over K.

Again we look at $c.s.a.$ over K with a Dedekind domain D in K such that $Q(D) = K$. Fix a nonzero prime ideal p of D and look at the set: $\mathcal{Q}(A) = (C, Q, \Lambda)$, Λ a subring of A such that $\Lambda \cap K = D$, Q a prime of Λ, $Q \cap K = p$.

We consider $\mathcal{Q}_f(A) \subset \mathcal{Q}(A)$ where $(Q, \Lambda) \in \mathcal{Q}_f(A)$ if Λ is a finitely generated R-module.

Both sets can be ordered in two ways: first by the domination relation: $(Q, \Lambda) < (Q_1, \Omega_1)$ if and only if $\Lambda \subset \Lambda_1, Q \subset Q_1, Q = Q_1 \cap \Lambda$, secondly by the inclusion on the Q in the (Q, Λ).

Maximal elements with respect to the domination relation are called **dominating pairs**, resp. **finite dominating pairs**, in $\mathcal{Q}(A)$, resp. $\mathcal{Q}_f(A)$. Clearly maximal (finite) pairs are also dominating (finite) pairs. There are maximal pairs that are not finite, for example in the case D is O_v with maximal ideal m_v, then $m_v + Ka$ for some $a \in A$ such that $a^2 = 0$ (exists if A is not a skewfield) is contained in a maximal pair which cannot be finite because $m_v + Ka$ is not a finitely generated O_v-module.

2.5.20 Lemma

Let (Q, Λ) be a finite dominating pair and Ω a maximal D-order containing Λ.

1. (Q, Λ) is a prime in Ω and Q is a maximal ideal in Λ.
2. Q contains a nonzero ideal of Ω.
3. Q and Λ are D-lattices in A (i.e. both contain a K-basis for A).

Proof. 1. Consider $\mathcal{Q}_f(\Omega)$ the pairs of $\mathcal{Q}_f(A)$ contained in Ω. If $(Q', \Lambda') \in \mathcal{Q}_f(\Omega)$ dominates (Q, Λ) then $Q' \cap \Lambda = Q$ yields that (Q', Λ') lies over (p, D). Since $\Lambda' \subset \Omega$ it is a finitely generated D-module, thus $(Q', \Lambda') \in \mathcal{Q}_f(A)$. Since (Q, Λ) was dominating, $Q = Q'$ and $\Lambda = \Lambda'$ follows and thus (Q, Λ) is dominating in $\mathcal{Q}_f(\Omega)$. As in the proof of Proposition 1.1.8 it follows that (Q, Λ) is a prime if Ω. Now Λ/Q is Artinian (finite dimensional over $k = D(p)$ hence Q is a maximal ideal in Λ.

2. For $n \in \mathbb{N}$, $p^n\Omega + Q$ is a finitely generated D-module. If for all $n \in \mathbb{N}$, $(p^n\Omega + Q) \cap \Lambda \neq Q$ then $1 \in p^n\Omega + Q$. Localizing centrally at p yields $1 \in p^n\Omega_p + Q_P$ for all n, and the latter is a finitely generated D_p-module. Since Ω_p is a maximal D_p-order, then $p\Omega_p \subset J(\Omega_p)$ and Nakayama's lemma entails $1 \in Q_p$. Hence there is a $d \in D$ such that $d \in D - p$, $d.1 \in Q$, but that would contradict $Q \cap D = p$. Thus for some n, $1 \notin p^n\Omega + Q$, i.e. $(p^n\Omega + Q) \cap \Lambda = Q$. Put $B = (p^n\Omega + Q) + \Lambda = p^n\Omega + \Lambda$. Then B is a ring and $p^n\Omega + Q$ is a maximal ideal of B since $Bp^n\Omega + Q = \Lambda/Q$ is simple (see (1)). Now $(p^n\Omega + Q, B)$ dominates (Q, Λ) in $\mathcal{Q}_f(A)$ thus $B = \Lambda, Q = p^n\Omega + Q$, yielding that $p^n\Omega \subset Q$ for some $n \in \mathbb{N}$.

3. Since Ω contains a K-basis, thus Q contains a K-basis by multiplying a suitable element of p^n (as $p^n\Omega \subset Q$). □

The following theorem tells us that maximal pairs (primes) in $\mathcal{Q}_f(A)$ are the indecomposable normal ideals of A lying over a fixed $p \subset D$. hence the maximal finite pairs are exactly the generators of the Brandt groupoid in classical maximal order theory [57].

2.5.21 Theorem

1. If (Q, Λ) is a dominating pair in $Q_f(A)$ then Q is a maximal ideal or a maximal one-sided ideal in a maximal order Ω of A containing a maximal ideal lying over $p \subset D$.
2. Every maximal one-sided ideal L in a maximal order Ω of A, lying over $p \subset D$, arises as a maximal finite pair (L, Ω^L).

Proof. 1. Look at a dominating pair (Q, Λ) in $\mathcal{Q}_f(A)$. By the lemma we find a maximal order Ω over D such that (Q, Λ) is a prime in Ω. Since (Q, Λ) is dominating we have $\Lambda = \Omega^Q$. Since Q contains a nonzero ideal of Ω, $Q^o \subset Q$ is a nonzero prime ideal of Ω, thus a maximal ideal of Ω. Now Q/Q^o is a prime in Ω/Q^o and the latter is a *c.s.a.* over $k = D/p$. The $\bar{Q} = Q/Q^o$ lies over the trivial prime (o, k) in k. We have $\Omega/Q^o = \mathrm{End}_\Delta V$ over some skewfield Δ. Now primes of $\mathrm{End}_\Delta V$ over (o, k) are characterized by $\bar{Q} = \varepsilon_1\bar{\Omega} + \bar{\Omega}\varepsilon_2$ where $\varepsilon_1, \varepsilon_2, \varepsilon_3$ are orthogonal idempotents in $\bar{\Omega} = \Omega/Q^o$ such that $\varepsilon_1 + \varepsilon_2 + \varepsilon_3 = 1$ (from the classification of k-primes in $\mathrm{End}_\Delta V$ [69]). Thus $Q = e_1\Omega + \Omega e_2 + Q^o$ with $\bar{e}_i = \varepsilon_i$. Observe that the characterization used here holds only if $\Omega/Q^o \not\cong \Delta$, but in the other case $\bar{Q} = 0$ is the only prime lying over o in k, then $Q = Q^o$ and (1) follows. Now consider $S = Q + Qe_1(Q^o)^{-1}e_2Q$. If $S \cap \Lambda \not\subset Q$, then by the lemma: $1 \in Q + Qe_1(Q^o)^{-1}e_2Q$. Since $e_1, e_2 \in Q \subset S$ we have $1 - e_1 - e_2 \in S$ or $e_3 \in S$. From $e_ie_j \in Q^o$ and $e_i^2 = e_i \bmod Q^o$ it follows that $e_3Se_3 \subset Q^o$. Then $e_3^3 \in Q^o$, yielding $\varepsilon_3 = \varepsilon_3^3 = e_3^3 + Q^o = 0$, a contradiction. Therefore $S \cap \Lambda \subset Q$ thus $S \cap \Lambda = Q$. As in the proof of the lemma, this is impossible unless $S = Q$ or $Qe_1(Q^o)^{-1}e_2Q \subset Q$. We calculate:

$$\begin{aligned} e_1e_1(Q^o)^{-1}e_2e_2 \subset Q &\Rightarrow e_1e_1(Q^o)^{-1}e_2e_2 \subset (e_1 + Q^o)(Q^o)^{-1}(e_2 + Q^o) \\ &\subset e_1(Q^o)^{-1}e_2 + Q^o(Q^o)^{-1}e_2 + e_1(Q^o)^{-1}Q^o + Q^o(Q^o)^{-1}Q^o \subset \\ &\subset e_1(Q^o)^{-1}e_2 + \Omega e_1 + e_1\Omega + Q \end{aligned}$$

Since $\Omega e_1 + e_1\Omega + Q^o \subset Q$ we deduce that $e_1(Q^o)^{-1}e_2 \subset Q$. If neither e_1 nor e_2 is in Q^o then $\Omega(e_i + Q^o)\Omega = \Omega$ implies:

$$(Q^o)^{-1} = \Omega(Q^o)^{-1}\Omega = \Omega(e_1 + Q^o)\Omega(Q^o)^{-1}\Omega(e_2 + Q^o)$$

a contradiction because $(Q^o)^{-1} \not\subset \Omega$. Thus e_1 or e_2 is in Q^o. Thus Q is either a left or right ideal in Ω. If Q is an ideal of Ω, then $Q = Q^o$. The first part is proved.

2. If L is a maximal one-sided, say left, ideal of Ω, then (L, Ω^L) is a localized pair (hence dominating hence a prime). That (L, Ω^L) is a maximal pair then follows from the maximality of L and part (1).

   Take $a, b \in \Omega^L - L$ and suppose $ab \in L$ then $1 = \lambda + ra$ for $\lambda \in L, r \in \Omega$. Then $1 - \lambda = ra$ and $(1 - \lambda)b = b - \lambda b = rab \in L$. But $\lambda b \in L$ as $b \in \Omega^L$, thus $b \in L$ follows, a contradiction. hence L is a prime ideal in Ω^L, or $(L, \Omega^L) \in \mathcal{Q}_f(A)$. To prove it is localized take $x \in \Omega - \Omega^L$, i.e. $Lx \not\subset L$. Then $1 = \lambda + \lambda' x$ for some $\lambda, \lambda' \in L$, or $\lambda' x \notin L$ with $L = L\lambda + L\lambda' x$, thus $L\lambda' x \subset L$ or $\lambda' x \in \Omega^L$ with $\lambda' x \notin L$. □

If we take (D, p) to be (O_v, m_v) a discrete valuation of K then Theorem 2.5.21 becomes the following

2.5.22 *Theorem*

With (D, p) being (O_v, m_v) in the foregoing theorem, the following statements are equivalent:

1. (P, Λ) is a maximal finite prime over (m_v, O_v).
2. (P, Λ) is a maximal finite pair over (m_v, O_v).
3. P is a generator for the Brandt groupoid over (m_v, O_v), that is a maximal one-sided ideal in a maximal order Ω over O_v in A.

Proof. Trivial from the foregoing theorem. □

2.6 The Riemann–Roch Theorem for Central Simple Algebras Over Function Fields of Curves

As an application of our theory or arithmetical pseudo valuations we will present a Riemann–Roch theorem for central simple algebras over curves. This goes back to some first approaches by Van Deuren and Van Oystaeyen [66] and Van Deuren, Van Oystaeyen, Van Geel [67], but the most elegant form was given in J. Van Geel's thesis, cf. [69]. In [76] it was considered as a part of noncommutative geometry. It will show that the *apv* allow a divisorial calculus on function algebras. A cohomological version of the Riemann–Roch theorem over higher dimensional varieties was developed by M. Van den Bergh, but here there is no relation with valuation theory. Probably the first approach to a kind of Riemann–Roch theorem is due to Witt based on a duality result, cf. [81], but the theory described hereafter is more general and more complete. Notwithstanding the fact that the results are now more than 30 years old, it seems that the tools for effective divisorial calculations in *c.s.a.* over function fields have not been used to the full. Perhaps the inclusion of this theory in this book may help to gain new attention to the subject.

In this section K is an algebraic function field in one variable over k, i.e. k is algebraically closed in K is assumed. A central simple algebra A over K is a **function algebra,** $A = M_r(\Delta)$ for some skewfield Δ.

$\mathcal{D}(K)$ is the set of k-valuation rings of K, these are discrete valuation rings. Fix for every valuation ring $O_v \subset \mathcal{D}(K)$ a maximal order of A over O_v, say Λ_v, this defines a set $\mathcal{D}(A)$. The latter may be viewed as a set of localized primes of A and $\mathcal{D}(A)$ is a divisorial set. For the choice of the maximal O_v-order Λ_v in A we may first choose a maximal Δ_v of Δ and put $\Lambda_v = M_r(\Delta_v)$. To every Λ_v there corresponds an $a.p.v.$ (we denote this again by v) via the set of fractional ideals of Λ_v. The complement of a finite set V_K in $\mathcal{D}(K)$ defines a Dedekind domain $D = \cap\{O_v, v \in \mathcal{D}(K) - V_K\}$. Then $R = \cap\{\Lambda_v, v \in \mathcal{D}(A) - V_A\}$ is a maximal D-order, where V_A is the set of Λ_v for $v \in V_K$; the set $\mathcal{D}(A) - V_A$ is a divisorial set satisfying condition (A). Hence for every finite set $V \subset \mathcal{D}(A)$ we have (A) and $(A1)$, $(A2)$. An apv defined on $\mathcal{F}(R)$ yields an apv defined on $\mathcal{F}(\Lambda_v)$ for some v (exactly the apv corresponding to v^* for the v defined over R). Indeed Λ_v for $v \in D(K) - V_K$ is the central localization of R at $Z(R) \cap P$ (as the latter is a maximal order contained in Λ_v) hence a fractional ideal I for R extends to a fractional ideal $\Lambda_v I$ (a Λ_v-bimodule!) for Λ_v and the associated value of the apv is in both cases given as the $\gamma \in \Gamma$ such that $I \subset F_\gamma A$ and $I \not\subset F_{\gamma'} A$ for $\gamma' < \gamma$; since $I \subset F_\gamma A$ if and only if $F_0 AI \subset F_\gamma A$, or $\Lambda_v I \subset F_\gamma A$.

We write $m_v \subset O_v$ for the maximal ideal and $P_v \subset \Lambda_v$ for the maximal ideal of Λ_v; we put $k_v : O_v/m_v$, $\bar{A}_v = \Lambda_v/P_v$. Then k_v is a finite extension of k and $\bar{A}_v$ is a central simple algebra with $Z(\bar{A}_v) \supset k_v$. Both $\bar{A}_v$ and $Z(\bar{A}_v)$ are finite dimensional over k. We put: $[A : K] = N$, $f_v = [\bar{A}_v : k]$ is the absolute residue class degree of v, $\rho_v = [\bar{A}_v : k_v]$ is the relative residue class degree of v. If v^c is the normalized restriction of v to K, i.e. $v^c(\pi) = 1$ for some uniformizing element π of O_v put $f_{v^c} = [k_v : k]$. For some $e_v \in \mathbb{N}$, $\Lambda_v m_v = P_v^{e_v}$; this number is the ramification index of v. If $e_v = 1$ then v is said to be **unramified** in A. The following equalities hold:

1. $e_v \rho_v = N$
2. $\rho_v f_{v^c} = f_v$
3. If Π is uniformizing for P_v then $v(\pi) = v(\Pi^{e_v}) = e_v$

Let $\mathrm{Div}(A)$ be the group of divisors generated by $\mathcal{D}(A)$, $\mathrm{Div}(K)$ the group of divisors generated by $\mathcal{D}(K)$.

For the $\delta \in \mathrm{Div}(A)$, $\deg\delta = \sum_v f_v \mathrm{ord}_v(\delta)$ is the **degree** of δ. To a divisor $d = \sum \gamma_v v^c$ in $\mathrm{Div}(K)$ we correspond $\delta_d = \sum e_v \gamma_v v$ in $\mathrm{Div}(A)$ and call it the **extended divisor** for d.

2.6.1 Observation

For d and δ_d as before: $\deg\delta_d = N \deg d$. Indeed, $\deg\delta_d = \sum f_v \mathrm{ord}_v \delta_d = \sum f_v e_v \mathrm{ord}_{v^c} d = \sum \rho_v f_{v^c} e_v \mathrm{ord}_{v^c} d = N \Sigma f_{v^c} \mathrm{ord}_v d = N.\deg d$.

2.6.2 Definition

The subring $l = \cap\{\Lambda_v, v \in \mathcal{D}(A)\}$ is called the **constant subring** of A.

2.6.3 Proposition

The constant subring l of A consists of k-algebraic elements and it is a central simple algebra finite dimensional as k-vectorspace.

Proof. It suffices to prove the statement in case $A = \Delta$ is a skewfield (by taking afterwards matrix rings over everything). Consider the reduced norm $\mathrm{Nr} : \Delta \to K$ (cf. Reiner, Sect. 9). If $\Delta_v \in \mathcal{D}(\Delta)$, P_v the maximal ideal of Δ_v then the ideal generated by $\{Nrx, x \in P_v\}$ is in m_v (see also Reiner [57] Theorem 24.13). If $a \in l$ then $a \in \Delta_v$ for all v and $Nra \in O_v$ for all $v \in \mathcal{D}(K)$, i.e. $\mathrm{Nr}a \in k$. Since Δ is a skewfield, $\mathrm{Nr}a = 0$ if and only if $a = 0$ thus $\mathrm{Nr}a \notin m_v$ for all v as $k \cap m_v = \{0\}$.

Hence $a \notin P_v$ for $v \in \mathcal{D}(\Delta)$. Then $l \hookrightarrow \Delta_v/P_v$, the latter being finite dimensional over k, thus $[l : k] < \infty$. Since l is an Artinian domain it is certainly simple. □

2.6.4 Corollary

With notation as before: $l = \{a \in A, a = 0 \text{ or } v(a) = 0 \text{ for all } v \in \mathcal{D}(A)\}$.

Proof. That $l \subset \{a \in A, a = 0 \text{ or } v(a) = 0 \text{ for all } v \in \mathcal{D}(A)\}$ is clear. If $a \neq 0$ in l has $v(a) > 0$ for some $v \in \mathcal{D}(A)$, then $P_v \cap l \neq 0$. Since l is simple and $P_v \cap l \neq l$ we arrive at a contradiction. □

With notation as introduced before, we fix a K-basis for A contained in R, say $\{u_1, \ldots, u_N\}$, hence $u_i \in \Lambda_v$ for almost all $v \in \mathcal{D}(A)$, in fact for all $v \in \mathcal{D}(A) - V_A$. The divisors of $\mathrm{Div}(A)$ generated by elements of $\mathcal{D}(A) - V_A$ are divisors of R. Since $\mathcal{D}(A) - V_A$ is divisorial and (A) holds, the ideal divisors of R are given exactly by the$\mathcal{F}(R)$, the fractional R-ideals.

Consider the completion $\hat{A}_v$ of A with respect to v, i.e. $\hat{K}_{v^c} \otimes_K A$. Let $\hat{U}(A) = \Pi_{v \in \mathcal{D}(A)} \hat{A}_v$, this is a k-algebra.

The algebra of **valuation vectors** $\hat{V}(A)$ is the k-subalgebra of $\hat{U}(A)$ given by $\{\xi \in \hat{U}(A), v(\xi) = v(\xi_v) \geq 0 \text{ for almost all } v \in \mathcal{D}(A)\}$. If $\xi \in \hat{V}(A)$ has $v(\xi) = 0$ for almost all $v \in \mathcal{D}(A)$ then it is called an **idèle** of A. Idèles form a multiplicative subgroup of $\hat{V}(A)$, written $I(A)$. To every idèle ξ of A there corresponds in a natural way a divisor of A, $\gamma_\xi = \sum_{v \in \mathcal{D}(A)} v(\xi)v$, and it is easily seen that all divisors are obtained this way. We embed A into $I(A)$ by $a \mapsto (a, a, \ldots)$, using this embedding A may be viewed as a subalgebra of $\hat{V}(A)$.

2.6.5 *Definition*

For $\delta \in \mathrm{Div}(A)$ we define the δ-parallelotope of $\hat{V}(A)$ to be the k-subspace of $\hat{V}(A)$ given as:

$$\Pi_\delta = \{\xi \in \hat{V}(A), v(\xi) \geq \mathrm{ord}_v\delta \text{ for all } v \in \mathcal{D}(A)\}$$

In case $a \in A, \alpha \in I(A)$ we mention a-or α-parallelotope when we mean Π_{γ_a}, resp, Π_{δ_α}.

Let $\hat{U}(K)$ be the space of valuation vectors for $\mathcal{D}(K)$. We have a well defined k-linear map: $\phi : \hat{V}(K)^N \to \hat{V}(A), (\xi_1, \ldots, \xi_N) \mapsto \xi_1 u_1 + \ldots \xi_N u_N$. The relation between $\phi(\hat{V}(K)^N), \Pi_\delta, \hat{V}(A)$ is expressed in the following lemma.

2.6.6 *Lemma*

1. For every $\delta \in \mathrm{Div}(A) : \phi(\hat{V}(K)^N + \Pi_\delta = \hat{V}(A)$.
2. There exists an $\alpha \in I(A)$ such that: $\Pi_\alpha + A = \hat{V}(A)$.

Proof. 1. For $\xi \in \hat{V}(A)$ define $Q = \{v \in \hat{V}(A), v(\xi) > 0, \mathrm{ord}_v(\delta) \neq 0\}$. In view of property A.1 there is an $a \in A$ such that: $v(a - \xi) \geq \mathrm{ord}_v(\delta)$ for all $v \in Q$. Write $a = \Sigma a_i u_i$ and define η_i by: $(\eta_i)_{v^c} = a_i$ if $v \in S, (\eta_i)_{v^c} = 0$ if $v \notin S$. Then it is clear that $\delta = \Sigma \eta_i u_i \in \phi(\hat{V}(K)^N)$ and $\xi - \delta \in \Pi_\gamma$.

2. From the commutative case we know there exists an $\alpha_K \in I(K)$ such that $\hat{V}(K) = K + \Pi_{\alpha_K}$, hence $\phi(\hat{V}(K)^N) = \phi(\Pi^N_{\alpha_K}) + A$. Since $v(\Sigma\xi_i u_i) \geq \min\{v(\xi_i) + v(u_i)\}$ and $v(u_i)$ depends only on u_i, there exists a lower bound for the values of elements in $\phi(\Pi^N_{\alpha_K})$ thus there exists a parallelotope Π_β such that $\phi(\Pi^N_{\alpha_K}) \subset \Pi_{\beta_K}$. This yields $\hat{V}(A) = \phi(\hat{V}(K)^N) + \Pi_\beta \subset \Pi_\beta + \Pi_\beta + A \subset \Pi_\alpha + A$ for some $\alpha \in I(A)$. □

For a divisor δ of A and finite subset S of $\hat{V}(A)$ we consider $L(\delta|S) = \{a \in A, v(a) \geq \mathrm{ord}_v(\delta), v \in S\}$. If $\delta_1|\delta_2$ then $L(\delta_1|S) \supset L(\delta_2|S)$ and also $\Pi_{\delta_1} \supset \Pi_{\delta_2}$.

2.6.7 *Lemma*

Let $\delta_1|\delta_2$ be divisors of A, $S = \{v, \mathrm{ord}_v(\delta_1) \neq 0 \text{ or } \mathrm{ord}_v(\delta_2) \neq 0\}$ then: $\Pi_{\delta_1}/\Pi_{\delta_2} = L(\delta_1|S)/L(\delta_2|S)$.

Proof. If $a \in L(\delta_1|S)$ put $\xi_a \in \Pi_{\delta_1}$ by $(\xi_a)_v = a$ if $v \in S$ and $(\xi_a)_v = 0$ if $v \notin S$. The map $a \mapsto \xi_a$ is k-linear $i : L(\delta_1|S) \to \Pi_{\delta_1}$. If $i(a) \in \Pi_{\delta_2}$ then $v(a) \geq \mathrm{ord}_v(\delta_2)$ for $v \in S$, i.e. $a \in L(\delta_2|S)$. Thus i defines a monomorphism $L(\delta_1|S)L(\delta_2|S) \to \Pi_{\delta_1}/\Pi_{\delta_2}$. For $\alpha \in \Pi_{\delta_1}$ there is an $a \in A$ such that $v(a - \alpha) \geq \mathrm{ord}_v(\delta_2)$ for all $v \in S$, by the A_1-property. Thus $\xi_a - \alpha \in \Pi_{\delta_2}$ and also $v(a) \geq \min\{\mathrm{ord}_v(\delta_2), v(\alpha)\}$, which is larger than $\mathrm{ord}_v(\delta_1)$, thus $\alpha \in L(\delta_1|S)$. This yields the desired isomorphism. □

2.6.8 Lemma

Let $\delta_1|\delta_2$ as in the foregoing lemma, then:

$$\dim_k(L(\delta_1|S)/L(\delta_2|S)) = \deg\delta_2 - \deg\delta_1$$

Proof. In view of the additivity property of the statement we may assume, without loss of generality, that $\mathrm{ord}_v(\delta_2) = 1 + \mathrm{ord}_v(\delta_1)$ and $\mathrm{ord}_{v'}(\delta_2) = \mathrm{ord}_v(\delta_1)$ for all $v' \neq v$, i.e. $\deg\delta_2 - \deg\delta_1 = f_v$. By the property A.3 (cf. Lemma 2.5.18) we may take $u \in A$ such: $v(u) = \mathrm{ord}_v(\delta_1)$ and $v'(u) \geq \mathrm{ord}_v(\delta_1)$ for all $v' \in S, v' \neq v$ where u is regular and right invariant for Λ_v. For $a_1, \ldots, a_{f_v+1}$ in $L(\delta_1|S)$ the $a_1u^{-1}, \ldots, a_{f_v+1}u^{-1}$ have positive value at v so they are in Λ_v. But $[\bar{A}_v : k] = f_v$, so there is a relation $\Sigma\lambda_i a_i u^{-1} \in m_v\Lambda_v$ with not all $\lambda_i \in K$ being zero. Thus $\Sigma\lambda_i a_i \in L(\delta_2|S)$ and it follows that: $\dim_k L(\delta_1|S)/L(\delta_2|S) \leq f_v$. If $x_1, \ldots, x_{f_v}$ in Λ_v are linearly independent modulo $m_v\Lambda_v$ then choose $x_i' \in A$ such that $v(x_i' - x_i) > 0, v'(x_i) > 0$ for all v' in S, $v' \neq v$. Then $x_i' = x_i \bmod \Lambda_v m_v$ and $x_i u \in L(\delta_1|S)$ and these are easily seen to be k-independent. So $\dim_k(L(\delta_1|S)/L(\delta_2|S)) \geq f_v$ and equality follows. □

2.6.9 Corollary

As before let $\delta_1|\delta_2$ then:

$$\dim_k \Pi_{\delta_1}/\Pi_{\delta_2} = (\Pi_{\delta_1} : \Pi_{\delta_2}) = \deg\delta_2 - \deg\delta_1$$

If S equals $\hat{V}(A)$ then we write $L(\delta|S) = L(\delta), L(\delta) = A \cap \Pi_\delta$

2.6.10 Proposition

With notation as above: $l(\delta) = \dim_k L(\delta)$ is finite and bounded.

Proof. Let ε be the unit divisor, i.e. $\mathrm{ord}_v(\varepsilon) = 0$ for $v \in \mathcal{D}(A)$. Then $L(\varepsilon) = l$ and Proposition 2.6.3 entails that $\dim_R L(\varepsilon) = n < \infty$. If $\delta_1|\delta_2$ then: $(\Pi_{\delta_1} : \Pi_{\delta_2}) = (\Pi_{\delta_1} \cap A) : (\Pi_{\delta_1} \cap A)$, thus:

$$\deg\delta_2 - \deg\delta_1 = l(\delta_1) - l((\delta_2) + (\Pi_{\delta_1} + A : \Pi_{\delta_2} + A) \qquad (*)$$

Put $m(\delta) = \hat{V}(A) : \Pi_\delta + A$ and chose α such that $\Pi_\alpha + A = \hat{V}(A)$ (see Lemma 2.5.8.(2)). Select a divisor δ_1 for which $\Pi_{\delta_1} = \Pi_\alpha + \Pi_\delta$. Then $\delta_1|\delta$ and $\deg\delta - \deg\delta_1 = l(\delta_1) - l(\delta) + m(\delta)$.

Thus $m(\delta)$ is finite and $\dim_k(L(\delta_1)/L(\delta)) \leq \dim_k(L(\delta_1|S)/L(\delta_2|S))$ where $S = \{v, \mathrm{ord}_v\delta_1 \neq 0 \text{ or } \mathrm{ord}_v\delta_2 \neq 0\}$ is a finite set. Thus $l(\delta)$ is finite because $l(\varepsilon) = l$ is finite dimensional over k.

We now may rephrase $(*)$ as follows:

$$\deg\delta_2 - \deg\delta_1 = l(\delta_1) - l(\delta_2) + m(\delta_2) - m(\delta_1) \text{ hence}$$

$$(\deg\delta_1 + l(\delta_1) - m(\delta_1) = \deg\delta_2 + l(\delta_2) - m(\delta_2)$$

Comparing any two divisors δ and δ' with their greatest common divisor δ'' defined by: $\mathrm{ord}_r(\delta'') = \min\{\mathrm{ord}_v(\delta), \mathrm{ord}_v(\delta')\}$ then yields that $\deg\delta + l(\delta) - m(\delta)$ is independent of δ and it is finite in view of the foregoing. □

2.6.11 Definition

The integer $1 - \deg\delta - l(\delta) + m(\delta) = g_A$ is called the genus of A (it is independent of δ). Note that g_A does depend on the choice of $\mathcal{D}(A)$ but we will show later that another admissible choice for $\mathcal{D}(A)$ will define the same genus.

Using the Π_δ as a fundamental system of neighbourhoods one may define a topology on $\hat{U}(A)$ and $\hat{V}(K)$ and put the product topology on $\hat{U}(K)^N$. Lemma 2.3.8 then yields that $\phi(\hat{V}(K)^N)$ is everywhere dense in $\hat{V}(A)$. We also may derive (as in the commutative case).

2.6.12 Corollary (of the Finite Dimensionality of $L(\delta)$)

The map $\phi : \hat{V}(K)^N \to \hat{V}(A)$ is bicontinuous k-linear and thus $\phi(\hat{V}(K)^N) = \hat{V}(A)$ i.e. $\hat{V}(A) \cong \hat{V}(K)^N$ as k-vectorspaces (topologically). Thus $\hat{V}(A)$ is complete like $\hat{V}(K)$.

To phrase the Riemann–Roch theorem in its classical form we have to introduce noncommutative differentials or valuation forms, this will allow to give other interpretations of the genus. The theory is now a modification of the commutative case using the extension of the reduced trace map.

2.6.13 Definition

A **valuation form** is an element of the dual $\hat{V}(A)^*$ of $\hat{V}(A)$ as a k vectorspace, vanishing on a subset $\Pi_\delta + A$ (i.e. it is continuous in the product topology of $\hat{V}(K)^N$).

If $\delta_1|\delta_2$ then $\Pi_{\delta_2} + A \subset \Pi_{\delta_1} + A$ so a valuation form vanishing on $\Pi_{\delta_1} + A$ also vanishes on $\Pi_{\delta_2} + A$. Put $M(\delta) = \{\text{all valuation forms vanishing on } \Pi_{\delta^{-1}} + A\}$.

Clearly $M(\delta)$ is a k-space which is dual to $\hat{V}(A)/\Pi_{\delta^{-1}} + A$, thus $\dim_k M(\delta) := m(\delta^{-1}) = \dim_k \hat{V}(A)/\Pi_{\delta^{-1}} + A$.

In the commutative case valuation forms have been studied extensively. We just recall the following: the valuation forms of a function field K form a one dimensional K-space and for every valuation form w there is an upper-bound $\Pi_{\delta^{-1}}$ such that $w \in M(\delta)$. Moreover, a fixed class i.e. the canonical class, is obtained as the class of divisors determining maximal parallelotopes in which valuation forms vanish and the degree of a divisor in the canonical class is $2-2g$ where g is the genus of the function field. We will now derive similar properties for function algebras A over K as before.

We let Tr be the reduced trace map of A (cf. [57]). Fix a K-basis $\{u_1, \dots, u_N\}$ of A and consider $Tr(u,-) : x \mapsto Tr(ux)$ for every $u \in A$. Then $Tr(u,-)$ is K-linear and $Tr(u,-), \dots, Tr(u_N,-)$ are K-linearly independent in $A^* = \mathrm{Hom}_K(A,K)$ because $Tr(ux) = 0$ for all $x \in A$ if and only if $u = 0$. We have a K-basis $\{Tr(u_1,-), \dots, Tr(u_N,-)\}$ for A^* and for $f \in A^* : f(-) = \sum_{i=1}^N a_i Tr(u_i,-) = Tr(\sum_{i=1}^N a_i u_i,-) = Tr(u,-)$ for $u = \Sigma a_i u_i \in A$; thus every $f \in A^*$ is of the form $Tr(u,-)$ for some $u \in A$. Any $f \in A^*$ extends to a K-linear $\hat{f} : \hat{V}(A) \to \hat{V}(K)$ as follows, if $f(u_i) = a_i$ then for $\xi = \Sigma \xi_{K,i} u_i \in \hat{V}(A)$ we put $f(\xi) = \Sigma \xi_{K,i} a_i$ with $a_i \in K$. Then $\hat{f}$ is K-linear and $\hat{V}(K)$-homogeneous: $f(\eta_K \xi) = \eta_K f(\xi)$ for every $\eta_K \in \hat{V}(K), \xi \in \hat{V}(A)$. It will not create ambiguity if we write f for $\hat{f}$ again.

2.6.14 Proposition

Every valuation form Ω on $\hat{V}(A)$ is of the form $w(Tr(a,-)$ for some $a \in A$ and w a fixed nontrivial valuation form on $\hat{V}(K)$.

Proof. Consider a valuation form Ω on $\hat{V}(A), \xi = \Sigma_i \xi_{K,i} u_i \in \hat{V}(A)$, then $\Omega(\xi) = \Sigma_i \Omega(\xi_{K,i} u_i)$ and the maps $\Omega(-u_i)$ define valuation forms on $\hat{V}(K)$. Let us fix a valuation form w on K then $\Omega(-u_i) = w(\lambda_i -)$ for some $\lambda_i \in K$ (one dimensionality of the valuation forms on K!)

Thus $\Omega(\xi) = w(\Sigma_i \lambda_i \xi_{K,i})$ and the map $\hat{V}(A) \to \hat{V}(K)$ given by $\Sigma \xi_{K,i} u_i \mapsto \Sigma_i \lambda_i \xi_{K,i}$ is obviously K-linear and $\hat{V}(K)$-homogeneous thus equal to $Tr(a,-)$ for $a \in A$. □

In the sequel we fix a nontrivial valuation form w for K, i.e. a generator for the space of valuation forms.

2.6.15 Proposition

There is an upperbound for the parallelotopes $\Pi_{\delta^{-1}}$ such that $w(Tr(-)) \in M(\delta)$.

Proof. To an $x \in L(\delta^{-1})$ there corresponds a valuation form $xw(Tr(-))$ defined by: $xw(Tr(\xi)) = w(Tr(x\xi))$, for $\xi \in \hat{V}(A)$. This yields a valuation form on $\Pi_\varepsilon + A$, where ε is the unit divisor, because we have $w(Tr(-)) \in M(\delta)$. The map $\Psi : L(\delta^{-1}) \to M(\varepsilon), x \mapsto xw(Tr(-))$ is k-linear and if $w(Tr(x\xi)) = 0$ for all ξ in $\hat{V}(A)$ then $Tr(x\xi) = 0$ for all ξ in $\hat{V}(A)$ (either $Tr(x-)$ is surjective or $Tr(x-) = 0$ as w is nontrivial). Thus then $x = 0$ or Ψ is injective. For the k-dimensions we get:

$$l(\delta^{-1}) \leq \dim M(\varepsilon) = l(\varepsilon) + \deg(\varepsilon) + g_A - 1 = n + g_A - 1, \text{and}$$

$$l(\delta^{-1}) + \deg(\delta^{-1}) = 1 - g_A + \dim M(\delta) \geq 1 - g_A$$

Thus $n + g_A - 1 + \deg(\delta^{-1}) \geq 1 - g_A$ or $\deg(\delta^{-1}) \geq 2 - 2g - n$. So if $wTr(-) \in M(\delta)$ then the degree of δ^{-1} is bounded from below. Hence there is an upper bound for strict chains $\ldots \subset \Pi_{\delta_i^{-1}} \subsetneqq \Pi_{\delta_{i+1}^{-1}} \subset \ldots$, where $w(Tr(-)) \in M(\delta_i)$, since $\deg(\delta_i^{-1}) > \deg(\delta_{i+1}^{-1})$.

An upper bound for all parallelotopes in which $w(Tr(-))$ vanishes follows from the fact that: if $w(Tr(-)) \in M(\delta_1), w(Tr(-)) \in M(\delta_2)$ then $w(Tr(-))$ is in $M(\gcd(\delta_1, \delta_2))$ where $\gcd(\delta_1, \delta_2)$ is the greatest common divisor of δ_1 and δ_2 as defined earlier. □

2.6.16 *Observation*

In case A is a skewfield then any valuation form Ω is of the type $w(Tr(a-))$ for some $a \in A$ and now a is not a zero-divisor. Thus the above proposition then holds for all valuation forms Ω on $\hat{V}(A)$.

We now establish the Riemann–Roch theorem, as before its proof is mainly linear algebraic and uses the form (A.3) of the approximation theorem.

2.6.17 *Theorem* Riemann–Roch

Let α be any divisor for A, then $\deg(\alpha) + l(\alpha) = l(\alpha^{-1}\delta^{-1}) + 1 - g_A$, where δ^{-1} is the canonical divisor.

Proof. From the proof of the foregoing proposition $\psi : L(\delta) \to M(\alpha\delta)$, $x \mapsto xw(Tr(-)$ is a k-linear injection. If $\Pi_{\delta^{-1}}$ is chosen to be maximal such that $w(Tr(-)) \in M(\delta)$ (defining the "canonical" divisor) then ψ is surjective. Indeed, if $\Omega \in M(\alpha\delta)$ then $\Omega = w(Tr(a-))$ for $a \in A$. If $a \notin L(\alpha)$ then $v(a) < \mathrm{ord}_v(\alpha)$ for some v. If we take $\xi \in \Pi_{\alpha^{-1}\delta^{-1}}$ then $w(Tr(a\xi)) = 0$, so the minimal parallelotope containing $a\Pi_{\alpha^{-1}\delta^{-1}}$ maps to zero, by continuity of A_a, and maximality of $\Pi_{\delta^{-1}}$ yields a $\Pi_{\alpha^{-1}\delta^{-1}} \subset \Pi_{\delta^{-1}}$. This however contradicts $a \notin L(\alpha)$ because $\xi + \Pi_{\alpha^{-1}\delta^{-1}}$

can be taken such that ξ_v is right invariant for $\cap_v$ and $v(\xi) = \text{ord}_v(\alpha^{-1}\delta^{-1})$ (using A.3, Lemma 2.5.18), hence $v(a\xi) < \text{ord}_v(\delta^{-1})$. Thus we may assume that $L(\alpha) \cong M(\alpha\delta)$ for every divisor α and a dimension calculation yields: $m(\alpha^{-1}\delta^{-1}) = l(\alpha)$ or $m(\alpha) = l(\alpha^{-1}\delta^{-1})$. As in the proof of Proposition 2.6.10, $\deg(\alpha) + l(\alpha) = m(\alpha) + 1 - g_A$, so the foregoing yields the desired formula. □

2.6.18 Remarks

1. Take $\alpha = \varepsilon$ then we obtain from the Riemann–Roch theorem: $\deg(\varepsilon) + l(\varepsilon) = l(\delta^{-1}) + 1 - g_A$, thus $l(\delta^{-1}) = n - 1 + g_A$.
2. Take $\alpha = \delta^{-1}$ he canonical divisor, then:

$$\deg(\delta^{-1}) + l(\delta^{-1}) = l(\varepsilon) + 1 - g_A, \text{ thus}$$

$$(\delta^{-1}) + n - 1 + g_A = n + 1 - g_A, \text{ or } d(\delta^{-1}) = 2 - 2g_A$$

To every v we may look at the completion $\hat{\Lambda}_v$ in $\hat{A}^v$, the v-completion of A; it is clear that $\hat{\Lambda}_v$ is an $\hat{O}_v$-maximal order in $\hat{A}^v$. Again let w be the fixed valuation form for K, $\Omega = w(Tr(-))$. The **local components** of w and Ω are defined by the valuation forms: $w_v(x) = w(x_v)$, $\Omega_v(\xi) = \Omega(\xi_v)$. let $\Pi_{\alpha^{-1}}$, resp. $\Pi_{\delta^{-1}}$, be maximal parallelotopes in $\hat{V}(K)$, $\hat{V}(A)$, so that w, resp. Ω, vanish on them. The v components of the divisors $\alpha\delta$ can be interpreted as $\hat{O}_v$-ideals resp. $\hat{\Lambda}_v$-ideals, say $(\widehat{m}_v)^{\gamma_v}$, resp. $(\hat{P}_v)^{\mu_v}$.

2.6.19 Definition

The inverse local different for $\hat{\Lambda}_v$ is the fractional $\hat{\Lambda}_v$-ideal $\mathcal{D}_v^{-1} = \{x \in A, Tr(xR) \subset \hat{O}_v\}$. The ideal $(\mathcal{D}_v^{-1})^{-1} = \mathcal{D}_v$ is the **local different** for $\hat{A}_v$.

From [57] we recall that $\mathcal{D}_v = (\hat{P}_v)^{e_v - 1}$ so D_v is the unit ideal if and only if v^c is unramified in A. The different of $\mathcal{D}(A)$ is given by $\mathcal{D} = \sum_{v \in \mathcal{D}(A)} (e_v - 1)v$, i.e. the different is the divisor for $\mathcal{D}(A)$ having for the local components exactly the D_v.

2.6.20 Corollaries

1. For α, δ as before: $\delta = \delta_\alpha \mathcal{D}$ where δ_d is the extension of α to $\mathcal{D}(A)$.
2. The genus g_A is an invariant of A.
3. We have: $g_A = Ng_K - N + 1 + \frac{1}{2}\Sigma f_v(e_{v-1})$.

Proof. 1. $\Omega_v(\Pi_{\delta^{-1}}) = 0$ is equivalent to $w_v(Tr(P_v^{-\mu_v})) = 0$, or $Tr(\hat{P}_v^{-\mu_v}) \subset m_v^{-\gamma_v}$, i.e. $Tr(\hat{P}_v^{-\mu_v} m_v^{\gamma_v}) \subset \hat{O}_v$. This yields that $\hat{P}_v^{-\mu_v} m_v^{\gamma_v}$ is the inverse local different of $\hat{A}_v$. By looking at all local components we find $\delta = \delta_\alpha \mathcal{D}$.
2. The ramification of v^c in A does not depend on the choice of the maximal O_v-orders Λ_v in A. The genus of K is an invariant of K, thus $\deg(\alpha^{-1}) = 2 - 2g_K$, $\deg(\delta^{-1}) = 2 - 2g_A$ together with the relation $\delta = \delta_\alpha \mathcal{D}$ yield that g_A is not depending on the choices involved.
3. From $\deg\delta_\alpha = N\deg\alpha$ and a straightforward calculation argument. □

2.6.21 Examples

1. If every v^c is unramified in A, e.g. if $A = M_n(K)$, then: $g_A = Ng_K - N + 1$.
2. Let A be $M_m(\Delta)$, Δ a skewfield. Then $e_{\Delta,v} = e_{A,v} = e_v$, thus $f_v = [A_v/P_v : \Delta_v/P_v \cap \Delta_v] f_{\Delta,v} = m^2 f_{\Delta,v}$. So putting $N = m^2v^2$, $r^2 = [\Delta : K]$ we obtain:

$$\begin{aligned} g_A &= Ng_K - N + 1 + \frac{1}{2}\Sigma f_v(e_v - 1) \\ &= m^2(r^2 g_K - r^2 + \frac{1}{2}\sum f_{\Delta,v}(e_v - 1) + 1 \\ &= m^2(r^2 g_K - r^2 + \frac{1}{2}\sum f_{\Delta,r}(e_v - 1 + 1) + 1 - m^2 \end{aligned}$$

Thus $g_A = m^2 g_\Delta - m^2 + 1$.

There are many other examples, see for example [69], but the calculations of the genus and related invariants for elements of the Brauer group $\mathrm{Br}(K)$ remain almost uninvestigated. Explicit calculations for higher dimensional function fields, e.g. with respect to noncommutative valuations of higher rank or even in terms of Dubrovin valuations also remains up to now a totally open area.

Chapter 3
Extensions of Valuations to Quantized Algebras

3.1 Extension of Central Valuations

We look at skewfields obtained as total quotient rings of algebras defined by generators and relations. It is of particular interest to consider so-called quantized algebras stemming from noncommutative geometry because we hope to use valuation theory in the construction of a kind of divisor theory in noncommutative geometry.

Consider a field K with valuation ring $O_v \subset K$ having maximal ideal $m_v \subset O_v$ and residue field $k_v = O_v/m_v$. Let A be a connected positively graded K-algebra, $A = K \oplus A_1 \oplus \ldots \oplus A_n \oplus \ldots$, where each A_i is a finite dimensional K-space and $A = K[A_1]$, $A_1 = \oplus_{i=1}^n K a_i$. We view A as an algebra given by generators and relations:

$$0 \to \mathcal{R} \to K < X_1, \ldots, X_n > \xrightarrow{\pi} A \to 0$$

where $K < X_1, \ldots, X_n >$ is the free K-algebra on $\{X_1, \ldots, X_n\}$ and π is given by $\pi(X_i) = a_i, i = 1, \ldots, n$. The ideal of relations $\mathcal{R}$ is homogeneous in the usual gradation of $K < X_1, \ldots, X_n >$. We can also consider the ungraded case where A is a finitely generated K-algebra with generators $a_1, \ldots, a_n$ and π defined as before but then $\mathcal{R}$ is not homogeneous is the usual gradation of $K < X_1, \ldots, X_n >$. Restriction of π to $O_v < \underline{X} >$ defines a graded subring Λ of A with $\Lambda_0 = O_v$

$$0 \to \mathcal{R} \cap O_v < \underline{X} > \to O_v < \underline{X} > \xrightarrow[\text{res}\pi]{} \Lambda \to 0$$

It is clear that π maps $w_v < \underline{X} >$ to $w_v \Lambda$ which is a graded ideal of Λ. We write: $\overline{\Lambda} = \Lambda / w_n \Lambda$ and $\overline{\mathcal{R}} = (\mathcal{R} \cap O_v < \underline{X} >) + w_v < \underline{X} > / w_v < \underline{X} >$, so we arrive at the following commutative diagram with exact rows:

Marubayashi and Van Oystaeyen, *Prime Divisors and Noncommutative Valuation Theory*, Lecture Notes in Mathematics 2059, DOI 10.1007/978-3-642-31152-9_3,

$$\begin{array}{ccccccccc}
0 & \longrightarrow & \overline{\mathcal{R}} & \longrightarrow & k_r < \underline{X} > & \xrightarrow{\overline{\pi}} & \overline{\Lambda} & \longrightarrow & 0 \\
 & & \uparrow & & \uparrow & & & & \\
0 & \longrightarrow & \mathcal{R} \cap O_v < \underline{X} > & \longrightarrow & O_v < \underline{X} > & \longrightarrow & \Lambda & \longrightarrow & 0 \\
 & & \downarrow & & \downarrow & & \downarrow & & \\
0 & \longrightarrow & \mathcal{R} & \longrightarrow & K < \underline{X} > & \xrightarrow[\pi]{} & A & \longrightarrow & 0
\end{array}$$

When $\mathcal{R}$ is generated by $p_1(\underline{X}), \ldots, p_d(\underline{X})$ as a two-sided ideal, then we may assume $p_i(\underline{X}) \in O_v < \underline{X} >$ up to multiplying by some constant but it does not follow that $\mathcal{R} \cap O_v < \underline{X} >$ is generated as a (two-sided) ideal by $\{p_1(\underline{X}), \ldots, p_d(\underline{X})\}$, nor that $\overline{\mathcal{R}}$ is generated by the reduced expressions $\overline{p}_1(\underline{X}), \ldots, \overline{p}_d(\underline{X})$, obtained by reducing coefficients at m_v.

3.1.1 Definition

We say that $\mathcal{R}$ (or A) **reduces well at** $O_v$or that Λ defines a **good reduction**, if $\overline{\mathcal{R}}$ is generated as an ideal by $\{\overline{p}_1(\underline{X}), \ldots, \overline{p}_d(\underline{X})\}$.

Let us write fK for the Γ-valuation filtration of K associated to v and define a Γ-filtration $fK < \underline{X} >$ by putting: for $\gamma \in \Gamma$, $f_\gamma K < \underline{X} >= (f_\gamma K) < \underline{X} >$. The latter is a strong filtration on $K < \underline{X} >$ with $f_0 K < \underline{X} >$ equal to $O_v < \underline{X} >$. A left ideal J of $O_v < \underline{X} >$ is said to be **v-comaximal** of for all $\gamma \in \Gamma$, $J \cap (f_\gamma K) < \underline{X} >= (f_\gamma K) J$.

3.1.2 Lemma

If the ideal L of $O_v < \underline{X} >$ generated by $p_1(\underline{X})$, $p_d(\underline{X})$ is v-comaximal then $\mathcal{R}$ reduces well at O_v.

Proof. Since $fK < \underline{X} >$ is a strong filtration and for any $r \in f_\gamma K < \underline{X} >$ for some $\gamma \in \Gamma$ yields $f_{\gamma^{-1}} K < \underline{X} > r \in \mathcal{R} \cap f_0 K < \underline{X} >$, we have that $\mathcal{R} = K < \underline{X} > (O_v < \underline{X} > \cap \mathcal{R})$. Let L' be the left ideal in $O_v < \underline{X} >$ generated by $\{p_1(\underline{X}), \ldots, p_d(\underline{X})\}$; then we have $L' K < \underline{X} >= \mathcal{R}$ since $L' K < \underline{X} >$ is the two-sided ideal generated by $\{p_1(\underline{X}), \ldots, p_d(\underline{X})\}$. For $x \in f_0(L' K < \underline{X} >)$ there is a $\gamma \in \Gamma$ such that $x f_{\gamma^{-1}} K < \underline{X} > \subset L$ as well as $x f_{\gamma^{-1}} K < \underline{X} > \subset f_{-\gamma} K < \underline{X} >$ since $x \in f_0 K < \underline{X} >$. Therefore we arrive at $x f_{\gamma^{-1}} K < \underline{X} > \subset L \cap f_{\gamma^{-1}} K < \underline{X} >= (f_{\gamma^{-1}} K) L$ by the v-comaximality of L. From this it follows that $f_\gamma K < \underline{X} > x f_{-\gamma} K < \underline{X} > \subset (f_\gamma K)(f_{\gamma^{-1}} K) L = L$ hence $x \in L$.

Then we obtain:

$$L \subset \mathcal{R} \cap O_v < \underline{X} >= f_0\mathcal{R} = f_0(L'K < \underline{X} >) \subset L$$

arriving at $\mathcal{R} \cap O_v < \underline{X} >= L$ being the two-sided ideal in $O_v < \underline{X} >$ generated by $\{p_1(X), \ldots, p_d(\underline{X})\}$; from this it follows easily that $\overline{\mathcal{R}}$ is the two-sided ideal generated by the reductions $\overline{p}_i(\underline{X})$ of $p_i(\underline{X})$. □

In case the reduced relations $\overline{p}_1(\underline{X}), \ldots, \overline{p}_d(\underline{X})$ determine a simple algebra then the O_v-reduction is necessarily a good reduction, indeed the ideal $(\overline{p}_1(\underline{X}), \ldots, \overline{p}_d(\underline{X}))$ is now maximal in $k_v < \underline{X} >$ hence $\overline{\mathcal{R}} = (\overline{p}_1(\underline{X}), \ldots, \overline{p}_d(\underline{X}))$.

3.1.3 *Corollary*

If $A = \mathbb{A}_n(K)$ is the n-th Weyl algebra defined as $K < X_i, Y_i, i = 1, \ldots, n > /(Y_iX_i - X_iY_i - 1, X_iX_j - X_jX_i, Y_iY_j - YiY_j)$ then the reduced relations define $\mathbb{A}_n(k_v)$ which is known to be a simple algebra (if char$(k_v) = 0$) so the reduction at O_v is good if char$(k_v) = 0$.

As we have already pointed out the results concerning good reduction are valid in the ungraded case, but it is interesting to look at positively filtered algebras since any finitely generated K-algebra inherits a standard filtration via $\pi : K < \underline{X} > \to A, X_i \mapsto a_i$, from the gradation filtration of the free algebra $K < X >$. So, let us assume again that the K-algebra A is given by generators and relation via

$$(*) : 0 \to \mathcal{R} \to K < \underline{X} > \to A \to 0$$

Let FA be the generator filtration of A induced by the gradation filtration of $K < \underline{X} >$ making $\pi : K < \underline{X} > \to A, X_i \mapsto a_i$, into a strict filtered morphism. On $\mathcal{R}$ we may consider the induced filtration $F\mathcal{R} = \mathcal{R} \cap FK < \underline{X} >$. Then (*) is a strict exact sequence that is to say that the image filtration on $\mathcal{R}$ is exactly the filtration induced by $FK < \underline{X} >$ and this yields exactness of $G(*) : 0 \to G(\mathcal{R}) \to G(K < \underline{X} >) \to G_{FA}(A) \to 0$. In fact we have the following:

3.1.4 *Lemma*

With notation as above, $G(A) = G_{FA}(A)$ is defined by:

$$0 \to \dot{\mathcal{R}} \to K < \underline{X} > \xrightarrow{G(\pi)} G(A) \to 0$$

where $\dot{\mathcal{R}}$ is the left ideal of $K < \underline{X} >$ generated by the $\dot{p}$ for all $p \in \mathcal{R}$ and $\dot{p}$ is the highest degree component of p in the decomposition of p in the gradation of $K < \underline{X} >$. For the Rees ring $\widetilde{A}$ with respect to FA we obtain:

$$0 \to \widetilde{\mathcal{R}} \to K < \underline{X} >^{\sim} \underset{\overline{\pi}}{\longrightarrow} \widetilde{A} \to 0$$

where $\overline{\pi}$ corresponds to $\pi : K < \underline{X} > \to A$ on the Rees object level.

Proof. See e.g. [40, Proposition 1.1.5. p. 10]. □

3.1.5 Theorem

If $G(A)$ reduces well with respect to O_v, say $\dot{\mathcal{R}}$ is generated as a two-sided ideal by $q_1(\underline{X}), \ldots, q_q(\underline{X})$ then there are $p_1(\underline{X}), \ldots, p_d(\underline{X})$ in $K < \underline{X} >$ such that $\mathcal{R} = (p_1(\underline{X}), \ldots p_d(\underline{X}))$ and $\dot{p}_i(\underline{X}) = q_i(X)$ for $i = 1, \ldots, d$, such that $\mathcal{R}$ (i.e. A) reduces well with respect to O_v.

Proof. Choose $p_i'(\underline{X}) \in \mathcal{R}$ such that $q_i(\underline{X})$ is the homogeneous part of highest degree in the decomposition of $p_i'(\underline{X})$, for $i = 1, \ldots, d$. Pick $\mu \in f_{\gamma^{-1}}K$ for $\gamma \in \Gamma_+$ large enough (how large will be clear in the sequel) and replace X_i by $\mu X_i, i = 1, \ldots, d$. Put $\deg q_i(\underline{X}) = m$. Then $\mu^m p_i'(\underline{X}) = q_i(\mu\underline{X}) + \mu\Psi(\mu\underline{X})$ where Ψ has degree lower than m; put this equal to $p_i(\mu\underline{X})$ for $i = 1, \ldots, d$. In the new variables $\mu X_i, i = 1, \ldots, d$, the homogeneous part of highest degree of $p_i(\mu\underline{X})$ is exactly $q_i(\mu\underline{X})$ and $p_i(\mu\underline{X})$ is in $\mathcal{R}$ because $\mu^m p_i'(\underline{X})$ is a relation for A. By choosing γ large enough we may assume that the coefficients appearing in $\mu\Psi(\mu\underline{X})$ are contained in O_v so that $p_i(\mu\underline{X}) \in O_v < \mu\underline{X} >$. Obviously $q_i(\mu\underline{X})$ viewed in $K < \mu\underline{X} >= K < \underline{X} >$ still generates the ideal of relations of $G(A)$. Now consider the two-sided ideal I in $K < \underline{X} >$ generated by $p_i(\underline{X})$, then $I \subset \mathcal{R}$. By construction we have $\dot{I} = \dot{\mathcal{R}}$ so $I \subset \mathcal{R}$ then yields $I = \mathcal{R}$ (for example see [51, 52]). Indeed if $r \in \mathcal{R} - I$ then $\dot{r} = \dot{\iota}$ for some $\iota \in I$ hence $r - \iota \in \mathcal{R}$ and in $F_m\mathcal{R}$ with $m < n$ where $r \in F_nR - F_{n-1}R$, thus $(r - \iota)^{\cdot} = \dot{\iota}_1$ with $\iota_1 \in F_{m_1}I$ then $r - \iota - \iota_1 \in \mathcal{R}$ and in $F_{m_i}\mathcal{R}$ with $m_1 < m$, and so on, leads to $r - \iota - \iota_1 - \ldots - \iota_t = 0$ since FR is a positive filtration, i.e. $r \in I$ as claimed. The good reduction assumption for $G(A)$ means that $\dot{\mathcal{R}} \cap O_v(\underline{X})$ is generated as a two-sided ideal by $q_1(\underline{X}), \ldots, q_d(\underline{X})$. Taking $(\mathcal{R} \cap O_v < \underline{X} >)^{\cdot}$ in $O_v < \underline{X} >$ we obtain:

$$(\mathcal{R} \cap O_v < \underline{X} >)^{\cdot} \subset \dot{\mathcal{R}} \cap O_v < \underline{X} >$$

Since $q_i(\underline{X})$ is the highest homogeneous part of $p_i(X) \in \mathcal{R} \cap O_v < \underline{X} >$ it is clear that $(\mathcal{R} \cap O_v(\underline{X}))^{\cdot}$ contains $q_i(\underline{X})$ and is an ideal of $O_v(\underline{X})$ (because if $\overline{h}(\underline{X})$ is a homogeneous element of $(\mathcal{R} \cap O_v < \underline{X} >)^{\cdot}$ then it is the leading term of some $h(\underline{X})$ in $R \cap O_v(\underline{X})$ and a nonzero $\overline{h}(\underline{X}).\underline{x}$ for some $\overline{x} \in O_v < \underline{X} >= G(O_v(\underline{X}))$ is the leading term of $h(X)x$ for some x with $\sigma(x) = \overline{x}$ and $h(x)x \in \mathcal{R} \cap O_v < \underline{X} >$). Hence we obtain: $(\mathcal{R} \cap O_v < \underline{X} >)^{\cdot} = \dot{\mathcal{R}} \cap O_v < X >$.

The $O_v < \underline{X} >$-ideal J generated by the $p_i(\underline{X})$ is in $\mathcal{R} \cap O_v < \underline{X} >$ and $q_i(\underline{X}) \in \dot{J} \subset (\mathcal{R} \cap O_v < X >)^{\cdot} = \dot{\mathcal{R}} \cap O_v < \underline{X} >$ yielding: $\dot{J} = (\mathcal{R} \cap O_v < \underline{X} >)^{\cdot}$. As before: $J = \mathcal{R} \cap O_v < \underline{X} >$ follows and this states exactly that $\mathcal{R}$ (hence A) reduces well at O_v.

The filtration fA defined by $f_\gamma A = (F_\gamma K)\Lambda$ will be used for extending the valuation v of K to some quotient ring of Λ.

3.1.6 Lemma

1. Let A be graded and $\Gamma = \mathbb{Z}$ and assume A is gr-simple, then the filtration fA is separated and $G_f(A)$ is strongly graded. If $\overline{\Lambda}$ is a domain then $G_f(A)$ is a domain and Λ is a domain.
2. If A is not graded but simple then the statement of 1 holds too.
3. For a non-discrete Γ assume that A has a PBW-basis $\{a_1, \ldots, a_d\}$ i.e. the $\{a_1, \ldots, a_d\}$ can be ordered such that elements of A have a unique expression as ordered polynomials in the generators $a_1, \ldots, a_d$. Then the statements of 1 are still true.

Proof. 1. Consider $I = \cap\{(f_{\gamma^{-1}}K)\Lambda, \gamma \in \Gamma_+\}$. Clearly $KI \subset I, IK \subset I$ hence $AI \subset I$ and $IA \subset I$ since $K\Lambda = A$. Thus I is a graded ideal of A hence $I = 0$. That $G_f(A)$ is strongly graded follows from fA being a strong filtration. If $\overline{\Lambda}$ is a domain, then from Lemma 1.8.9 it follows that $G_f(A)$ is a domain and then A is domain too.
2. In the ungraded situation but with A simple the statements of 1 follow in an almost identical way.
3. If we can establish that fA is Γ-separated then it is again a strong filtration and the statements in (1) follow in the same way. Suppose fA is not separated, that is there in an $x \in A$ such that for every $\gamma \in \Gamma$ such that $x \in F_\gamma A$ there is a $\delta < \gamma$ in Γ such that $x \in f_\delta A$ too! So for $x \in (f_\gamma K)\Lambda$ this means $x \in (f_\delta K)\Lambda$ for some $\delta < \gamma$. Assume that $\{a_1, \ldots, a_d\}$ is a PBW-basis for A in the ordering given by the indices . Then $x = \sum \xi_{\underline{i}} \underline{a}^{\underline{i}} = \sum \eta_{\underline{i}} \underline{a}^{\underline{i}}$ with $\xi_{\underline{i}} \in f_\gamma K, \eta_{\underline{i}} \in f_\delta K$. Pick $c \in f_{\gamma^{-1}}K$ such that $c\xi_{\underline{i}} \in O_v$ but not all in m_v and $c\eta_{\underline{i}} \in m_v$ (because $\delta < \gamma$). Adapting a common multi-index notation (i.e. inserting some zero-coefficients $\xi_{\underline{i}}$ or $\eta_{\underline{i}}$ when necessary) we obtain $\sum(c\xi_{\underline{i}} - c\eta_{\underline{i}})\underline{a}^{\underline{i}} = 0$. This relation is non-trivial since not all coefficients are in m_v, but that contradicts the PWB-basis property of $\{a_1, \ldots, a_d\}$. Hence such x does not exist so for every $z \in A$ there exists a $\gamma \in \Gamma$ such that $z \in f_\gamma A$ and $z \notin f_\mu A$ with $\mu < \gamma$, or fA is separated. □

Since we consider Γ-valuations on K the Γ-filtration defined on a K-algebra A is not Zariskian i.e. $\widetilde{A}$ need not be Noetherian, so we cannot use results or Zariskian filtration here. We consider a separated Γ-filtration fA on a ring A and S an Ore set of A such that $\sigma(S)$ consists of regular elements of $G_F(A) = G(A)$. We define the localized filtration $FS^{-1}A$ by putting $x \in F_\gamma S^{-1}A$ if there exists an $s \in S$, s of $\deg s = \tau \in \Gamma$, such that $sx \in f_{\tau\gamma}A$.

3.1.7 Proposition

With notation as before, $FS^{-1}A$ is a Γ-filtration of $S^{-1}A$, Γ-separated, inducing fA on A.

Proof. Since $\sigma(S)$ consists of regular elements of $G(A)$ also S consists of regular elements of A. If $x \in F_\gamma S^{-1}A$ then $sx \in f_{\tau\gamma}A$ for some $s \in S$ with $\deg\sigma(s) = \tau$; there is a $\delta \in \Gamma$ such that $sx \in f_\delta A$ but $sx \notin f_{\delta'}A$ for $\delta' < \delta$. Hence $x \in F_{\tau^{-1}\delta}A$ and $x \notin F_{\delta'}A$ with $\delta' < \tau^{-1}\delta$. This follows from the uniqueness of γ, suppose $s_\delta x \in f_{\delta\gamma}A$ and $s_\sigma x \in f_{\sigma\tau}A$ with $\tau \neq \gamma$, say $\tau < \gamma$ in Γ. By the Ore condition there is an s_α such that $s_\alpha s_\delta = a s_\sigma$ with $a \in A$, where the index of the $s's$ refers to the degree of the $\sigma(s)$. Since $\sigma(s_\sigma)$ is regular in $G(A)$ we must have that $\deg\sigma(a) = \alpha\delta\sigma^{-1}$ in Γ. Then $0 \neq s_\alpha s_\delta x = a_{\alpha\delta\sigma^{-1}} s_\sigma x \in a_{\alpha\delta\sigma^{-1}} f_{\sigma\tau}A \subset f_{\alpha\delta\tau}A$; on the other hand we also have that $s_\alpha s_\delta x \in s_\alpha f_{\delta\gamma}A \subset f_{\alpha\delta\gamma}A$, so if we assume $\delta\gamma$ to be the lowest in Γ such that $s_\delta x \in f_{\delta\gamma}A$ then from $\tau < \gamma$ we reach a contradiction because $\alpha\delta\gamma$ is then the lowest containing $s_\alpha s_\delta x$ (as $\sigma(s_\alpha s_\delta x) = \sigma(s_\alpha)\sigma(s_\delta x)$. If $x, y \in F_\gamma S^{-1}A$ then $s_\delta x \in f_{\delta\gamma}A$, $s_\rho y \in f_{\rho\gamma}A$ for some $s_\delta, s_\rho \in S$; then $s_\delta y \in F_{\delta\gamma}S^{-1}A$ for $s_\tau s_\delta = a_{\tau\delta\rho^{-1}} s_\rho$ for some $s_\tau \in S, a_{\tau\delta\rho^{-1}} \in A$ yields $s_\tau s_\delta y = a_{\tau\delta\rho^{-1}} s_\rho y \in f_{\tau\delta\rho^{-1}} A f_{\rho\gamma} A$ hence $s_\tau s_\delta y \in f_{\tau\delta\gamma}A$, consequently: $s_\tau s_\delta(x + y) = s_\tau(s_\delta x) + s_\tau s_\delta y \in f_{\tau\delta\gamma}A$, or $x + y \in F_\gamma S^{-1}A$, proving that the $F_\gamma S^{-1}A$ are additive subgroups. Now for $x \in F_\gamma S^{-1}A, y \in F_\tau S^{-1}A$ we have $s_\alpha x \in f_{\alpha\gamma}A, s_\beta y \in f_{\beta A}$. Write $a_{\alpha\gamma}$ for $s_\alpha x$ and pick $s_\mu \in S$ such that $s_\mu a_{\alpha\gamma} = a' s_\beta$ where $a' \in f_{\mu\alpha\gamma\beta^{-1}}A$ follows from $\sigma(a')\sigma(s_\beta) = \sigma(s_\mu)\sigma(a_{\alpha\gamma})$ and $\deg a_{\alpha\gamma} \le \alpha\gamma$, hence $\deg\sigma(a') \le \mu\alpha\gamma\beta^{-1}$. Now $s_\mu s_\alpha xy = s_\mu a_{\alpha\gamma} y = a' s_\beta y$ with $s_\beta y \in f_{\beta\tau}A$ yields $s_\mu s_\alpha xy \in f_{\mu\alpha\gamma\beta^{-1}} A f_{\beta\tau} A \subset f_{\mu\alpha\gamma\tau}A$. Putting $s_\mu s_\alpha = s_{\mu\alpha}$ yields $xy \in F_{\gamma\tau}S^{-1}A$, so $FS^{-1}A$ is a filtration. The filtration is separated because for $x \in S^{-1}A$ there is an $s_\delta \in S$ such that $s_\delta x \in f_{\delta\gamma}A$ and if $\delta\gamma$ is such that $s_\delta x \notin f_{\gamma'}A$ for $\gamma' < \delta\gamma$ then $x \notin F_\tau S^{-1}A$ for $\tau < \gamma$ (observe that $F_\tau S^{-1}A \cap A = f_\tau A$ because for $a \in A \cap F_\tau S^{-1}A$ some $s_\delta a \in f_{\delta\tau}A$ so $\deg\sigma(a) \le \tau$, hence $FS^{-1}A$ induces fA on A). □

Next we look at the Weyl skewfield $D_1(K)$ and a Γ-valuation O_v in K.

3.1.8 Theorem

Every Γ-valuation O_v of K extends to a noncommutative valuation ring Λ_v of $D_1(K)$.

Proof. In view of Proposition 1.8.10.3 it suffices to construct a separated Γ-filtration on $D_1(K)$ extending the valuation filtration of K such that the associated graded ring is a domain. In fact we only have to construct a Γ-separated filtration on

$\mathbb{A}_1(K)$ extending v on K such that the associated graded ring is a domain because by Proposition 3.1.7 we can extend this to the localized filtration at the Ore set $\mathbb{A}_1(K)^*$ (the Weyl algebra is an Ore domain) provided $\sigma(\mathbb{A}_1(K)^*)$ consists or regular elements. Now $\Lambda = \mathbb{A}_1(O_v)$ defines a good reduction of $\mathbb{A}_1(K)$ at O_v and $\overline{\Lambda} = \mathbb{A}_1(k_v)$ is a Weyl algebra over the residue field k_v, hence a domain. Thus the filtration $f^v\mathbb{A}_{,}(K)$ defined by $f^v_\gamma\mathbb{A}_1(K) = (f_\gamma K)\mathbb{A}_1(O_v)$ has the properties mentioned in (3) of Lemma 3.1.6 and the elements of $\sigma(\mathbb{A}_1(K)^*)$ form exactly the set of homogeneous elements of $G_f(\mathbb{A}_1(K)) = \mathbb{A}_1(k_v)\Gamma$ where $G_f(K) = k_v\Gamma$ and these form even an Ore set because $\mathbb{A}_1(k_v)$ is an Ore domain (and $k_v\Gamma$ is central in $G_f\mathbb{A}_1$ and they are certainly regular in $G_f(\mathbb{A}_1(K))$. For the localized filtration $FD_1(K)$ of $f\mathbb{A}_1(K)$ the associated graded $G_F D_1(K)$ is the graded quotient ring of $\mathbb{A}_1(k_v)\Gamma$ which is $\mathbb{D}_1(k_v)\Gamma$ and a domain! □

3.1.9 Observation

In the foregoing Γ is abelian because it comes from O_v on the commutative K. We shall see later that any valuation on $D_n(K)$ is in fact abelian!

We can extend the foregoing theorem to K-algebras with a PBW-basis as follows.

3.1.10 Proposition

Let A be a K-algebra with PBW-basis $\{a_1, \ldots, a_d\}$ and $\Lambda = O_v < a_1, \ldots, a_d >$. Suppose that A is an Ore domain with skew field of fractions $Q(A)$ and that $\overline{\Lambda} = \Lambda/m_v\Lambda$ is a domain then v extends to a noncommutative valuation of $Q(A)$.

Proof. Define fA by $f_\gamma A = (f_\gamma K)\Lambda$ for every $\gamma \in \Gamma$. Statement (3) from Lemma 3.1.6 yields that fA is a Γ-separated filtration and $G_f(A)$ is a domain. We have that $\sigma(A^*)$ is a graded Ore set of $G_f(A)$ in fact $\sigma(A^*)$ is the set of homogeneous elements (nonzero) of $G_f(A)$; indeed if $\overline{a}, \overline{b} \in h(G_f(A))^*$ then there are $a', b' \in A^*$ such that $a'b = b'a$ by the Ore condition for A^* and since $G_f(A)$ is a domain $\sigma(a'b) = \sigma(a')\sigma(b) = \sigma(b')\sigma(a) = \sigma(b'a)$, or $\sigma(a')\overline{b} = \sigma(b')\overline{a}$. For $x \in G_f(A)$ say $x = x_{\gamma_1} + \ldots + x_{\gamma_n}$ with $x_{\gamma_i} \in hG_f(A)^*$. There is an $s_1 \in hG_f(A)^*$, $s_1 x_{\gamma_1} = y_1\overline{a}$, hence $s_1 x = y_1\overline{a} + s_1 x_{\gamma_2} + \ldots + s_1 x_{\gamma_n}$ with $\overline{a} \in hG_f(A)^*$ and $y_1 \in hG_f(A)^*$.

Then take $s_2 \in hG_f(A)^*$ such that $s_2 s_1 x_{\gamma_2} = y_2\overline{a}$ with $y_2 \in hG_f(A)^*$, then $s_2 s_1 x = s_2 y_1 \overline{a} + y_2\overline{a} + s_2 s_1 x_{\gamma_3} + \ldots + s_2 s_1 x_{\gamma_n}$. Repeating this n times we arrive at $s_1, \ldots, s_n \in hG_f(A)^*$ such that $s_n \ldots s_1 x = y\overline{a}$ with $y \in G_f(A)^*$, so $hG_g(A)^*$ is an Ore set in $G_f(A)$. Thus fA defines $FQ(A)$ by localization and the associated graded ring of $Q(A)$, $G_F(Q)$ is the localization of $G_f(A)$ at $hG_f(A)^*$ which is a domain (and in fact a gr-skewfield $Q_{cl}(\overline{\Lambda})\Gamma$). Therefore $F_0Q(A)$ is a Γ-valuation ring extending v on K to $Q(A)$. □

We have a similar result for Dubrovin valuations using now Theorem 1.8.11.

3.1.11 Proposition

Let A be a K-algebra with PBW-basis $\{a_1, \ldots, a_d\}$ and put $\Lambda = O_v\langle a_1, \ldots, a_d\rangle$. If A is a prime Goldie ring such that Λ_{reg} maps to regular elements of $\overline{\Lambda} = \Lambda/m_v$ and $\overline{\Lambda}$ is a prime Goldie ring than v extends to a Dubrovin valuation on the simple Artinian $Q_{cl}(A)$.

Proof. The filtration fA defined by $f_\gamma A = (f_\gamma K)\Lambda$ is again separated and strong, hence $G_f(A)$ is strongly graded by Γ over $G_f(A)_0 = \overline{\Lambda}$. The homogeneous elements of $G_f(K)$ are central units in $G_f(A)$ and $G_f(A) = G_f(A)_0 G_f(K)$, hence $G_f(A)$ is also a prime Goldie ring. A regular element of A, x say, may be multiplied by a $\lambda \in K$ to a regular element λx of Λ, such that $\lambda x \notin m_v\Lambda$. Hence $\sigma(\lambda x)$ is regular in $\overline{\Lambda}$ hence in $G_f(A)$, since $\sigma(\lambda)$ is regular in $G_f(A), \sigma(\lambda x) = \sigma(\lambda)\sigma(x)$ hence $\sigma(x)$ is regular in $G_f(A)$. Then fA extends to the localized filtration $FS^{-1}A$, where $S = A_{reg}$ and $S^{-1}A$ is a simple Artinian ring. The associated graded ring of $S^{-1}A$ is $G(S)^{-1}G_f(A)$ which is again a prime Goldie ring as it is an order in the simple Artinian ring $T^{-1}G_f(A)$ where $T = G_f(A)_{reg}$ ($G_f(A)$ is prime Goldie). In fact $\sigma(S)^{-1}G_f(A) = Q_{cl}(G_f(A)_0)G_f(K)$ where $Q_{cl}(G_f(A)_0)$ is simple Artinian. In view of Theorem 1.8.11 we obtain that $F_0S^{-1}A$ is a Dubrovin valuation ring. □

The extension problem for valuations of K to K-algebra appearing as simple Artinian or skewfield quotient rings of algebras given by generators and relations has now been reduced to finding "good reductions" or more directly to the existence of an O_v-order Λ defining a suitable filtration on A that extends well to a localized filtration of $Q_{cl}(A)$. This comes down to the verification of domain or prime Goldie properties of the associated graded ring. This method applied to the Weyl field, but also to other interesting examples.

3.1.12 Observation

In all of the following situation the extension result for valuations $O_v \subset K$ to the quotient ring of the K-algebra is valid.

(a) The quantum plane $A = K < X, Y > /(XY - qYX)$ and $O_v \subset K$ such that q is a unit in O_v.
(b) The quantized Weyl algebra $A_1(K, q)$ defined as $K < X, Y > /(XY - qYX - 1)$ and $O_v \subset K$ containing q as a unit.

(c) The enveloping algebra $U(g)$ for a finite dimensional Lie algebra g over K.
(d) Quantum 2×2-matrices defined as $K < a, b, c, d >$ with relations: $ba = q^{-2}ab, ca = q^{-2}ac, bc = cb, db = q^{-2}bd, dc = q^{-2}cd, ad - da = (q^2 - q^{-2})bcx$.
(e) The conformal sl_2-enveloping algebra in the sense of L. Le Bruyn given as $K < X, Y, Z >$ modulo the relations:

$$\begin{cases} XY - aYX = Y, ZX - aXZ = Z \\ YZ - cZY = bX^2 + Y \end{cases}$$

at O_v containing a, b, c as units.
(f) Let A be as in Proposition 3.1.10 but assuming now that $\overline{\Lambda}$ is Auslander regular (cf. [40]) and positively graded over a field of characteristic zero. Then it is known that $\overline{\Lambda}$ is a domain and the extension result follows.

The results in this section open the possibility for developing a valuation and divisor theory on quantized algebras, these are deformations of classical algebras depending on certain parameters (as in Observation 3.1.12 above).

3.2 Discrete Valuations on the Weyl Skewfield

In this section K is a field of characteristic zero and $\mathbb{A}_1(K)$ is the first Weyl algebra, $\mathbb{A}_1(K) = K < x, y >= K < X, Y > /(YX - XY - 1)$. We know that $A_1(K)$ is a simple Noetherian non-Artinian, Ore domain and it has a skewfield of fractions $\mathbb{D}_1(K)$ called the first Weyl field. For $A_n(K) = \mathbb{A}_1(K) \otimes \ldots \otimes \mathbb{A}_1(K)$ we have a skewfield of functions $\mathbb{D}_n(K)$.

The Bernstein filtration of $\mathbb{A}_1(K)$ is defined by putting $\deg x = \deg y = 1$, i.e. $F_0\mathbb{A}_1(K), F_n\mathbb{A}_1(K) = K \oplus Kx \oplus Ky, \ldots, F_n\mathbb{A}_n(K) = (F_1\mathbb{A}_1(K))^n, \ldots$. It is a separated $\mathbb{Z}$-filtration with $G_F\mathbb{A}_1(K) \cong K[X, Y]$, we let σ be the principal symbol map of F. On $\mathbb{D}_1(K)$ we consider the quotient filtration $F\mathbb{D}_1(K)$, then $G_F\mathbb{D}_1(K) = Q^g(K[X, Y])$, the graded quotient field of $K[X, Y]$. Since the latter is a domain we know that $F_0\mathbb{D}_1(K)$ is a valuation ring of $\mathbb{D}_1(K)$ and the valuation filtration of it coincides with $F\mathbb{D}_1(K)$; the corresponding valuation v_B is called the Bernstein valuation ring of $\mathbb{D}_1(K)$. To a discrete valuation v of $\mathbb{D}_1(K)$ there corresponds a noncommutative valuation ring Λ_v and a valuation filtration $f_v\mathbb{D}_1(K)$ with $(f_v\mathbb{D}_1(K))_0 = \Lambda_v$.

3.2.1 *Observation*

If Λ_v is a valuation ring of a skewfield Δ then the following statements are equivalent for $a, b \in \Delta^*$.

1. $\Lambda_v a \subset \Lambda_v b$.
2. $v(a) \geq v(b)$.
3. $a\Lambda_v \subset b\Lambda_v$.

Proof. (an expansion of Lemma 1.3.2.8).

1. $\Rightarrow$ 2 If $\Lambda_v a \subset \Lambda_v b$ than $a = \lambda b$ for some $\lambda \in \Lambda_v$ then $v(a) = v(\lambda) + v(b)$ yields $v(a) \geq v(b)$.
2. $\Rightarrow$ 3. From $v(a) \geq v(b)$ it follows that $v(b^{-1}a) \geq 0$ or $b^{-1}a \in \Lambda_v$ and $a\Lambda_v \subset b\Lambda_v$ follows.
3. $\Rightarrow$ 2. and 3. $\Rightarrow$ 1. follow by symmetry from the foregoing. □

Recall that two discrete valuations v_1 and v_2 are said to be equivalent if there exist n and m in $\mathbb{Z}$ such that $nv_1(x) = mv_2(x)$ for every $x \in \Delta$.

Let us recall how the valuation function $v : \Delta^* \to \Gamma$ is constructed from a valuation ring Λ of Δ. Put $P \subset \Lambda$ equal to the ideal $P = \{x \in \Lambda, x^{-1} \notin \Lambda\}$. For $\lambda \in \Lambda$ define $(P : \lambda) = \{(a, b) \in \Delta \times \Delta, a\lambda b \in P\}$ and call $\lambda_1 \simeq \lambda_2$ if $(P : \lambda_1) = (P : \lambda_2)$. Let $[P : \lambda]$ denote the class of $(P : \lambda)$ with respect to the foregoing equivalence relation. On the set of equivalence classes Γ introduce the total order induced by the inclusion ordering on the set of $(P : \lambda), \lambda \in \Delta$. The function $v : \Delta^* \to \Gamma, x \mapsto [P : \lambda]$ is well-defined. Multiplication of Δ induces a multiplication in Γ making Γ into a totally ordered group. The valuation ring Λ_v coincides with Λ and $P = w_v$.

3.2.2 Proposition

A valuation v on a skewfield Δ has rank one exactly when Λ_v is maximal as a proper subring of Δ (this extends Proposition 1.2.12 to the noncommutative case).

Proof. If v has rank (1) then Γ is Archimedean (cf. Proposition 1.3.1.4). If Λ_v is not maximal let $\Lambda' \supsetneqq \Lambda_v$ be a proper subring of Δ, suppose $a \in \Lambda' - \Lambda_v$ and consider $b \in \Delta^* - \Lambda'$. Since $v(a), v(b) < 0$, the Archimedean property yields that there is an $n \in \mathbb{N}$ such that $v(a^{-n}) \geq v(b^{-1})$, or $ba^{-n} \in \Lambda_v$ with $a^n \in \Lambda'$ and $b \in \Lambda_v a^n \subset \Lambda'$, contradiction. Conversely, if Λ_v is maximal then $htw_v = 1$. Indeed, any nontrivial prime $P \subsetneqq w_v$ is a completely prime ideal (since left ideals of Λ_v are idreals!). Moreover $S = \Lambda_v - P$ is an Ore set of Λ_v since for given $s \in D, \Lambda \in \lambda_v$ we have $s\lambda \in s\Lambda_v = \Lambda_{v^2}$ or $s\lambda = \lambda' s$ for some $\lambda' \in \Lambda_v$. Now $\Lambda_v \subsetneqq S^{-1}\Lambda_v$. It is clear that $S^{-1}\Lambda_v \neq \Delta$ since $(S^{-1}\Lambda_v)P$ is a proper ideal of $S^{-1}\Lambda_v$. Maximality of Λ_v thus entails $ht(w_v) = 1$. If $rk\Gamma > 1$ then Γ contains a convex subgroup C. Put $P = \{x \in \Lambda, v(a) \in C\}$. It is easily verified that P is a prime ideal of Λ_v and also $P \subsetneqq w_v$ because $C \neq \Gamma^+$, this would contradict $ht(w_v) = 1$. □

A slight extension of the final part of the foregoing proof yields also a proof of the following.

3.2.3 Observation

Nonzero prime ideals P of Λ_v correspond bijectively to nontrivial convex subgroups of Γ. Normal convex subgroups of Γ correspond to prime ideals of Λ_v which are invariant under inner automorphisms of Δ.

When studying valuations on $\mathbb{D}_n(K)$ one may restrict to abelian Γ. It is known that every valuation on a finite dimensional skewfield is abelian but for $\mathbb{D}_n(K)$ this result is somewhat surprising. They are in some sense very noncommutative rings, in fact they even contain free subalgebras of any countable rank! The result is due to J. Shtipel'man but we follow L. Makar-Limanov's proof.

3.2.4 Theorem

Let v be a Γ-valuation on $\mathbb{D}_1(K)$ then Γ is abelian.

Proof. Write $\mathbb{A}_1(K) = K < x, y > \subset \mathbb{D}_1(K) = K(< x, y >)$. Take $r \neq 0$ in $\mathbb{A}_1(K)$ and suppose that $v(xr) \neq v(rx)$, say $v(rx) > v(xr)$ (in the other case the proof is formally similar). Then for $[x, r] = xr - rx$ we have $v([x, r]) = v(xr)$. By an easy induction argument we then obtain: $v([x, -]^n(r)) = v(x^n r)$. For every $r \in \mathbb{A}_1(K)$ there is an $e = e(r)$ such that $[x, -]^e(r) = 0$ (because every $r \in \mathbb{A}_n(K)$ has a unique finite polynomial expression in x and y with powers in x before powers in y and $[x-]$ lowers the y-degree because $xy - yx = -1$). Thus we obtain $v(0) = v([x, -]^e r) = v(x^e r)$ but that is a contradiction since $x^e r \neq 0$. Since Γ is generated as a group by the semigroup $v(\mathbb{A}_1(K))$ it follows from $v(x) + v(r) = v(r) + v(x)$ that $v(x) \in Z(\Gamma)$. In fact the foregoing establishes that $v(f) \in Z(\Gamma)$ for every f such that every $r \in \mathbb{A}_1(K)$ is annihilated by some power of $[f, -]$, in particular this holds for all $f \in K[x]$. Since Γ is totally ordered $\gamma^m \sigma = \sigma \gamma^m$ for some m entails $\gamma\sigma = \sigma\gamma$ hence $Z(\Gamma)$ is root-closed in Γ. Now assume $r \in \mathbb{A}_1(K)$ is such that $v(r) \notin Z(\Gamma)$. Since $GK\dim(\mathbb{A}_1(K)) = 2$ it follows that for any $s \in \mathbb{A}_1(K)$ we have a relation: $\Sigma x_{ij} r^i s^j = 0$ with $x_{ij} \in K[x]$ (the GK dimension bounds the transcendence of the ring, so the r and s cannot be algebraically independent over $K[x]$). At least two monomials in this relation have the same valuation, otherwise $v(\Sigma x_{ij} r^i s^j)$ would necessarily be the valuation of the unique monomial in it having minimal valuation but that could not be equal to $-\infty$. Say $v(x_{i_0 j_0} r^{i_0} s^{j_0}) = v(x_{i_1 j_1} r^{i_s} s^{j_1})$, then either some $v(s^k) \in Z(\Gamma) < v(r) >$ or $v(r^l) \in Z(\Gamma) < v(s) >$ with k, l larger than zero, because $v(x_{ij}) \in Z(\Gamma)$ by foregoing remarks. In either case we obtain that some power of $v(r)$ commutes with some power of $v(s)$. Since Γ is totally ordered ($\gamma^n \sigma = \sigma \gamma^n$ entails $\gamma\sigma = \sigma\gamma$) it then follows that $v(r)$ and $v(s)$ commute. This holds for arbitrary $s \in \mathbb{A}_1(K)$, hence it contradicts $v(r) \notin Z(\Gamma)$. Consequently $Z(\Gamma) = \Gamma$ or Γ is abelian. □

3.2.5 *Corollary*

Every valuation of $\mathbb{D}_n(K)$ is abelian.

Proof. $\mathbb{D}_n(K)$ is the n-fold tensor product of copies of $\mathbb{D}_1(K)$, its value group is a subgroup of a product of the value groups $v(\mathbb{D}_1(K))$ which is an abelian group. □

3.2.6 *Remark and Project*

The above proof is elementary except for the key result about the GK-dimension. For the general theory about GKdim we may refer to G. Krause and T. Lenagan, [34] or C. Năstăsescu and F. Van Oystaeyen [53]. It would be an interesting project to relate GKdim and valuation theory further, or perhaps the GKtd (Gelfand–Kirrilov transcendence degree) could be used instead of GKdim. The driving conjecture could be that for a skewfield of $GK\text{-dim}\Delta = n$ and a valuation v of Δ of rank m we would have $GK\text{dim}\overline{\Delta} = n-m$, $\overline{\Delta}$, the residue skewfield of v. Also it seems possible to extend the foregoing theorem to skewfields obtained as skewfields of fractions of enveloping algebras of nilpotent Lie algebras.

3.2.7 *Lemma*

There are no discrete K-valuations of $D_1(K)$ with residue field K.

Proof. Suppose v is a discrete valuation of $\mathbb{D}_1(K)$ with valuation ring Λ_v and $\Lambda_v/w_v = K$. Write $w_v = (\pi)$. If $a, b \in \Lambda_v$ then for each $n \in \mathbb{N}$ there are polynomials $f(\pi)$ and $g(\pi)$ with coefficients in K such that:

$$v(a - f(\pi)) \geq n \text{ and } v(b - g(\pi)) \geq n$$

Since $f(\pi)$ and $g(\pi)$ commute we obtain that $ab - ba$ is in $f^v_{-n}\mathbb{D}_1(K)$ and this holds for all $n \in \mathbb{N}$. Since Λ_v cannot be commutative as it has $\mathbb{D}_1(K)$ for its quotient skewfield $ab \neq ba$ for some $a, b \in \Lambda_v$ and then $ab - ba$ is not in $f^v_{-N}\mathbb{D}_1(K)$ for some $N \in \mathbb{N}$. □

3.2.8 *Lemma*

For any Γ-valuation v on $\mathbb{D}_1(K)$ we have that $v([x, y]) > v(xy) = v(x) + v(y)$.

Proof. Since Γ is abelian $v(xy) = v(yx)$. Hence for the valuation filtration degree: $\deg\sigma_v(xy - yx) < \deg\sigma_v(xy)$. Therefore $v(xy - yx) > v(xy) = v(x) + v(y)$. □

3.2.9 *Corollary*

From $xy - yx = -1$ we obtain that $v(x) + v(y) < 0$ (see the foregoing lemma). If $v(x) > v(y)$, then $v(x+y) = v(y)$ and we may generate $\mathbb{A}_1(K)$ as $K < x+y, y >$. In other words we may assume that $\mathbb{A}_1(K)$ is generated by x and y with $[y,x] = 1$ and $v(x) = v(y) < 0$.

The valuation v_B corresponding to the Bernstein filtration factors over the principal symbol map $\sigma_B : \mathbb{D}_1(K) \to Q_g(K[X,Y])$. In fact there is only one such discrete valuation up to equivalence.

3.2.10 *Proposition*

If $\overline{v}$ is a discrete K-valuation of $K(X,Y)$ such that $\overline{v}\sigma_B$ is a nontrivial discrete K-valuation of $\mathbb{D}_1(K)$ then $\overline{v}\sigma_B$ is equivalent to the valuation v_B induced by the Bernstein filtration.

Proof. Suppose $a, b \in \mathbb{D}_1(K)$ are such that $\deg\sigma_B(b) < \deg\sigma_B(a)$. Then $\overline{v}(\sigma_B(a)) = \overline{v}(\sigma_B(a+b)) = \overline{v}\sigma_B(a+b) \geq \min\{\overline{v}\sigma_B(a), \overline{v}\sigma_B(b)\}$. If $\overline{v}\sigma_B(a) \neq \overline{v}\sigma_B(b)$, then equality holds, thus $\overline{v}\sigma_B(a) < \overline{v}\sigma_B(b)$. In particular when $\sigma_B(b)$ is homogeneous in $K(X,Y)$ of strictly negative degree, then $\overline{v}\sigma_B(b) \geq 0$. For every homogeneous element x of $K(X,Y)$ of degree zero (this is always σ_B of some element of $\mathbb{D}_1(K)$) in $Q^g(K[X,Y])$ we thus have $\overline{v}(x) = 0$. For $a \in \mathbb{D}_1(K)$ we have $\deg\sigma_B(a) = n$, then $\overline{v}\sigma_B(a) = \overline{v}(\sigma_B(ay^{-n})\sigma_B(y^n)) = 0 + n\overline{v}\sigma_B(y)$ as σ_B is multiplicative. Consequently $\overline{v}\sigma_B = (\overline{v}\sigma_B(y)).\deg\sigma_B$ hence $\overline{v}\sigma_B$ is equivalent to the Bernstein valuation defined by $\deg\sigma_B$. □

We write $K(\frac{X}{Y})$ for $G_B(\mathbb{D}_1(K))_0$ and let v be a discrete K-valuation on $\mathbb{D}_1(K)$ with $v(x) = v(y) < 0$. Let Λ_v be the valuation ring of v with maximal ideal w_v and residue skewfield Δ_v. From Lemma 3.2.8 we may derive that Δ_v is commutative; indeed if $a, b \in \Lambda_v - w_v$ then $v([a,b]) > v(a) + v(b) = 0$ or $ab - ba \in w_v$ and thus $\Delta_v = \Lambda_v / w_v$ is commutative. Consequently $G_v(\mathbb{D}_1(K))$ is commutative and of the form $\Delta_v[T, T^{-1}] \cong \Delta_v\mathbb{Z}$.

The valuation filtration $f^v\mathbb{D}_1(K)$ induces a filtration on $G_B(\mathbb{D}_1(K))_0 = K(\frac{X}{Y})$, $f_i^v(K(\frac{X}{Y})) = f_i^v\mathbb{D}_1(K) \cap \Lambda_B / (f_i^v\mathbb{D}_1(K) \cap w_B)$. This is an exhaustive filtration but it need not be separated.

3.2.11 *Lemma*

With notation as above, $f^vK(\frac{X}{Y})$ is separated if and only if $\sigma_B(z) = 1$ entails $v(z) \leq 0$. In case $f^vK(\frac{X}{Y})$ is not separated then $\cap_i f_i^vK(\frac{X}{Y}) = K(\frac{X}{Y})$.

Proof. Put $I = \cap_i f_i^v K(\frac{X}{Y})$. If $\sigma_B(x) \neq 0$ is in I then there exists a $y \in \mathbb{D}_1(K)$ such that: $\sigma_B(z) = \sigma_B(y)$ and $v(y) > v(z)$ (choose y in $f_i^v \mathbb{D}_1(K)$ for appropriate i.) Then $\sigma_B(yx^{-1}) = 1$ but $v(yz^{-1}) > 0$. Conversely if $z \in \mathbb{D}_1(K)$ is such that $\sigma_B(z) = 1$ and $v(z) > 0$ then for all $n \in \mathbb{N}$ we have: $1 = (1 - z^n) + z^n$. Thus $1 \in f_{-n}^v K(\frac{X}{Y})$, then for all $n \geq 0$ we obtain that $f_{-n}^v K(\frac{X}{Y}) = K(\frac{X}{Y}))$ as desired. □

From here on we assume that K is algebraically closed. We say that v is F^B-**compatible** if $f^v K(\frac{X}{Y})$ is a separated filtration, i.e. $\sigma_B(z) = 1$ yields $v(z) \geq 0$.

3.2.12 Proposition

A discrete K-valuation on $\mathbb{D}_1(K)$ that is F^B-compatible is determined by its restriction to the subfield $K(\frac{X}{Y})$ in $\mathbb{D}_1(K)$.

Proof. A pseudo-homogeneous element of $\mathbb{A}_1(K)$ is one having a homogeneous expressions in x and y (this is not unique since $yx = xy + 1$ but that is harmless here). Consider $f \in \mathbb{A}(K)$ and write $f = f_1 + f_2$ where f_1 is pseudo-homogeneous of degree $\deg\sigma_B(f)$ and $\deg\sigma_B(f_2) < \deg a_{\sigma_B}(f)$. From $\sigma_B(f/f_1) = 1$ it follows that $v(f) \leq v(f_1)$ since v is assumed to be F^B-compatible. Thus, $v(f) \geq \min\{v(f_1), v(f_2)$ with equality whenever $v(f_1) \neq v(f_2)$, yields $v(f) = \min\{v(f_1), v(f_2)\}$.

In case f is pseudo-homogeneous of degree n, say $f = \sum_{i=0}^n a_i x^i y^{n-i}$, then $v(fy^{-n}) = v(g)$ where we put $g = \sum_{i=0}^n a_i (xy^{-1})^i$. If for example $v(fy^{-n}) > v(g)$ then $\sigma_B(fy^{-n}/g) = 1$ and $v(fy^{-n}/g) > 0$ leads to a contradiction. Otherwise look at $\sigma_B(g/fy^{-n}) = 1$. Then $v(f) = v(g) + nv(y)$ and the proof is finished. □

The discrete K-valuations of $K(T), T = \frac{X}{Y}$ are well-known i.e. v is either trivial or v corresponds to an $\alpha \in K$. In the first case v is equivalent to the valuation of the Bernstein filtration, meaning that for all $z \in \mathbb{A}_1(K)$, $v(z) = -\deg\sigma_B(z)v(y)$. By the foregoing proposition the F^B-compatible valuations are determined by three parameters $p = v(x) = v(y) \in \mathbb{Z} - \mathbb{N}$, $q = v(\frac{x}{y} - a) \in \mathbb{N} - \{0\}, a \in K^*$. Given p, q, a then there is at most one discrete K-valuation of $\mathbb{D}_1(K)$ compatible with the Bernstein filtration such that $v(x) = v(y) = p$ and $v(\frac{x}{y} - a) = q$. From foregoing remarks it follows that v, if it exists, may be calculated in the following way. To calculate v of $f \in \mathbb{A}_1(K)$ at first decompose f into pseudo-homogeneous elements, say $f = \sum_{i=0}^n f_i$ and put $v(f) = \min\{v(f_i), i = 0, \ldots, n\}$. To calculate v on a pseudo-homogeneous f_n, of degree n, put $v(f_n) = np + v_a(\sigma_B(f_n))q$ where v_a is the graded K=valuation on $K(\frac{X}{Y})[Y, Y^{-1}]$ such that $v_a(Y) = 0, v_a(\frac{X}{Y} - a) = 1$. It is also possible to define $v(f_n) = v'_a(\sigma_B(f_n))$ where v'_a is he graded valuation of $K(\frac{X}{Y})[Y, Y^{-1}]$ such that $v'_a(\frac{X}{Y} - a) = q, v'_a(Y) = p$. We observe that $K(\frac{X}{Y})[Y, Y^{-1}]$ is a graded field (gr-field) in the sense that every homogeneous element different from 0 is invertible; a gr-valuation is associated to a gr-valuation ring, being a graded subring such that for every homogeneous element of the gr-field, say z, either z

or z^{-1} is in the subring. A gr-valuation of $K(\frac{X}{Y})[Y, Y^{-1}]$ is always induced by a valuation of $K(X, Y)$ which is a graded valuation in the sense that $v(z) \leq 0$ if and only if $v(z_n) \leq 0$ for every homogeneous component n. Graded valuations are not studied in detail in this work, we refer to [38, 66].
We are now ready to prove the existence theorem.

3.2.13 Theorem

Let v be defined by p, q, a as above and assume that $q \leq -p$ then v is a discrete K-valuation of the Weyl field $\mathbb{D}_1, (K)$.

Proof. It suffices to verify the valuation properties on elements of $\mathbb{A}_1(K)$. If f, g are pseudo homogeneous of different degree then: $v(f+g) = \min\{v(f), v(g)\}$ holds by definition; if f, g have the same degree but $f+g \neq 0$, then $v(f+g) = np + v_a(\sigma_B(f+g))q \geq np + \min\{v_a(\sigma_B(f)), v_a(\sigma_B(g))\}q = \min(v(f), v(g)\}$. In the situation that f, g are not pseudo-homogeneous the relation follows by decomposition into pseudo-homogeneous elements and the definition of v.
For pseudo-homogeneous f and g we write:

$$f = \sum_{i=0}^{n} a_i x^i y^{n-i} \text{ and } \sigma_B(f) = \overline{f}$$

$$g = \sum_{i=0}^{m} b_j x^j y^{m-j} \text{ and } \sigma_B(g) = \overline{g}$$

$$fg = \sum_{i=0}^{m+n} \sum_{j=0}^{i} c_{ji} x^j y^{i-j}, h = \sum_{i=0}^{m+n} \sum_{j=0}^{i} c_{ji} X^j Y^{i-j}$$

For the Weyl algebras it is a well-known fact that:

$$h = \sum_{k=0}^{r} (-1)^k \frac{1}{k!} \frac{\partial^k \overline{f} \partial^k \overline{g}}{\partial X^k \partial Y^k}$$

where r is the integral part of $\frac{n+m}{2}$. It is now straightforward to calculate

$$v(fg) = \min\{v \sum_{j=0}^{i} c_{ji} x^j y^{i-j}, i = 0, \ldots, m+n\}$$

$$= \min\{ip + v_a(\sigma_B(\sum_{j=0}^{i} c_{ji} x^j y^{i-j}))q, i = 0, \ldots, m+n\}$$

$$= \min\{(n+m-2k)p + v_a\left(\frac{\partial^k \overline{f} \partial^k \overline{g}}{\partial Y^k \partial X^k}\right) q, k = 0, \ldots, r\}$$

$$= \min\{(n+m+2k)p + (v_a(\overline{f}) - k + v_a(g) - k)q, k = 0, \ldots, r\}$$

$$= (n+m)p + v_a(\overline{f}\overline{g})q + \min\{k(-2p-2q), k = 0, \ldots, r\}$$

$$= v(f) + v(g)$$

The last equality follows from $v_a\left(\frac{\partial^k f \partial^k g}{\partial y^k \partial x^k}\right) \geq v_a(f) - k + v_a(g) - k$. Now more generally, if $f = \sum_{i=0}^{n} f_i, g = \sum_{i=0}^{m} g_i$, then let k and l be maximal such that we have $v(f) = v(f_k), v(f_k), v(g) = r(g_l)$. Then $v(fg) \leq v(\sum_{j=0}^{k+l} f_j g_{k+l-j}) = v(f_k g_l) = v(f_k) + v(g_l) = v(f) + v(g)$.

The other inequality follows from: $v(\sum_{j=0}^{i} f_j g_{i-j}) \geq v(f) + v(g)$ for all $i = 0, \ldots, k+l$. □

It is clear from foregoing proof that v will not be a valuation if $q > -p$, so we suppose hereafter that $q \leq -p$. The valuation filtration f^v defines a commutative associated graded ring and we can calculate this explicitly. For f and g as in the proof we have:

$$fg - gf = \sum_{i,j} d_{ij} x^i y^j$$

Again it follows that:

$$h = \sum_{k=1}^{r} (-1)^{k+1} \frac{1}{k!} \left(\frac{\partial^k \overline{f} \partial^k \overline{g}}{\partial X^k \partial Y^k} - \frac{\partial^k \overline{f} \partial^k \overline{g}}{\partial Y^k \partial X^k}\right)$$

r being the integral part of $\frac{n+m}{2}$. Hence we obtain: $v(fg - gf) = \min\{(n+m-2k)p + v_a(\frac{\partial^k \overline{f}}{\partial X^k} \frac{\partial \overline{g}}{\partial Y^k} - \frac{\partial^k \overline{f}}{\partial Y^k} \frac{\partial^k \overline{g}}{\partial X^k})q, k = 1, \ldots, r\}$.

Now: $v_a\left(\frac{\partial^k \overline{f}}{\partial X^k} \frac{\partial \overline{g}}{\partial Y^k} - \frac{\partial^k \overline{g}}{\partial Y^k} \frac{\partial^k \overline{g}}{\partial X^k}\right) > v_a(\overline{f}) + v_a(\overline{g}) - 2k$, thus $v(fg - gf) > v(f) + v(g)$!

The residual field of an F^B-compatible K-valuation does not only turn out to be commutative, it is actually a purely transcendent extension of degree one of K. From a transcendence (Gelfand–Kirillov) argument it would follow that the transcendent degree is one but for pure transcendence some arithmetical information is needed in the proof.

3.2.14 Theorem

Let $q \leq -p$ and v given by p, q, a. The residue field $\Delta_v \cong K(t)$ with $t = (\frac{x}{y} - a)^l y^k$, where $-kp = lq$ is the least common multiple of $-p$ and q.

Proof. First we observe that v is a $K(t)$ valuation because $v(t-\beta)=0$ for all $\beta \in K$ and K is algebraically closed. Indeed, it is clear that $v(t-\beta) \geq 0$, the leading pseudo homogeneous term of $t-\beta$ is equal to the one of t which is equal to: $\sum_{i=0}^{l}\binom{l}{i}(-a)^{l-i}x^i y^{k-i}$. It follows that at least one of the pseudo-homogeneous terms in the decomposition of $t-\beta$ must have v-value equal to zero, hence $v(t-\beta)=0$. Next we show for $f,g \in \mathbb{A}_1(K)$ such that $v(f)=v(g)$ there is an $h \in K(t)$ such that $v(g^{-1}f-h)>0$. Since $v(f)$ is obtained as $v(f_i)$ for some pseudo homogeneous part f_i of f we may assume that f is pseudo-homogenous. Take F,G in $\mathbb{A}_1(K)$ such that $fG=gF$; there is a pseudo-homogenous $A \in \mathbb{A}_1(K)$ such that $v(A)=v(G)=v(F)$. If there are F_1, G_1 in $K(t)$ such that $v(FA^{-1}-F_1)>0$ and $v(GA^{-1}-G_1)>0$ then $g^{-1}f-F_1G_1^{-1}= g^{-1}(fG_1-gF_1)G_1^{-1}=g^{-1}(g(FA^{-1}-F_1)-f(GA^{-1}-G_1))G_1^{-1}$. Consequently: $v(g^{-1}f-F_1G_1^{-1})>-v(g)+v(g)-v(G_1)=0$. From the foregoing it follows that we may select h such that $v(g^{-1}f-h)>0$ with $v(f)=v(g)$ and both f,g are assumed to be pseudo-homogenous. We arrive at:

$$v(f)=p\deg\sigma_B(f)+qv_a(\sigma_B(f))$$

$$v(g)=p\deg\sigma_B(g)+qv_a(\sigma_B(g))$$

The integer $(\deg\sigma_B(f)-\deg\sigma_B(g))(-p)=v_a(\sigma_B(f(g))q$ is a common multiple of $-p$ and q. Thus we obtained an $n \in \mathbb{Z}$ such that: $nk=\deg\sigma_B(f)-\deg\sigma_B(g), nl= v_a(\sigma_B(f/g))$.

We now calculate $\sigma_B(\frac{f}{g}y^{-nk})=(T-a)^{nl}h(T)$ with $h(T) \in K(T), T=\frac{X}{Y}$, and $\gamma=h(a)\neq 0$.

If $\sigma_B(f/g)\neq\sigma_B(\gamma t^n)$, then $v(\frac{f}{g}-\gamma t^n)=nkp+v_a((h(T)-\gamma)(T-a)^{nl})q$, and the latter is strictly bigger than $nkp+nlq=0$. In case $\sigma_B(f/g)=\sigma_B(\gamma t^n)$ we may assume that $\gamma=1$ and $n\in\mathbb{N}$. Clearly, then $t^n=\sum_{i=0}^{ln}\binom{ln}{i}(-a)^{ln-i}x^i y^{kn-i}$ + terms having a strictly positive value.

Since both f and the term $\sum_{i=0}^{ln}\binom{ln}{i}(-a)^{ln-i}x^i g y^{kn-i}$ are pseudo homogeneous and have the same image under the principal symbol map they must be equal! (in the Weyl algebra every element has a unique pseudo-homogeneous decomposition with powers of x preceding powers of y). Hence, modulo terms with value strictly larger than $v(f)=v(g)$ we obtain:

$$f-gt^n=\sum_{i=0}^{ln}\binom{ln}{i}(-a)^{ln-i}[x^i,g]y^{kn-i}$$

If $g=\sum_{j=0}^{m}a_j x^{m-j}y^j$ then $[x^i,g]$ is equal to

$$\sum_{j=0}^{m}a_j\left(\sum_{k=0}^{\min(i,j)}(-1)^k\frac{i!}{(i-k)!}\binom{j}{k}x^{m-j+i-k}y^{j-k}\right)$$

Therefore: $v(f - gt^n - \sum_{k=1}^{\min(lnm)} R_k > v(f) = v(g)$, where

$$R_k = (-1)^k \sum_{i=k}^{l_n} \binom{ln}{i} (a)^{ln-i} \frac{i!}{(i-k)!} \left(\sum_{j=k}^{m} a_j \binom{j}{k} x^{m-j+i-k} y^{kn-i+j-k} \right)$$

Calculate:

$$v(R_k) = (kn + m - 2k)p + v_a \left(\sum_{i=k}^{ln} \binom{ln}{i} (-a)^{ln-i} \frac{i!}{(i-k)!} T^{i-k} \right) q$$

$$+ v_a \left((-1)^k \sum_{j=k}^{m} a_j \binom{j}{k} T^{m-j} \right) q$$

The first value we need to know is $ln - k$ since we recognize the kth-derivative of $(T-a)^{ln}$. The second value equals:

$$v_a \left(\frac{1}{k!} (-1)^k \sum_{j=k}^{m} a_j \frac{j!}{(j-k)!} T^{-j-1} \right)$$

$$= v_a \left(\sum_{j=k}^{m} a_j (k - j - 1) \ldots (-j+1)(-j) T^{-j-1} \right)$$

$$= v_a \left(\frac{a^k}{dT^k} \left(\sum_{i=0}^{m} a_j T^{k-j-1} \right) \right)$$

$$\geq v_a \left(\sum_{j=0}^{m} a_j T^{-j} \right) - k = v_a(\sigma_B(g)) - k$$

So we obtain for all k that: $v(R_k) = v(g) - 2k(p+q)$ which is strictly larger than $v(g)$ if $p + q < 0$. Thus if $p + q < 0$ then $v(\frac{f}{g} - G^n) > 0$.

For the case where $p + q = 0$ and $\sigma_B(f/g) = \sigma_B(t^n)$ we finish the proof by induction on n. Observe that $t = x - ay$ and $k = l - 1$. If $n = 0$ then $\sigma_B(f) = \sigma_B(g)$ and thus $f = g$ since both f and g are pseudo homogeneous. Then suppose $\sigma_B(f/g) = \sigma_B(t^n)$. Previous calculation establishes that: $\frac{f}{g} - t^n = g^{-1} \left(\sum_{k=1}^{\min(n,m)} R_k \right)$ plus terms of strictly positive value.

For each k there is a $\gamma_k \in K(t)$ such that: $v(g^{-1} R_k - \gamma_k) > 0$ either by induction in case $\sigma_B(g^{-1} R_k) = \sigma_B(\beta t^{n-2k})$ for some $\beta \in K^*$ or by the first part of the proof in case $\sigma_B(g^{-1}R_k) \neq \sigma_B(\beta t^{n-2k})$ for any $\beta \in K^*$. Finally we arrive at:

$$\left(t^n + \sum_{k=1}^{\min(n,m)} \gamma_k\right) \in K(t) \text{ and } v(\frac{f}{g} - (t^n + \sum_{k>1}^{\min(n,m)} \gamma_k)) > 0$$

□

3.3 Some Divisor Theory for Weyl Fields Over Function Fields

In this section we let K be an algebraic function field of degree one over an algebraically closed $k \subset K$ of characteristic zero, i.e. K is the function field of a nonsingular projective curve C over k.

Points on the curve C correspond bijectively to the discrete k-valuations of K and each such valuation induces a valuation filtration f^vK on K. This filtration extends to $\mathbb{A}_1(K)$ and to $f^v\mathbb{D}_1(K)$ as observed earlier. The associated graded ring of $f^v\mathbb{D}_1(K)$ is exactly $\mathbb{D}(k)[T, T^{-1}]$, T a central variable. Hence $f^v\mathbb{D}_1(K)$ is a valuation filtration corresponding to a discrete noncommutative valuation ring $f_0^v\mathbb{D}_1(K)$. In a sense the constant field k is now replaced by $\mathbb{D}_1(k)$ but we will point out some essential new features related to this "skewfield of constants".

3.3.1 Proposition

If v is a $\mathbb{D}_1(k)$-valuation of $\mathbb{D}_1(K)$, then:

for $a_{ij} \in K, i = 0\dots, n, j = 0, \dots, m$:

$$v\left(\sum_{i,j=0}^{n,m} a_{ij}x^iy^j\right) = \min\{v(a_{ij}); i = 0, \dots, n, j = 0, \dots, m\}$$

Proof. Write $\underline{a} = \sum_{i=0}^n \sum_{j=0}^m a_{ij}x^iy^j, p = \min\{v(a_{ij}), i = 0, \dots, n, j = 0, \dots, m\}$.

Since $x, y \in \mathbb{D}_1(k)$ the equality $v(\underline{a}) \geq p$ is obvious. Let $\delta(\underline{a}) \in \mathbb{N}$ be the filtration degree of $\underline{a}$ in the Bernstein filtration of $\mathbb{A}_1(K)$. If $\delta(\underline{a}) = 0$ then the claim holds. So we assume that the claim holds for $\underline{b} \in \mathbb{A}_1(K)$ with $\delta(\underline{b}) < d$. We calculate:

$$\begin{aligned} v(x\underline{a} - \underline{a}x) &= v\left(\sum_{i=0}^{n}\sum_{j=0}^{m} ja_{ij}xy^{j-1}\right) \\ &= \min\{v(a_{ij}), j \neq 0\} \end{aligned}$$

$$v(y\underline{a} - \underline{a}y) = v\left(-\sum_{i=0}^{n}\sum_{j=0}^{m} i a_{ij} x^{i-1} y^j\right)$$
$$= \min\{v(a_{ij}, i \neq 0\}$$

Now $v(x\underline{a} - \underline{a}x) \geq v(\underline{a}) \geq p$ and similarly $v(v\underline{a} - \underline{a}y) \geq p$. The foregoing implies that $v(\underline{a})$ is smaller than $\min\{v(y\underline{a} - \underline{a}y), v/(\underline{a} - \underline{a}y)\}$, hence $v(\underline{a}) \leq p$ or $v(\underline{a}) = p$ follows. □

Since we assume that k is algebraically closed in K we have that $k = \cap_v\{O_v, O_v$ a discrete valuation ring of $K\}$. We may look at the ring R obtained as the intersection $\cap_v \Lambda_v$ for all discrete $\mathbb{D}_1(k)$-valuation rings of $\mathbb{D}_1(K)$. For a k-valuation O_v of K we have a nontrivial $\mathbb{D}_1(k)$-valuation Λ_v extending v to $\mathbb{D}_1(K)$ it is given by v on $\mathbb{A}_1(K)$ as in the foregoing proposition.

We have that $\mathbb{D}_1(k) \subset R$ but R is not equal to $\mathbb{D}_1(k)$. Look at $(X + a)^{-1}$ since $v(X + a)$ is at most zero we have that $v(X + a)^{-1}$ is at least zero, hence $(X + a)^{-1} \in R$ for every $a \in K - k$. Clearly $(X + a)^{-1} \in R$ is not invertible in R so R is not a skewfield, yet it has many invertible elements e.g. $(ax + y)(x + ay)^{-1}$. As an intersection of valuation rings R has the property that every one-sided ideal of R is an ideal of R, moreover R is invariant under inner automorphism of $\mathbb{D}_1(K)$. A formal sum $D = \sum'_{v \in C} n_v v$ with $n_v \in \mathbb{Z}$ almost all being zero, is called a **divisor** of C. When we chose Λ_v to represent v of C (C the curve of $K(k)$) we define a **divisor for** $\mathbb{D}_1(K)$ as $\sum'_v n_v v$. To an element $q \in \mathbb{D}_1(K)$ we associate a principal divisor: $\mathrm{div}(q) = \sum_v v(q)v$. Divisors for $\mathbb{D}_1(K)$ are partially ordered by: $\sum'_v n_v v > \sum'_v m_v r$ if and only if $n_v \geq m_v$ for all v. So the ring R may be obtained by putting $R = \{q \in \mathbb{D}_1(K), \mathrm{div}(q) \geq 0\}$. A divisor for $\mathbb{D}_1(K), D$ is said to be **positive** if $D > 0$.

3.3.2 *Lemma*

A positive divisor is principal, i.e. if $D > 0$ then $D = \mathrm{div}(r)$ for some $r \in R$.

Proof. It will be sufficient to establish that for each $v \in C$ there is an $r \in R$ such that $\mathrm{div}(r) = v$, that is $v(r) = 1$ and $w(r) = 0$ for every $w \neq v$ in C. Take $a \in K$ such that $v(a) = -1$ and write $\mathrm{div}(a) = D_1 - D_2$ with D_1 and D_2 being both positive divisors. The Riemann–Roch theorem on C yields the existence of an integer N such that for any divisor D on C of degree (being the sum of the n_v appearing in the divisor) larger than N we have: $\dim_k L(D) = \deg D + 1 - g$, where $L(D) = \{f \in K, \mathrm{div}(f) + D > 0\}$ and g is the genus of the curve C. Fix a positive divisor D_3 of degree larger than N such that every valuation with nonzero coefficient in D_2 appears with zero coefficient in D_3 (note: v appears in D_2). Then it is easily checked that:

$$\dim_k L(D_3 + v) = 1 + \dim_k L(D_3)$$

So there must be a nonzero b in $L(D_3+v)-L(D_3)$. Consequently $v(b)=-1=v(a)$; moreover for all $w \in C-\{v\}$ we have that $\max\{w(a),w(b)\} \geq 0$. Now define: $r=(a^{-1}x+b^{-1})(a^{-1}x+b^{-1}+y)$ in $\mathbb{D}_1(K)$. If w is a k-valuation of K different from v, then:

$$w(a^{-1}x+b^{-1}) = \min\{w(a^{-1}),w(b^{-1})\}$$
$$= -\max\{w(a),w(b)\}$$
$$w(a^{-1}X+b^{-1}+y) = \min\{w(a^{-1}),w(b^{-1}),0\}$$
$$= -\max\{w(a),w(b)\}$$

Thus from $w \neq v$ we obtain $w(r)=0$. On the other hand

$v(a^{-1}x+b^{-1}) = \min\{1,1\} = 1$
$v(a^{-1}x+b^{-1}+y) = \min\{1,1,0\} = 0$

Hence $v(r)=1$ and $\mathrm{div}(r)=v$ as desired. □

This leads to a rather beautiful structure result on R.

3.3.3 Theorem

The ring R is a principal ideal domain.

Proof. As a first step we establish that the sum of two cyclic ideals is again cyclic. Hence consider Ra and Rb. For all $v \in C$ and $r,r' \in R$ we have: $v(ra+r'b) \geq \min\{v(ra),v(r'b)\} \geq \min\{v(a),v(b)\} = v(c)$ for some $c \in R$. Thus $ra+r'b \in Rc$ or $Ra+Rb \subset Rc$. Write $a=fg^{-1}, b=f'g^{-1}$ with f,f',g in $\mathbb{A}_1(K)$.

Let us fix an integer n larger than the Bernstein filtration degree of f and put: $b'=x^n f'g^{-1}$. Then $Rb=Rb'$ and

$$v(a+b') = v(f+x^n f') - v(g)$$
$$= \min\{v(f),v(f')\} - v(g)$$
$$= \min\{v(a),v(b)\} = v(c)$$

Consequently: $Rc \subset R(a+b') \subset Ra+Rb' \subset Ra+Rb$, hence $Ra+Rb=Rc$. From the first step it follows that every finitely generated ideal is cyclic and for every $b \notin Ra$ we have $Ra+Rb=Rc$ with $0<\mathrm{div}(c)<\mathrm{div}(a)$. Consequently any ascending chain of (left, right) ideals must terminate thus R is Noetherian and then every ideal is finitely generated (on the left) hence principal.

3.3.4 Corollary

The ring R is a Noetherian domain with quotient division ring $\mathbb{D}_1(K)$.

Proof. As observed in the theorem R is a Noetherian domain hence it has a classical ring of fractions which is a skewfield. If $q \in \mathbb{D}_1(K)$, then $\mathrm{div}(q) = D_1 - D_2$ for positive divisors D_1, D_2. From the lemma it follows that there are r_1 and r_2 in R such that $\mathrm{div}(r_1) = D_1, \mathrm{div}(r_2) = D_2$. It follows that $qr_2r_1^{-1} \in R$ is invertible in R as it has the zero divisor for its divisor. Therefore $q \in \mathbb{Q}_{\mathrm{cl}}(R)$ and $\mathbb{D}_1(K) = \mathbb{Q}_{\mathrm{cl}}(R)$ follows. □

To a divisor D on $\mathbb{D}_1(K)$ we associate a space $\mathcal{L}(D) = \{q \in \mathbb{D}_1(K), \mathrm{div}(q) + D > 0\}$. In particular $\mathcal{L}(0) = R$ and each $\mathcal{L}(D)$ is an R-bimodule. The ideals of R are exactly the $\mathcal{L}(D)$ with $D < 0$. The theorem states that any $\mathcal{L}(D) \cong R$ as an R-bimodule. For $D_1 < D_2$ we have $\mathcal{L}(D_1) \subset \mathcal{L}(D_2)$. We let $\Lambda_v = F_0^v\mathbb{D}_1(K)$ be the discrete valuation ring of $\mathbb{D}_1(K)$ corresponding to $F^v\mathbb{D}_1(K)$ and we write $\pi_v \subset \Lambda_v$ for its unique maximal ideal, $m_v = R \cap \pi_v$.

Observe that $m_v = \mathcal{L}(-v)$ is a maximal ideal of R and the correspondence $v \in C \to m_v \subset R$ defines a bijective correspondence between points of C and the set of maximal ideals $\Omega(R)$ of R. The following expands on this relation.

3.3.5 Proposition

With notation as above: $\Lambda_v = R_{m_v}$, the localization of R at the maximal ideal m_v.

Proof. For $q \in \Lambda_v$ write $\mathrm{div}(q) = D_1 - D_2$ where D_1 and D_2 are positive divisors having disjoint supports (no valuation appears with a nonzero coefficient in both D_1 and D_2, this is of course always possible). Again we find r_1 and r_2 in R such that $\mathrm{div}(r_1) = D_1, \mathrm{div}(r_2) = D_2$. Consequently $q = ur_1r_2^{-1}$ for some unit u of R. Since $v(q) \geq 0$ we cannot have $r_2 \in m_v$, hence $q \in R_{m_v}$. On the other hand the $R_{m_v} \subset \Lambda_v$ is obvious so equality follows: □

3.3.6 Theorem

If $D_1 < D_2$ then $\dim_{\mathbb{D},(k)}(\mathcal{L}(D_2)/\mathcal{L}(D_1))$ equals the degree of the divisor $D_2 - D_1$.

Proof. Form the foregoing proposition it follows that π_v is the extension of $m_v \subset R$ to Λ_v and $R/m_v \to \Lambda_v/\pi_v$ is an isomorphism. In particular $R/m_v \cong \mathbb{D}_1(k)$ and $R = \mathbb{D}_1(k) + m_v$. Let us write F^vR for the filtration induced on R by $F^v\mathbb{D}_1(K)$. For the associated graded rings we have: $G_v(\mathbb{D}_1(K)) \cong \mathbb{D}_1(k)[T, T^{-1}]$ and we may restrict this isomorphism to the associated graded ring of F^vR and obtain $G_v(R) \cong \mathbb{D}_1(k)[T^{-1}]$. It is straightforward to verify for every $v \in C$ and every divisor D

on $\mathbb{D}_1(K)$ we have: $\mathcal{L}(D) = m_v\mathcal{L}(D+v)$. Then it follows from this that $\mathcal{L}(D+v)/\mathcal{L}(D) \cong R/m_v \otimes_R \mathcal{L}(D+v) \cong R/m_v \cong \mathbb{D}_1(k)$. For $D_1 < D_2$ the left dimension over $\mathbb{D}_1(k)$ of the space $\mathcal{L}(D_2)/\mathcal{L}(D_1)$ may thus be counted as the degree of the divisor $D_2 - D_1$ (note that in a similar way this degree also equals the right $\mathbb{D}_1(k)$-dimension, so that left and right dimension of the bimodule $\mathcal{L}(D_2)/\mathcal{L}(D_1)$ are actually equal in this situation). □

The foregoing Riemann–Roch type theorem is independent of the genus of C, the formula proved actually corresponds to stating that this is a "genus-less" situation, a remark that may be related to the noncommutative geometry of $\mathbb{A}_1(K)$.

If we consider the Bernstein filtration then $\mathbb{D}_1(k)$ is not in $F_0^B R = F_0^B\, \mathbb{D}_1(K) \cap R$. We may compare the Bernstein filtrations on $\mathbb{D}_1(K)$ and R; in some sense R is a rather big subring of $\mathbb{D}_1(K)$, perhaps unexpected for the intersection of all $\mathbb{D}_1(k)$-valuation (discrete) rings of $\mathbb{D}_1(K)$.

3.3.7 *Proposition*

With respect to the Bernstein filtrations: $G_B(R) = G_B(\mathbb{D}_1(K))$.

Proof. It is known that $G_B(\mathbb{D}_1(K))$ is just the graded quotient field of $G_B(\mathbb{A}_1(K)) = K[X,Y]$. Consider p in $K[X,Y]_m$, $p = \sum_{i=0}^m a_i XY^{m-i}$ and look at $p^{-1} \in G_B(\mathbb{D}_1(K))$. Consider $q = \sum_{i=0}^m a_i xy^{m-i} \in \mathbb{A}_1(K)$ and write $\mathrm{div}(q) = D_1 - D$ with D_1 and D_z positive. Pick $r \in R$ such that $\mathrm{div}(r) = D_2$, say $r = fg^{-1}$ with $f, g \in \mathbb{A}_1(K)$ and let $n \in \mathbb{N}$ be larger than the degree of g in the Bernstein filtration. Now put $a = (rx^n q))(rx^n + 1)^{-1}$. Obviously $\sigma(a) = p$. For all $v \in C$ we obtain: $v(rx^n + 1) = \min\{v(r), 0\}$ because of the choice of n. Then we obtain $\mathrm{div}(a) = D_1$ and thus $a \in R$. Finally, if $m > 0$ then $r' = (q+1)^{-1}$ is in R and we have $\sigma(r') = p^{-1}$. Thus $Q^g_{\mathrm{Cl}}(K[X,Y]) \subset G(R)$ and also $G(\mathbb{D}_1(K)) = Q^g_{\mathrm{Cl}}(K[X,Y])$ entail $G(R) = G(\mathbb{D}_1(K))$. □

The results may be generalized to $\mathbb{D}_n(k)$-valuations of $\mathbb{D}_n(K)$ but we do not go into this here leaving it as an exercise for the zealous reader.

3.4 Hopf Valuation Filtration

The guiding principle in foregoing sections is that an extension of valuation theory to K-algebras can be obtained from a value function on A extending a valuation v of K with corresponding filtrations FA, resp. fK. The ring $F_o A$, where 0 is the neutral element of the value group Γ, is an order in A over O_v, and it enjoys certain properties like being a separated prime, a Dubrovin valuation, a noncommutative valuation ring, depending on properties of the value function. However we may look at a strong filtration FA and ask other structural properties of A possibly in combination with some properties of the value function, e.g. we may look at Hopf algebras or quantum groups A and ask F_0A to be also a Hopf algebra over O_v. In this way we shall define Hopf valuations and their filtrations.

3.4.1 Definition. Good Γ-Filtrations on K-Vector Spaces

Consider a field K with a separated Γ-filtration fK for some totally ordered group Γ. The category of Γ-filtered vector spaces over K is denoted by K-filt. A Γ-filtration on a K-vector space V, FV is a **good filtration** if there exist sets $\{v_\alpha, \alpha \in \mathcal{A}\}, \{\gamma_\alpha \in \Gamma, \alpha \in \mathcal{A}\}$ such that for $\gamma \in \Gamma$ we have: $F_\gamma V = \sum_{\alpha\in\mathcal{A}} f_{\gamma-\gamma_\alpha} K v_\alpha$. It is clear that $\{v_\alpha, \alpha \in \mathcal{A}\}$ is a set of K-generators for V. In the sequel fK will be a strong filtration, in fact a valuation filtration. Then from $f_{\gamma-\gamma_\alpha} K v_\alpha \in F_\gamma V$ it follows that $v_\alpha \in F_{\gamma_\alpha} V$ for all $\alpha \in \mathcal{A}$; moreover, for every $\gamma \in \Gamma$ we also have that $F_\gamma V = f_\gamma K F_0 V$. Hence, if FV is a good filtration then without loss of generality we may assume that $\{v_\alpha, \alpha \in \mathcal{A}\}$ is taken in F_0V and for $\gamma \in \Gamma$, $F_\gamma V = \sum_{\alpha\in\mathcal{A}} f_\gamma K v_\alpha$. If F_0V is free over F_0K with basis $\{w_i, i \in \mathcal{J}\}$ then FV may be given by $F_\gamma V = \sum_{i\in\mathcal{J}} f_\gamma K w_i$. If fK is a valuation filtration then it is strong and $f_0K = O_v$ is a valuation ring so torsion free finitely generated O_v-modules will be free.

For detail on Hopf algebras we refer to [20, 31]. We let H be a K-Hopf algebra with counit $\varepsilon : H \to K$, comultiplication $\Delta : H \to H \otimes H$ and antipode $S : H \to H$. A **Γ-filtered Hopf algebra** is a K-Hopf-algebra H with a filtration FH such that ε, Δ, S are filtered morphisms, e.g.

1. $\varepsilon(F_\gamma H) \subset F_\gamma K = f_\gamma K$, for all $\gamma \in \Gamma$.
2. $S(F_\gamma H) \subset F_\gamma H$, for all $\gamma \in \Gamma$.
3. $\Delta(F_\gamma H) \subset \sum_{\sigma+\tau=\gamma} F_\sigma H \otimes F_\tau H$, for all $\gamma \in \Gamma$.

The condition (3) just expresses that Δ is a filtered morphism if $H \otimes H$ is equipped with the tensor filtration defined by putting

$$F_\gamma(H\otimes H) = \sum_{\sigma+\tau=\gamma} F_\sigma H \otimes F_\tau H, \gamma \in \Gamma.$$

We write $F_0^0 H = \sum_{\gamma<0} F_\gamma H$, $F_\tau^0 H = \sum_{\gamma<\tau} F_\gamma H$ for $\tau \in \Gamma$.

3.4.2 Proposition

Let H be a Hopf algebra over K with Hopf filtration FH, then $G(H)$ is a Γ-graded Hopf algebra. If FH extends fK then $G(H) = k\Gamma \otimes_k F_0H$ with Hopf structure deriving from F_0H (via F_0H/F_0^0H) making it into a graded Hopf algebra over the gr-field $k\Gamma$ (where k is the residue field of K).

Proof. Observe that F_0H is a sub-Hopf algebra of H over $f_0K = O_v$. Indeed, $\Delta(F_0H) \subset \sum_{\gamma\in\Gamma} F_\gamma H \otimes F_{-\gamma}H$ but since $F_\gamma H = f_\gamma K \otimes F_0H$ for all $\gamma \in \Gamma$, it follows that $\Delta(F_0H) \subset F_0H \otimes F_0H$; the restriction of ε to F_0H defines the O_v-linear $\varepsilon|F_0H$ and the $k\Gamma$-linear $\overline{\varepsilon}$ on $G(H)$ extending $F_0H/F_0^0H \to K$. Since $S|F_0H$ defines the $k\Gamma$-linear $\overline{S} : G(H) \to G(H)$, all claims in the proposition follows easily. □

Note that for a Hopf filtration FH the inclusion $K \hookrightarrow H$ is a filtered morphism; for a strong Hopf filtration FH the condition of extending fK is equivalent to $F_0H \cap K = f_0K$.

Now we consider a Hopf algebra H over K with a Γ-valuation ring $O_v = D$ in K; the valuation of O_v is $v : K^* \to \Gamma$ and we also write $v : K \to \Gamma \cup \{\infty\}$ by putting $v(0) = \infty$. The residue field of v will be denoted by k.

A **Hopf valuation function** extending v is a function $-\xi : H \twoheadrightarrow \Gamma \cup \{\infty\}$, usually viewed as a **Hopf valuation filtration function** $\xi : H \twoheadrightarrow \Gamma \cup \{-\infty\}$, satisfying:

HV.1 We have $\xi(h) = -\infty$ if and only if $h = 0$.
HV.2 We have $\xi(1) = 0$.
HV.3 For $h \in H, \lambda \in K, \xi(\lambda h) = \xi(\lambda h) = \xi(h) - v(\lambda)$.
HV.4 For $g, h \in H, \xi(gh) \leq \xi(g) + \xi(h)$.
HV.5 For $g, h \in H, \xi(g + h) \leq \max\{\xi(g), \xi(h)\}$.
HV.6 For $h \in H, \xi(S(h)) \leq \xi(h), \xi(\varepsilon(h)) \leq \xi(h)$.
HV.7 For $h \in H, \Delta(h) = \Sigma h_1 \otimes h_2$ (Sweedler notation) $\xi(h) \geq \inf\{\max_\Sigma\{\xi(h_1) + \xi(h_2)\}\}$, where $\max_\Sigma$ is taken over the terms in a fixed expression of $\Delta(h)$ while inf is over all possible decompositions of $\Delta(h)$. By $\xi(h) \geq \inf\{\gamma, \gamma \in \mathcal{A} \subset \Gamma\}$ we just mean that if $\sigma \leq \gamma$ for $\gamma \in \mathcal{A}$ then $\xi(h) \geq \sigma$.

3.4.3 Proposition

If ξ is a Hopf valuation function then we have equality in HV.7, in fact:

$$\begin{aligned}\xi(h) &= \inf\{\max_\Sigma\{\xi(h_1) + \xi(h_2)\}\} = \inf\{\max_\Sigma\{\xi(\varepsilon(h_1)) + \xi(h_2)\}\} \\ &= \inf\{\max_\Sigma|\xi(h_1) + \xi(\varepsilon(h_2))\}\end{aligned}$$

Proof. Indeed, from $\Delta(h) = \Sigma h_1 \otimes h_2$ we may derive: $h = \Sigma\varepsilon(h_1)h_2 = \Sigma h_1\varepsilon(h_2)$. Applying HV.3 leads to:

$$\xi(h) \leq \max_\Sigma\{\xi(\varepsilon(h_1)h_2)\} = \max_\Sigma\{\xi(\varepsilon(h_1)) + \xi(h_2)\} \qquad (*)$$

Since (*) holds for any decomposition of $\Delta(h)$ we obtain:

$$\xi(h) \leq \inf\{\max_\Sigma\{\xi(\varepsilon(h_1)) + \xi(h_2)\}\}$$

and by this we just mean that $\xi(h) \le \max_\Sigma(\ldots)$ for all possible decompositions of $\Delta(h)$. Now using HV.6 this leads to: $\xi(\varepsilon(h_1)) + \xi(h_2) \le \xi(h_1) + \xi(h_2)$, or $\xi(h) \le \inf\{\max_\Sigma(\xi(\varepsilon(h_1)) + \xi(h_2)\}$. This proves:

$$\begin{aligned}\xi(h) &= \inf\{\max_\Sigma\{\xi(h_1) + \xi(h_2)\}\}\\ &= \inf\{\max_\Sigma\{\xi(\varepsilon(h_1)) + \xi(h_2)\}\}\\ &= \inf\{\max_\Sigma\{\xi(h_1) + \xi(\varepsilon(h_2))\}\}\end{aligned}$$

where now we may interpret inf in the classical way, because $\xi(h)$ is smaller than all $\max\{\ldots\}$ and bigger than all $\sigma \in \Gamma$ smaller than all $\max_\Sigma\{\ldots\}$. Hence HV.7 entails the existence of the inf as defined. The last equality following by using $h = \Sigma h_1 \varepsilon(h_2)$. □

3.4.4 Theorem

Hopf filtration functions $\xi : H \to \Gamma \cup \{-\infty\}$, satisfying HV.1,...,HV.7, correspond bijectively to the separated Hopf filtrations FH extending the valuation filtration fK of the valuation v.

Proof. Start from a Hopf valuation filtration function $\xi : H \twoheadrightarrow \Gamma \cup \{-\infty\}$ satisfying HV.1,...,HV.7. For $\gamma \in \Gamma$ put $F_\gamma H = \{h \in H, \xi(h) \le \gamma\}$ Properties HV.5 and HV.3 entail that $F_\gamma H$ is an additive subgroup of H, containing 0 because of HV.1. From HV.2, HV.4 and HV.5 it follows that FH is a filtration of the ring H. Putting $h = 1$ in HV.3 entails $\xi(\lambda) = -v(\lambda)$ for $\lambda \in K$, hence FH extends the valuation filtration fK corresponding to v. For $h \in H$ we have $h \in F_{\xi(h)} H$, hence FH defines an exhaustive filtration of H. From HV.6 we obtain that S and ε are filtered maps of degree zero with respect to FH. In case FH would not be separated, then there is a nonzero $z \in H$ such that for every $\gamma \in \Gamma$ such that $z \in F_\gamma H$ we have $z \in F_\gamma^0 H$. In any case $z \in F_{\xi(z)} H$ but if $z \in F_\gamma H$ with $\gamma < \xi(z)$, then by definition of $F_\gamma H$ it means that $\xi(z) \le \gamma$, contradiction. Thus FH is separated. Now consider $h \in H, \Delta(h) = \Sigma h_1 \otimes h_2$, then Proposition 3.4.3 entails: $\xi(h) = \inf\{\max_\Sigma\{\xi(h_1) + \xi(h_2)\}\}$. For $\Delta(F_\gamma H) \subset \sum_{\tau\in\Gamma} F_\tau H \otimes F_{\gamma-\tau} H$ it suffices to establish this for $\gamma = \xi(h)$, any $h \in H$ (indeed ξ like v is assumed to be surjective). If $\delta > \xi(h) = \gamma$, then for some decomposition $\Delta(h) = \sum h_1 \otimes h_2$ we have $\delta \ge \max_\Sigma\{\xi(h_1) + \xi(h_2)\}$, and this comes down to:

$$\Delta(h) \in \sum_{\tau\in\Gamma} F_\tau H \otimes F_{\delta-\tau} H \qquad (*)$$

Recall that FH is a strong filtration hence for every $\gamma \in \Gamma$, $F_\gamma H = F_\gamma K F_0 H = f_\gamma K F_0 H$. For the tensor filtration on $H \otimes H$ defined by FH we have:

$$F_\gamma(H \otimes H) = \sum_{\sigma\in\Gamma} F_{\gamma-\sigma} H \otimes F_\sigma H = \sum_{\sigma\in\Gamma} F_{\gamma-\sigma} K F_\sigma K F_0 H \otimes F_0 H = F_\gamma H \otimes F_0 H$$

From $(*)$ we obtain $\Delta(h) \in F_\delta H \otimes F_0 H$ for every $\delta > \xi(h)$. As a D-module $F_0 H$ is flat, indeed over a valuation domain every finitely generated torsion free modules is projective and every projective is free. Thus $F_0 H$ is the direct limit of free modules of finite rank, thus flat. Consequently: $\cap_\delta (F_\delta H \otimes_D F_0 H) = (\cap_\delta F_\delta) \otimes_D F_0 H$. Either $\xi(h)$ is equal to some $\max_\Sigma\{\xi(h_1) + \xi(h_2)\}$ for a certain decomposition $\Delta(h) = \Sigma h_1 \otimes H_2$ in which case $(*)$ applies with $\delta = \xi(h)$ and there is nothing left to prove, or else $\xi(h)$ appears as the inf of elements $\max_\Sigma\{\xi(h_1) + \xi(h_2)\} \in \Gamma$ (observe that $\max_\Sigma$ is over a finite set). In view of the remark after this proof, separatedness of FH yields $F_{\xi(h)} = \cap_{\delta > \xi(h)} F_\delta H$. Therefore we obtain $\Delta(h) \in F_{\xi(h)} H \otimes F_0 H$ and it follows that FH is a Hopf filtration. Conversely if FH is a separated Hopf filtration extending fK_0, then for $x \neq 0$ in H there is a unique $\gamma \in \Gamma$ such that $x \in F_\gamma H - F_\gamma^0 H$. The function $\xi : H \to \Gamma \cup \{-\infty\}$ defined by $\xi(0) = -\infty$ and for $x \neq 0$, $\xi(x) = \inf\{\tau \in \Gamma, \inf_\tau H\}$ is well-defined and surjective since FH extends fH and v is surjective. We may view $\xi(x)$ as the "filtration degree" with respect to FH. Verifying the properties HV.1,…,HV.7 is easy enough. □

3.4.5 Remark

If FR is a separated filtration on a ring R and $\mathcal{A}$ is a subset of Γ such that $\gamma = \inf\{\alpha, \alpha \in \mathcal{A}\} \in \Gamma \cup \{-\infty\}$, then $F_\gamma R = \cap_{\alpha \in \mathcal{A}} F_\alpha R$. Indeed if $x \notin F_\gamma R$ there is a unique $\delta \in \Gamma$ such that $x \in F_\delta R$ but $x \notin F_{\delta'} R$ for any $\delta' < \delta$ in view of the separatedness. Then $\gamma < \delta$ because $\delta \leq \gamma$ and $x \in F_\delta R$ leads to $x \in F_\gamma R$ which is excluded. Hence $\alpha_0 < \delta$ for sone $\alpha_0 \in \mathcal{A}$ and thus $x \notin F_{\alpha_0} R$ yields $x \notin \cap_{\alpha \in \mathcal{A}} F_\alpha R$.

From now on we consider a separated Hopf filtration FH extending fK associated to a valuation v of R with associated Hopf valuation function ξ. We have seen that $F_0 H$ is a Hopf algebra over $D = f_0 K$ and it is an order of H in the sense that $KF_0 H = K \otimes_D F_0 H = H$. For $h \in H$ we define $I_h \subset K$ by putting $I_h = \{\lambda \in K, \lambda h \in F_0 H\}$. The next proposition establishes that ξ may be calculated from data in K.

3.4.6 Proposition

For $h \in H$, $\xi(h) = v(I_h)$, in particular for $h = \lambda \in K, \xi(\lambda) = -v(\lambda)$. If ξ_1, ξ_2 correspond to Hopf filtrations $F^1 H$ resp. $F^2 H$, then $F_0^1 H \subset F_0^2 H$ is equivalent to $\xi_2 \leq \xi_1$.

Proof. Let FH be the Hopf filtration corresponding to ξ; note that surjectivity of ξ implies $F_\gamma H \neq F_\tau H$ for $\gamma \neq \tau$ in Γ. Indeed if $\sigma \in \Gamma$ hen $\sigma = \xi(z)$ for some $z \in H$ and $z \in F_{\xi(z)} H$ but $z \notin F_{\xi(h)} H$, i.e. $F_\sigma H \neq F_\tau H$ for all $\tau < \sigma$. Now take $h \in H$ then $h \in F_\gamma H - F_\gamma^0 H$ for some $\gamma \in \Gamma$, in fact $\gamma = \delta(h)$. If $\lambda \in K$ is such that $\lambda h \in F_0 H$ then $\xi(\lambda h) = \xi(h) - v(\lambda) \leq 0$. From $\gamma = \xi(h) \leq v(\lambda)$ it follows that $\xi(\lambda) \leq -\gamma$ or $I_h \subset f_{-\gamma}K$. Since $f_{-\gamma}K$ is obviously in $I_h = f_{-\gamma}K$, so $v(I_h) = \gamma = \delta(h)$ follows (in general we may define $v(L)$ for an O_v-submodule L of K as $-\max\{\sigma_v(\lambda), \lambda \in L\}$ if the maximum exists in Γ where σ_v is the principal symbol map for fK). For the second statement observe that F^1H and F^2H are strong filtrations hance $F_0^1H \subset F_0^2H$ entails $F_\gamma^1H \subset F_\gamma^2H$ for all $\gamma \in \Gamma$ and $\xi_2 \leq \xi_1$ follows. Conversely from $\xi_2 \leq \xi_1$ it follows that $F_0^1H = \{h \in H, \xi_1(h) \leq 0\} \subset F_0^2H = \{h \in H = \{h \in H, \xi_2(h) \leq 0\}$. □

For $h \in F_0H$ we have $\varepsilon(h) \in D$ and if $v(\varepsilon(h)) = \gamma$ then we may divide h by $\lambda \in K$ with $v(\lambda) = \gamma$ and we still have that $\varepsilon(\lambda^{-1}) \in D$ but we do not know whether $\lambda^{-1}h \in F_0H$. If a suitable set of K-generators for H, B say, can be selected such that the D-module generated by $\{\lambda_i^{-1}h_i, h_i \in B\}$ is a D-ring then we may obtain a method to construct D-orders in H. Elements of H which are candidates for the ones with best divisibility properties are those $h \in H$ such that $\varepsilon(h) = 0$, i.e. the elements of the augmentation ideal. The advantage of Theorem 3.4.4 is that ξ is known if we know the Hopf order F_0H and conversely. In case $H = KG$ for a finite group G the knowledge of a Zassenhaus valuation on G does not determine unambiguously a Hopf valuation on KG nor a Hopf order of KG. Indeed the construction of Larson orders different from DG in KG, cf.[35], exactly shows that different orders may be constructed from the same Zassenhaus valuation of G. Theorem 3.4.4 applied to $H = KG$ entails that DG and a nontrivial Larson order $\mathcal{L}$, $DG \subsetneq \mathcal{L} \subsetneq KG$, correspond to different Hopf valuation functions on KG but these take the same values on some K-basis of the Hopf algebra, e.g. G in KG. It turns out that some basis is better than another! In the sequel we obtain a construction method for (maximal) orders of Larson-type in any finite dimensional (semisimple) Hopf algebra, and a description in terms of some suitably selected basis; we include some examples with number theoretical flavour.

Consider the K-space H/K and define $d_\xi : H/K \to \Gamma \cup \{-\infty\}$ by putting $d_\xi(\overline{h}) = \xi(h - \varepsilon(h))$, where $\overline{h}$ is the class of h in H/K. We may define d_ξ on H by putting $d_\xi(h) = \xi(h - \varepsilon(h))$. Taking into account that $d_\xi(\lambda) = -\infty$ for every $\lambda \in K$. We call d_ξ the **derived valuation function** of ξ.

3.4.7 *Lemma*

With notation as above, either $\xi(h) = \xi(\varepsilon(h))$ or $\xi(h) = d_\xi(h)$, in other words $d_\xi(h) < \xi(h)$ only if $\xi(h) = \xi(\varepsilon(h))$ and $\varepsilon(h) \neq 0$.

Proof. From $h = (h - \varepsilon(h)) + \varepsilon(h)$ if $\varepsilon(h) \neq 0$, we obtain: $\xi(h) \leq \max\{d_\xi(h), \xi(\varepsilon(h))\}$. By definition of d_ξ we also have: $d_\xi(h) \leq \max\{\xi(h), \xi(\varepsilon(h))\}$. Combination of these inequalities yields either $\xi(h) = \xi(\varepsilon(h))$ or else $\xi(h) = d_\xi(h)$. The second statement in the lemma is now clear. Recall that ε is a filtered morphism, hence $\xi(\varepsilon(h)) \leq \xi(h)$. □

The properties of d_ξ are modifications of those of ξ.

3.4.8 Proposition

With notation and conventions as before the function d_ξ satisfies the following properties.

DV.1 For $h \in H$, $d_\xi(h) = -\infty$ if and only if $h \in K$.

DV.2 For $\lambda \in K, h \in H$, $d_\xi(\lambda h) = d_\xi(h) - v(\lambda)$.

DV.3 For $g, h \in H$, $d_\xi(gh) \leq \max\{d_\xi(g) + \xi(h), \xi(g) + d_\xi(h)\}$. In case $\varepsilon(g) = \varepsilon(h) = 0$, hen $d_\xi(gh) \leq d_\xi(g) + d_\xi(h)$.

DV.4 For $g, h \in H$, $d_\xi(g + h) \leq \max\{d_\xi(g), d_\xi(h)\}$.

DV.5 For $h \in H$, $d_\xi(Sh) \leq d_\xi(h)$.

DV.6 For $h \in H - K$ with $\varepsilon(h) = 0$ and $\Delta(h) = \Sigma h_1 \otimes h_h$, $d_\xi(h) \geq \inf\{\max_\Sigma\{d_\xi(h_1) + d_\xi(h_2)\}$.

Proof. The proof of DV.1, DV.2, DV.4 is straightforward.

DV.3 The first statement follows from:

$$gh - \varepsilon(gh) = (g - \varepsilon(g))h + \varepsilon(g)(h - \varepsilon(h))$$

Hence: $\xi(gh) \leq \max\{\xi(g - \varepsilon(g)h), \xi(\varepsilon(g))(h - \varepsilon(h))\}$. Now HV.4 and $\xi(\varepsilon(g)) \leq \xi(g)$ yields the statement. In case $\varepsilon(g) = \varepsilon(h) = 0$ we may assume $g \notin K$ and we have $d_\xi(g) = \xi(g)$, $d_\xi(h) = \xi(h)$ and HV.4 applies.

DV.5 Follows from $d_\xi(Sh) = \xi(Sh - \varepsilon(Sh)) \leq \xi(h - \varepsilon(h)) = d_\xi(h)$.

DV.6 Since $\varepsilon(h) = 0$, $d_\xi(h) = \xi(h)$ and the claim follows from [HV.7]. □

To a Hopf valuation filtration function δ there corresponds a Hopf order $H(\xi) = \{h \in H, \xi(h) \leq 0\} = F_0 H$, to d_ξ we may associate $H(d_\xi) = D \oplus \{h \in \varepsilon^{-1}(D) : -\infty < d_\xi(h) \leq 0\} \cup \{0\}$.

3.4.9 Observation

With notation as above: $H(d_\xi) = H(\xi)$. Indeed it is clear if $h \in H(d_\xi)$ with $h \in$ Kerε then $\xi(h) \leq 0$, hence $h \in H(\xi)$; if $h \notin$ kerε then $\xi(h) = \xi(\varepsilon(h))$ entails $\xi(h) \leq 0$ because $\varepsilon(h) \in D$. hence $H(d_\xi) \subset H(\xi)$. Conversely if $\xi(h) \leq 0$ then $\xi(h - \varepsilon(h)) \leq \max\{\xi(h), \xi(\varepsilon(h)\} = \xi(h) \leq 0$, thus $H(\xi) \subset H(d_\xi)$ follows, hence $H(\xi) = H(d_\xi)$.

3.4.10 Definition

A D-order in a finite dimensional K-algebra A, Λ say, is called a moderate order if it is integral over D and the prime radical rad(Λ) is a finite D-module.

Note

In the definition of moderate order given in [2] it is forgotten to remind that orders are assumed to be integral! The term "Hopf order" used before only refers to the property $KH(\xi) = H$ but from now on we shall restrict attention to ξ **corresponding to orders $H(\xi)$ which are integral over D.**

So we look at Hopf valuation functions $\xi \leq \xi_0$ with associated $H(\xi_0) \subset H(\xi)$ and assuming $H(\xi)$ is a moderated order. Let us write $m = m_v$ for the maximal ideal of $D = O_v$. Pick a K-basis $B = \{b_1 = 1, b_2, \ldots, b_n\}$ in $H(\xi_0)$ and we may assume without loss of generality that $\varepsilon(b_2) = \ldots = \varepsilon(b_n) = 0$.

Define $H_B(\xi)$ to be the D-algebra generated by $1, m^{d_\xi(b_i)} b_i$, $i = 1, \ldots, n$. Since $\xi(m^{d_\xi(b_i)} b_i) = 0$, it follows that $H_B(\xi) \subset H(\xi)$ hence $H_B(\xi)$ is integral over D. Since 1 and the $m^{d_\xi(b_i)} b_i$ for $i = 2, \ldots, n$ are again a K-basis for H it follows that $KH_B(\xi) = H$, hence $H_B(\xi)$ is a D-order of H.

We are now interested in the discrete case, so we suppose D is a discrete valuation ring of K from hereon, we may also assume $ch(K) = 0$ but that is not really necessary.

3.4.11 *Proposition*

With notation as before, if D is a discrete valuation ring of K then $H(\xi_0)$, $H_B(\xi)$, $H(\xi)$ are finite D-modules.

Proof. Since $K\mathrm{rad}H(\xi)$ is a nil ideal of H it is in $\mathrm{rad}H$ and therefore nilpotent hence $\mathrm{rad}H(\xi) = H(\xi) \cap \mathrm{rad}H$ and similar for $\mathrm{rad}H(\xi_0)$, $\mathrm{rad}H_B(\xi)$. Thus $\mathrm{rad}(\xi_0) = H(\xi_0) \cap \mathrm{rad}H$, $\mathrm{rad}H_B(\xi) = H_B(\xi) \cap \mathrm{rad}H(\xi)$ and we obtain $\overline{H(\xi_0)} \subset \overline{H(\xi)}$ and $\overline{H_B(\xi)} \subset \overline{H(\xi)} \subset \overline{H}$, where we denoted $\overline{R}$ for $R/\mathrm{rad}R$. Now $\overline{H}$ is semisimple Artinian with center $L_1 \oplus \ldots \oplus L_d$ say, each L_i the center of a simple component $\overline{S}_i$ of $\overline{H}$. The center of $\overline{H(\xi)}$, also for $\overline{H(\xi_0)}$ and $\overline{H_B(\xi)}$ is integral over D hence in the integral closure of D in $L_1 \oplus \ldots L_d$ and therefore it is a finitely generated D-module, thus also Noetherian. In any case for all three D-orders it follows by a result of G. Cauchon that they are finitely generated D-modules because they are finite over their center since it is Noetherian (P.I. rings with Noetherian center) and the center is a finite D-module as observed above. Since $\mathrm{rad}(\xi)$ is a finite D-module, so are $\mathrm{rad}H(\xi_0)$ and $\mathrm{rad}H_B(\xi)$, consequently the statement of the proposition follows. □

3.4.12 *Corollary*

There is a K-basis B in $H(\xi_0)$ such that $H_B(\xi) = H(\xi)$.

Proof. Since $H(\xi)$ is finitely generated as a D-module it is free of finite rank over D and it has a basis, B say. Without loss of generality we may assume $B = \{b_0 = 1, b_2, \dots, b_n\}$ with $\varepsilon(b_2) = \varepsilon(b_3) = \dots = \varepsilon(b_n) = 0$. Now (writing $m = (\pi) \subset D$) look at $\pi^{\beta_i} b_i$ where $\beta_i = \xi_0(b_i)$. This yields a K-basis in $H(\xi_0)$, say B_0, such that $H_{B_0}(\xi)$ contains B hence $D[B]$, consequently $H_{B_0}(\xi) = H(\xi)$. □

From the foregoing it follows that if there is a (finitely generated) Hopf order over D, say H', containing $H(\xi_0)$ then we may construct it from some K-basis contained in $H(\xi_0)$. Now let us forget $H(\xi)$, but start from a D-basis B of $H(\xi_0)$, this is also a K-basis for H, and make the D-order $H_B(\xi_0)$ and then try to check we obtain a finitely generated D-module which is a Hopf-order. In case $H(\xi_0)$ is not a maximal Hopf order, the foregoing corollary suggested we can find a maximal one by the H_B-construction but for some K-basis B in $H(\xi_0)$, not necessarily for a D-basis of $H(\xi_0)$! Nevertheless we shall show in a series of examples that this method leads to new Hopf orders in many cases.

3.4.13 Example. The Sweedler Hopf Algebra

Consider the Sweedler Hopf algebra over the rational fields Q, say $H = Q[x, y]$ with relations $x^2 = 1$, $y^2 = 0$ and $xy + yx = 0$, and put:

$$\varepsilon(x) = 1, S(x) = x, \Delta(x) = x \otimes x,$$
$$\varepsilon(y) = 0, S(y) = xy, \Delta(y) = 1 \otimes y + y \otimes x$$

Let $D \doteq \mathbb{Z}_p$ the localization of $\mathbb{Z}$ at the prime ideal (p) and $m = (p)$. Define $H(m) = \mathbb{Z}_p + \mathbb{Z}_p(x-1) + n^{-n}y + m^{-n}xy$, for every $n \geq 0$, $H(0) = H_{\mathbb{Z}_p}$. It is easily verified that $H(n)$ is a Hopf order in H. If $H_{\mathbb{Z}_p} = \mathbb{Z}_p[x, y]$ is in some Hopf order $H(\xi)$ then $\xi(x-1) \leq 0$ and from $\xi(S(h)) \leq \xi(h)$ it follows that $\xi(xy) \leq \xi(y)$ and $\xi(y) \leq \xi(xy)$ by taking $h = y$ resp. $h = xy$. Thus $\xi(xy) = \xi(y)$. So we put $\xi(y) = \xi(xy) = -n$ with $n \geq 0$ and $\xi(x-1) = 0$ is then the $\mathbb{Z}_p$-subalgebra of H generated by $b^{\xi(b)}(b - \varepsilon(b))$ for b in the chosen basis for $H_{\mathbb{Z}_p}$, has $\mathbb{Z}_p$-basis $a, x-1, \pi^{-n}y, \pi^{-n}xy$ and it is exactly the $H(n)$ we defined.

3.4.14 Example. The Taft Algebras Over $\mathbb{Q}$

Let $H_T(n)$ be the Taft algebra over $K = \mathbb{Q}$, i.e. $H_T(n) = \mathbb{Q}[x, y]$ with $x^n = 1$, $y^n = 0$ and $xy + yx = 0$, where we put:

$$\varepsilon(x) = 1, S(x) = x^{n-1}, \Delta(x) = x \otimes x,$$
$$\varepsilon(y) = 0, S(y) = -x^{n-1}, \Delta(y) = 1 \otimes y + y \otimes x$$

Then $H_T(n)(-1) = \mathbb{Z}_p + \sum_{i=1}^{n-1} \mathbb{Z}_p(x-1) + \sum_{i=0,j=1}^{n-1} \mathbb{Z}_p \pi^{-j} xy^j$ is a Hopf order on $H_T(n)$ with $\mathbb{Z}_p$-basis $\{x(\pi^{-1}y)^j, i, j = 0, \dots, n-1\}$. We can construct

$H_T(n)(-1)$ from $H_T(n)$ by constructing ξ by putting $\xi(x^{n-1}-1) = 0$, $i = 1,\ldots,n-1$, $\xi(y^j) = 1$, $j = 1,\ldots,n-1$. In checking that $H_T(n)(-1)$ is an Hopf order only $\Delta(\pi^{-j}x^j y^j) \in H_T(n)(-1) \otimes H_T(n)(-1)$ needs some work; this follows from the following lemma.

3.4.15 Lemma

In $H_T(n)$, from all $0 \leq i, j < n$ we have: $\Delta(x^i, y^i) = x^i y^i \otimes (x^i - 1) + x^i y^i \otimes 1 + \sum_{\tau=1}^{j} \alpha_\pi \binom{j}{\pi} x^{i+r+j-\nu} \otimes x^i y^r + 1 \otimes x^i y^j + (x^{i+j} - 1) \otimes x^i y^j$, where: for j even, $\alpha_r = 0$ when r is odd and $\alpha_\tau = \frac{(r-1)(r-3)\ldots}{(j-1)(j-3)(j-r+1)}$ when r is even, for j odd, $\alpha_r = \frac{r(r-2)\ldots}{j(j-2)\ldots(j-r+1)}$ if r is odd and $\alpha_r = \frac{(r-1)(r-3)\ldots}{j(j-2)\ldots(j-1+2)}$ if r is even.

Proof. Since Δ is an algebra morphism $\Delta(x^i y^j) = \Delta(x)^i \Delta(y)^j = (x^i \otimes x^i)(y \otimes 1 - x \otimes y)^j$. Applying the binomial formula and taking into account that $yx = -xy$, yields the result in a straightforward way. □

The group-like elements we have to consider are just the $g-1$ ($\varepsilon(g) = 1$). Therefore the following numerical lemma will be useful.

3.4.16 Lemma

Let g be any element in a Q-algebra A, then for any natural number n we have the equality:

$$\begin{aligned}(g-1)^n = g^n &+ \alpha_{n-1}(-1)(g-1)^{n-1} + \alpha_{n-2}(-1)^2(g-1)^{n-2} \\ &+ \ldots + \alpha_{n-i}(-1)^i(g-1)^{n-i} + \ldots \\ &+ \alpha_2(-1)^{n-2}(g-1)^2 + \alpha_1(-1)^{n-1}(g-1) + c_n(-1)^n\end{aligned}$$

where

$$\begin{aligned}\alpha_{n-1} &= \binom{n}{1}, \alpha_{n-2} = \binom{n}{2} - \alpha^{n-1}\binom{n}{1}, \ldots \\ \alpha_{n-i} &= \binom{n}{i} - \alpha_{n-1}\binom{n-1}{i-1} - \alpha_{n-2}\binom{n-2}{i-2} - \ldots \\ &\quad -\alpha_{n-(i-1)}\binom{n-(i-1)}{1} \text{ and } c_n = 1 - \alpha_{n-1} - \alpha_{n-2} - \alpha_{n-3} - \ldots - \alpha_1\end{aligned}$$

Proof. The proof is by induction on n. If $n = 2$, then

$$(g-1)^2 = g^2 + \alpha_1(-1)(g-1) + c_2(-1)^2$$

$\alpha_1 = \binom{2}{1}$ and $c_2 = 1 - \alpha_1 = 1 - 2 = -1$.

Suppose now that it is true for all $n \leq k$, it is true for $n = k+1$. Since it is true for $n = k$, then we can write:

$$\begin{aligned}(g-1)^{k+1} = {} & g^{k+1} + (\alpha'_{k-1}+1)(-1)(g-1)^k + \\ & (\alpha'_{k-2} - \alpha'_{k-1})(-1)^2(g-1)^{k-1} + \ldots + \\ & (\alpha'_{k-(i+1)} - \alpha'_{k-1})(-1^{i+1})(g-1)^{k-i} + \ldots + \\ & \alpha_2(-1)^{n-2}(g-1)^2 + (\alpha'_1 - \alpha'_2)(-1)^{k-i}(g-1)^2 + \\ & (c_k - \alpha'_1)(-1)^k(g-1) - c_k(-1)^{k+1}\end{aligned}$$

where

$$\alpha'_{k-1} = \binom{k}{1}, \alpha'_{k-2} = \binom{k}{2} - \alpha'_{k-1}\binom{k}{1}, \ldots,$$

$$\alpha'_{k-i} = \binom{k}{i} - \alpha'_{k-1}\binom{k-1}{i-1} - \alpha'_{k-2}\binom{k-2}{i-2} - \ldots - \alpha'_{k-(i-1)}\binom{k-(i-1)}{1}$$

and $c_k = 1 - \alpha'_{k-1} - \alpha'_{k-1} - \alpha'_{k-2} - \alpha'_{k-3} - \ldots - \alpha'_1$. To complete the proof we prove $\alpha_{k-i} - \alpha'_{k-(i+1)} - \alpha'_{k-1}, \alpha_k = \alpha'_{k-1} + 1$ and $\alpha_1 = c_k - \alpha'_1$. First we prove by induction that $\alpha_{k-i} = \alpha'_{k-(i+1)} - \alpha'_{k-i}$. Since $\alpha_k = \binom{k+1}{1}, \alpha'_{k-1} = \binom{k}{1}$, then $\alpha_k = \alpha'_{k-1} + 1$. Since

$$\alpha_{k-1} = \binom{k+1}{2} - \alpha_k\binom{k}{1} = \binom{k+1}{2} - \alpha'_{k-1}\binom{k}{1} - \binom{k}{1} = \binom{k}{2} - \alpha'_{k-1}\binom{k}{1}$$

and

$$\alpha'_{k-2} - \alpha'_{k-1} = \binom{k}{2} - \alpha'_{k-1}\binom{k}{1}$$

then $\alpha_{k-2} - \alpha'_{k-1}$. Suppose now this is true for all $j < i$, then

$$\begin{aligned}\alpha_{k-1} &= \binom{k+1}{i+1} - \alpha_k\binom{k}{i} - \alpha_{k-1}\binom{k-1}{i-1} - \ldots - \alpha_{k-(i-1)}\binom{k-(i-1)}{1} \\ &= \binom{k+1}{i+1} - \binom{k}{i} - \alpha'_{k-1}\left[\binom{k}{i} - \binom{k-1}{i-1}\right] \\ &\quad - \alpha'_{k-2}\left[\binom{k-1}{i-1} - \binom{k-2}{i-2}\right] - \ldots \\ &\quad - \alpha'_{k-(i-1)}\left[\binom{k-i}{2} - \binom{k-(i-1)}{1}\right] - \alpha'_{k-i}\binom{k-(i-1)}{1} \\ \alpha_{k-i} &= \binom{1}{i+1} - \alpha'_{k-1}\binom{k-1}{i} - \alpha'_{k-2}\binom{k-2}{i-1} - \ldots\end{aligned}$$

$$- \alpha'_{k-(i-1)} \binom{k-(i-1)}{2} - \alpha'_{k-i} \binom{k-(i-1)}{1}$$
$$= \alpha'_{k-(i-1)} - \alpha'_{k-i}$$

Using $\alpha_{k-i} = \alpha'_{k-(i+1)} - \alpha'_{k-i}$ we can prove that $\alpha_1 = c_k - \alpha'$, then

$$\begin{aligned}\alpha &= \binom{k+1}{k} - \alpha_k \binom{k}{k-1} - \alpha_{k-1} \binom{k-1}{k-2} - \dots \\ &\quad - \alpha_{k-1} \binom{k-i}{k-(i-1)} - \dots - \alpha_2 \binom{2}{1} \\ &= 1 - \alpha'_{k-1} - \alpha'_{k-2} - \alpha'_{k-3} - \dots - \alpha'_{k-i} - \dots - 2\alpha'_1 \\ &= c_k - \alpha'_1\end{aligned}$$

Using $\alpha_{k-1} = \alpha'_{k-(i+1)} - \alpha'_{k-i}$ we find that $\alpha_1 = c_k - \alpha'_1$, then $c_{k+1} = 1 - \alpha_k - \alpha_{k-1} - \alpha_{k-2} - \dots - \alpha_1 = c_k$.

3.4.17 Remark

1. $\alpha_{n-i} = (-1)^{i+1} \binom{n}{i}$.
2. $\alpha_{n-i} = (-1)^{i+1} \alpha_i$.
3. If $n = p_1^{i_1} p_2^{i_2} \dots p_s^{i_s}$ where $p_1, \dots, p_s$ are nonequal prime numbers, then $p_j^{i_j} | \alpha_{p_l^{i_l}}, 0 \le j, l \le s$ and $j \ne l$ and $p_j^{i_j} \not| \alpha_{p_j^{i_j}}$.
4. If $n = p^s, s \ge 1$ then $p^{s-i} | \alpha_{p^i}, \alpha_{p^i(p^{s-i}-1)}$ and $p^{s-i+1} \not| \alpha_{p^i}, \alpha_{p^i(p^{s-i}-1)}$.

Proof. 1. We prove it by induction.

If $i = 1$, then $\alpha_{n-1} = \binom{n}{1} = (-1)^2 \binom{n}{1}$. Suppose it is true for all $i \le k-1$, then

$$\begin{aligned}\alpha_{n-k} &= \tbinom{n}{k} - \alpha_{n-1} \tbinom{n-1}{k-1} - \alpha_{n-2} \tbinom{n-2}{k-2} - \dots \\ &\quad - \alpha_{n-(k-1)} \tbinom{n-(k-1)}{1} \\ \alpha_{n-k} &= \tbinom{n}{k} - k \tbinom{n}{k} + k(k-1)/2 \tbinom{n}{k} - \dots \\ &\quad \dots - (-1)^k k \tbinom{n}{k}\end{aligned}$$

If k is odd, then: $\alpha_{n-k} = (-1)^{k+1} \binom{n}{k}$. If $k = 2s$ is even, then:

$$\begin{aligned}\alpha_{n-k} &= \tbinom{n}{k} - 2k \tbinom{n}{k} + 2k(k-1)/2 \tbinom{n}{k} - \dots \\ &\quad + 2k(k-1)(k-2)\dots(k/2)/((k/2)-1)! \tbinom{n}{k} \\ &\quad - \ 2k(k-1)(k-2)\dots((k/2)+1)/(k/2)! \tbinom{n}{k} \\ &= \tbinom{n}{k} \left(1 - 2\tbinom{k}{1} + 2\tbinom{k}{2} - \dots + 2\tbinom{k}{s-1} - \tbinom{k}{s}\right)\end{aligned}$$

2. If $n = pq$ then $\alpha_p = (-1)^{n-p+1}pq(pq-1)\dots(pq-p+1)/p! = qz$. In fact, no factor $pq-i, 1 \le i \le p-1$ can be divided by p, otherwise $i = sp$, a contradiction.
3. If $n = p^s$ then $\alpha_{p^i} = (-1)^{n-p^i+1}p^s(p^s-1)\dots(p^s-p^i+1)/p^i! = p^{s-i}z$. In fact, no factor $p^i - t$, $1 \le t \le p^i - 1$ can be divided by p^l with $l \le i$ without reducing p^l by a factor of $(p^i-1)!$, otherwise $i = kp^i$, a contradiction.
4. Similar to (3). □

3.4.18 Proposition

Consider a numberfield $K/\mathbb{Q}$ and let D be a discrete valuation ring of K extending $\mathbb{Z}_p \subset Q$ (here $\mathbb{Z}_p$ is the location of $\mathbb{Z}$ with respect to the prime p). Let $e = v(p)$ be the absolute ramification index of K. Consider a finite dimensional Hopf algebra H over K and let $G = G(H)$ be its finite group of group-like elements. If ξ is any Hopf valuation filtration function on H extending v then we have:

1. $\xi(g) = 0$ for $g \in G$ such that the order $0(g)$ of g in G is not a power of p.
2. $\xi(g) \le e(p^s - p^{s-1})^{-1}$ if the order of g is p^s.

Proof. This follows from (3) and (4) in Remark 3.4.17. Indeed if $0(g) \ne p^s$ then it is a multiple of at least two different primes and none of these can divide all α's in the formula given in Lemma 3.4.16, hence in that case $\xi(g) = 0$. On the other hand if $n = p^s$ then p will be a divisor of all α's in the formula in Lemma 3.4.16 but p^2 will not divide α_{p^s-1}. From Lemma 3.4.16 we may obtain an expression for $(\pi^{\xi(g)}(g-1))^n$, i.e.: with $n = p^s$:

$$(\pi^{\xi(g)})^n = \alpha_{n-1}(-1)\pi^{\xi(g)}(\pi^{\xi(g)}(g-1))^{n-1} + \dots$$
$$\dots + \alpha_{p^s-i}\pi^{(n-p^{s-i})\xi(g)}(-1)^{p^{s-i}}(\pi^{\xi(g)}(g-1))^{p^{s-i}} + \dots$$

It follows from this that $p = d\pi^{(n-p^{s-1})\xi(g)}$ for some $d \in D$, therefore $e \ge (p^s - p^{s-1})\xi(g)$. □

3.4.19 Remark

1. For $H = KG$, G a finite group, and ξ corresponding to a Larson order (i.e. a Hopf order $H_B(\xi)$ corresponding to the basis $\{1, 1-g, g \in G\}$), then the conditions in Proposition 3.4.17 do reduce to the conditions also found by Larson in [35]. Note that in [35] the author proves that the constructed orders (of Larson-type) in RG are in fact finitely generated D-modules without the assumption that

they are constructed in an integral D-order. The proof is more combinatorial in nature.
2. The conditions (2) in Proposition 3.4.18 make it clear that the realization of a certain ξ forces rather demanding ramification properties of v, e.g. for $\xi(g) = 1$ one needs $e \geq p^s - p^{s-1}, p^s = 0(g)$.

Recall the definition of the generalized Taft algebra with respect to a root of unity ρ, say $\rho^n = 1$. Put $H_T(n)$ equal to the K-algebra generated by x and y satisfying $x^n = 1$, $y^n = 0$ and $xy = \rho yx$, with Hopf algebra structure given by:

$$\varepsilon(x) = 1, S(x) = x^{n-1}, \Delta(x) = x \otimes x$$
$$\varepsilon(y) = 1, S(y) = -\rho^{-1}x^{n-1}y, \Delta(y) = 1 \otimes y \otimes y \otimes x$$

3.4.20 Example

Let D be a discrete valuation ring of $K, \rho \in D$.

In case $n \neq p^s$ for $x \geq 1$, then

$$H_T(-n) = D + \sum_{i=1}^{n-1} D(x-1)^i + \sum_{i=0,j=1}^{n-1} D\pi^{-jn}(x-1)^i y^j$$

is a Hopf order.

In case $n = p^s$, $\pi^{m(p^s - p^{s-1})} | p$ and $\pi^m | (\rho - 1)$ in D, then

$$H_T(-p^s) = D + \sum_{i=1}^{n-1} D\pi^{-im}(x-1)^i + \sum_{i=0,j=1}^{n-1} D\pi^{-im-jn}(x-1)^i y^j$$

is a Hopf order.

Observe that $(\rho - 1)^{p^s} = p\sum_{i=1}^{p^s-1}(\alpha_i/p)(\rho - 1)^i$ (Lemma 3.4.16) and then $(\rho - 1)^{p^s} \in (p) \subset \pi^{(p^s - p^{s-1})m}$, where $p^s m < e + p^{s-1}m, e = v(p)$.

A full proof of the claims can be obtained via the quantum binomial formula (see [31]) applied to $f = \pi^{-m}(g-1)$, and via a careful coefficient calculation in an expression for $X^t f^s$, $t, s \in \mathbb{N}$. We omit these technical details here. Let us just provide a concrete case where all of the above phenomena are clear.

3.4.21 Example

Take $p = 2$ and look at the localization of $\mathbb{Z}_2[\rho]$ at $\rho - 1$ where $\rho = \sqrt[4]{-1} = \sqrt{i}$. For D we take $\mathbb{Z}[(\rho - 1)^{1/5}]_{((\rho-1)^{1/5})}$. In this case $\pi = (\rho - 1)^{1/5}$, $(2) = (\pi^5)$,

$v(\pi) = 1$, $e = v(2) = 5$, $v(\rho - 1) = 5$, $\rho^8 = 1$. So we have $(\rho - 1)^8 \in (2)$ and $(2) \subset (\rho - 1) \subset (\pi) \subset D$. The constructions in Example 3.4.20 apply in this case.

Let us conclude with an example showing the effect of base change.

3.4.22 *Example*

We start with the situation of Example 3.4.13 but with K a number field such that $\pi^m|2$ in D. Put $H(m,n) = D[f, \chi]$, $f = \pi^{-m}(g - 1)$, $\chi = \pi^{-n}h$. Then $H(m,n)$ is a Hopf algebra of rank 4 over D with:

$$\Delta(f) = f \otimes g + 1 \otimes f$$

$$f\chi + \chi f = \nu\chi, \text{ where } 2 = \nu\chi^m \text{ with } \nu \in D$$

$$\Delta(f\chi) = f\chi \otimes 1 + 1 \otimes f\chi + f \otimes g\chi + \chi \otimes fg$$

$$\Delta(\chi) = 1 \otimes \chi + \chi \otimes g$$

$$f\chi - \chi = \nu g\chi$$

We may also define $H(n)$ as in Example 3.4.13, it is of rank 4 over D with basis $\{1, g-1, \pi^{-n}h, \pi^{-n}gh\}$. Both $H(m,n)$, $H(n)$ contain the Hopf order $D[g,h]$ (viewed as $H(\xi_0)$ in Proof of 3.4.12).

Now $H(n)$ is of the form $H_B|\xi|$ with respect to $B = \{1, g-1, h, gh\}$ and $H(m,n)$ with respect to $B' = \{1, g-1, h, (g-1)h\}$, and these orders are obviously different.

Bibliography

1. G.Q. Abbasi, S. Kobayashi, H. Marubayashi, A. Ueda, Noncommutative unique factorization rings. Comm. Algebra **19**(1), 167–198 (1991)
2. F. Aly, F. Van Oystaeyen, Hopf filtrations and Larson-type orders in Hopf algebras. J. Algebra **267**, 756–772 (2004)
3. C. Baetica, F. Van Oystaeyen, Valuation extensions of filtered and graded algebras. Comm. Algebra **34**, 829–840 (2006)
4. K. Brown, H. Marubayashi, P.F. Smith, Group rings which are v-HC orders and Krull orders. Proc. Edinb. Math. Soc. **34**, 217–228 (1991)
5. H. Brungs, N. Dubrovin, A Classification and examples of rank one chain domains. Trans. Am. Math. Soc. **335**(7), 2733–2753 (2003)
6. H. Brungs, H. Marubayashi, E. Osmanagic, A classification of prime segments in simple Artinian rings. Proc. Am. Math. Soc. **128**(11), 3167–3175 (2000)
7. H. Brungs, H. Marubayashi, E. Osmanagic, Gauss extensions and total graded subrings for crossed product algebras. J. Algebra **316**, 189–205 (2007)
8. H. Brungs, H. Marubayashi, E. Osmanagic, Primes of Gauss extensions over crossed product algebras. Comm. Algebra **40**(6), (2012)
9. H. Brungs, H. Marubayashi, A. Ueda, A Classification of primary ideals of Dubrovin valuation rings. Houston J. Math. **29**(3), 595–608 (2003)
10. H. Brungs, G. Törmer, Extensions of chain rings. Math. Z. **185**, 93–104 (1984)
11. S. Caenepeel, M. Van den Bergh, F. Van Oystaeyen, Generalized crossed products applied to maximal orders, Brauer groups and related exact sequences. J. Pure Appl. Algebra **33**(2), 123–149 (1984)
12. G. Cauchon, Les T-anneaux et les anneaux à identités polynomials Noethé riens, Thèse de doctorat, Université Paris XI, 1977
13. M. Chamarie, Anneaux de Krull non commutatifs, Thèse, Univ. de Lyon, 1981
14. M. Chamarie, Anneaux de Krull non commutatifs. J. Algebra **72**, 210–222 (1981)
15. M. Chamarie, *Sur les orders maximaux au sense d'Asano*, Vorlesungen aus dem Fachbereich Mathematik der Univ. Essen, Heft 3 (1979)
16. A. Chatters, C. Hajarnavis, *Rings with Chain Conditions*. Research Notes in Mathematics, vol. 44 (Pitman, London, 1980)
17. C. Chevalley, *Introduction to the Theory of Algebraic Functions of One Variable*. Mathematical Surveys, vol. VI (American Mathematical Society, New York, 1951)
18. P.M. Cohn, *Skew Field Constructions*. London Mathematical Society, vol. 27 (Cambridge University Press, London, 1977)
19. I. Connell, Natural transform of the spec functor. J. Algebra **10**, 69–81 (1968)

Marubayashi and Van Oystaeyen, *Prime Divisors and Noncommutative Valuation Theory*, Lecture Notes in Mathematics 2059, DOI 10.1007/978-3-642-31152-9,

20. S. Dascalescu, C. Năstăsescu, S. Raianu, *Hopf Algebras*. Pure and Applied Mathematics, vol. 235 (M. Dekker, New York, 2001)
21. N. Dubrovin, Noncommutative valuation rings. Trans. Moscow Math. Soc. **45**, 273–287 (1984)
22. O. Endler, *Valuation Theory* (Springer, Berlin, 1972)
23. L. Fuchs, *Teilweise Geordnete Algebraische Strukturen*, Vandenhoeck and Ruprecht, Göttingen, 1966
24. P. Gabriel, R. Rentschler, Sur la dimension des anneaux et ensembles ordonnées. C. R. Acad. Paris **265**, 712–715 (1967)
25. R. Gilmer, *Multiplicative Ideal Theory*. Queens' Papers in Pure and Applied Mathematics, vol. 90 (Queens' University, Kingston, 1992)
26. A. Goldie, *The Structure of Noetherian Rings*. Lecture Notes in Mathematics, vol. 246 (Springer, Berlin, 1972)
27. R. Gordon, J.C. Robson, *Krull Dimension*. Memoirs American Mathematical Society, vol. 133 (1973)
28. D. Haile, P. Morandi, On Dubrovin valuation rings in crossed product algebras. Trans. Am. Math. Soc. **338**, 723–751 (1993)
29. C. Hajarnavis, T. Lenagan, Localization in Asano orders. J. Algebra **21**, 441–449 (1972)
30. D. Harrisson, *Finite and Infinite Primes for Rings and Fields*. Memoirs of the American Mathematical Society, vol. 68 (1968)
31. C. Kassel, *Quantum Groups*. Graduate Texts in Mathematics, vol. 155 (Springer, Berlin, 1995)
32. J. Kauta, H. Marubayashi, H. Miyamoto, Crossed product orders over valuation rings II. Bull. Lond. Math. Soc. **35**, 541–552 (2003)
33. S. Kobayashi, H. Marubayashi, N. Popescu, C. Vracin, G. Xie, Noncommutative valuation rings of the quotient Artinian ring of a skew polynomial ring. Algebra Rep. Theor. **8**, 57–68 (2005)
34. G. Krause, T. Lenagan, *Growth of Algebras and Gelfand-Kirillov Dimension*. Research Notes in Mathematics, vol. 116 (Pitman, London, 1985)
35. R. Larsson, Hopf algebra orders determined by group valuations. J. Algebra **38**, 414–452 (1976)
36. L. Le Bruyn, F. Van Oystaeyen, A note on noncommutative Krull domains. Comm. Algebra **14**(8), 1457–1472 (1986)
37. L. Le Bruyn, F. Van Oystaeyen, Generalized Rees rings and relative maximal orders satisfying polynomial identities. J. Algebra **83**(2), 409–436 (1993)
38. L. Le Bruyn, M. Van den Bergh, F. Van Oystaeyen, *Graded Orders* (Birkhauser, Boston, 1988)
39. H. Li, F. Van Oystaeyen, Filtrations on simple Artinian rings. J. Algebra **232**, 361–371 (1990)
40. H. Li, F. Van Oystaeyen, *Zariskian Filtration*. K-Monographs in Mathematics, vol. 2 (Kluwer Academic, Dordrecht, 1996)
41. H. Marubayashi, H. Miyamoto, A. Ueda, *Noncommutative Valuation Rings and Semi-hereditary Orders*. K-Monographs in Mathematics, vol. 3 (Kluwer Academic, Dordrecht, 1997)
42. H. Marubayashi, E. Nauwelaerts, F. Van Oystaeyen, Graded rings over arithmatical orders. Comm. Algebra **12**, 745–775 (1984)
43. H. Marubayashi, G. Xie, A classification of graded extensions in a skew Laurent polynomial ring. J. Math. Soc. Jpn. **60**(2), 423–443 (2008)
44. H. Marubayashi, G. Xie, A classification of graded extensions in a skew Laurent polynomial ring, II. J. Math. Soc. Jpn. **61**(4), 1111–1130 (2009)
45. K. Mathiak, Bewertungen nicht kommutativer Körper. J. Algebra **48**, 217–235 (1977)
46. G. Maury, J. Raynand, *Ordres maximaux au sense de K. Asano*. Lecture Notes in Mathematics, vol. 808 (Springer, Berlin, 1980)
47. J. McConnell, J.C. Robson, *Noncommutative Noetherian Rings* (Wiley-Interscience, New York, 1987)
48. G. Michler, Asano orders. Proc. Lond. Math. Soc. **3**, 421–443 (1969)
49. H. Moawad, F. Van Oystaeyen, Discrete valuations extend to certain algebras of quantum type. Comm. Algebra **24**(8), 2551–2566 (1996)

50. C. Năstăsescu, E. Nauwelaerts, F. Van Oystaeyen, Arithmetically graded rings revisited. Comm. Algebra **14**(10), 1991–2007 (1986)
51. C. Năstăsescu, F. Van Oystaeyen, *Graded and Filtered Rings and Modules*. Lecture Notes in Mathematics, vol. 758 (Springer, Berlin 1979)
52. C. Năstăsescu, F. Van Oystaeyen, *Graded Ring Theory* (North Holland, Dordrecht, 1982)
53. C. Năstăsescu, F. Van Oystaeyen, *Dimensions of Ring Theory*. Mathematics and Its Applications, vol. 36 (D. Reidel Publ. Co., Dordrecht, 1987)
54. E. Nauwelaerts, F. Van Oystaeyen, *Localization at Primes in Algebras over Fields*. Indag. Math. **39**, 233–242 (1977)
55. E. Nauwelaerts, F. Van Oystaeyen, Finite generalized crossed products over Tame and maximal orders. J. Algebra **101**(1), 61–68 (1986)
56. D. Passman, *Infinite Crossed Products*. Pure and Applied Mathematics, vol. 135 (Academic, New York, 1987)
57. I. Reiner, *Maximal Orders*. London Mathematical Society, Monographs Series, vol. 5 (Academic, New York, 1975)
58. J.C. Robson, Noncommutative Dedekind rings. J. Algebra **9**, 249–265 (1968)
59. J. Rotman, *Notes on Homological Algebra* (Van Nostrand Reinhold, New York, 1970)
60. H. Rutherford, Characterizing primes in some noncommutative rings. Pac. J. Math. **27**, 387–392 (1968)
61. O. Schilling, *The Theory of Valuations*. Mathematical Surveys and Monographs, vol. 4 (American Mathematical Society, Providence, 1950)
62. Ya.I. Shtipel'man, Valuations on the quotient field of the ring of quantum mechanics. Funkts. Anal. Pnlozh. **71**(1), 56–63 (1973)
63. B. Stenström, *Rings of Quotients* (Springer, Berlin, 1975)
64. B. Torrecillas, F. Van Oystaeyen, Divisorially graded rings, related groups and sequences. J. Algebra **105**(2), 411–428 (1978)
65. M. Van den Bergh, F. Van Oystaeyen, Lifting maximal orders. Comm. Algebra **17**(2), 341–349 (1989)
66. J.P. Van Deuren, F. Van Oystaeyen, *Arithmetically Graded Rings I*. Lecture Notes in Mathematics, vol. 825 (Springer, Berlin, 1980), pp. 130–152
67. J.P. Van Deuren, J. Van Geel, F. Van Oystaeyen, *Genus and a Riemann-Roch Theorem for Noncommutative Function Fields in Ore Variable*. Lecture Notes in Mathematics, vol. 867 (Springer, Berlin, 1981), pp. 295–318
68. J. Van Geel, *A Noncommutative Theory for Primes*. Lecture Notes in Pure and Applied Mathematics, vol. 51 (M. Dekker, New York, 1979), pp. 767–781
69. J. Van Geel, *Places and Valuations in Noncommutative Ring Theory*. Lecture Notes in Pure and Applied Mathematics, vol. 71 (M. Dekker, New York, 1981)
70. F. Van Oystaeyen, On pseudo-places of algebras. Vull. Soc. Math. Belg. **25**, 139–159 (1973)
71. F. Van Oystaeyen, Primes in algebras over fields. Pure Appl. Algebra **5**, 239–252 (1974)
72. F. Van Oystaeyen, *Prime Spectra in Noncommutative Algebra*. Lecture Notes in Mathematics, vol. 444 (Springer, Berlin, 1975)
73. F. Van Oystaeyen, Crossed products over arithmetically graded rings. J. Algebra **80**(2), 537–551 (1983)
74. F. Van Oystaeyen, On orders over graded Krull domains. Osaka J. Math. **20**(4), 757–765 (1983)
75. F. Van Oystaeyen, *Algebraic Geometry for Associative Algebras*. Monographs in Pure and Applied Mathematics, vol. 227 (M. Dekker, New York, 2000)
76. F. Van Oystaeyen, A. Verschoren, *Noncommutative Algebraic Geometry, an Introduction*. Lecture Notes in Mathematics, vol. 887 (Springer, Berlin, 1981)
77. F. Van Oystaeyen, A. Verschoren, *Relative Invariants of Rings; The Noncommutative Theory*. Monographs in Pure and Applied Mathematics, vol. 86 (M. Dekker, New York, 1984)
78. F. Van Oystaeyen, L. Willaert, Valuations on extensions of Weyl skew fields. J. Algebra **183**, 359–364 (1996)
79. H. Warner, Finite primes in simple algebras. Pac. J. Math. **36**, 245–265 (1971)

80. L. Willaert, Discrete valuations on Weyl skew fields. J. Algebra **187**(2), 537–547 (1997)
81. E. Witt, Riemann-Rochser Satz und z-Funktionen in Hyperkomplexen. Math. Ann. **110**, 12–28 (1934)
82. O. Zariski, P. Samuel, *Commutative Algebra I*. Graduate Texts in Mathematics, vol. 28 (Springer, Berlin, 1958)

Index

Abelian valuation, 25
Archimedean, 20, 49
Arithmetical (a.p.v.), 153
Arithmetical order, 153
Arithmetical prime, 22

Bezout, 36

Class group, 143
δ-Class group, 143
σ-Class group, 143
Classical D-order, 127
Compatible, 81
Completely prime, 1
Crossed product, 70

Derived valuation function, 202
Differential polynomial ring, 132
Discrete rank one valuation ring, 12
Divisor, 159
  of poles, 160
  of zeros, 160
Divisorial, 46
Divisorially simple, 124
D-lattice, 128
Dominating pair, 3
Domination relation, 3
Dubrovin valuation ring, 33

Exceptional, 49
Exhaustive, 31

Γ-Filtration, 31
Fractional, 2
F-reductor, 97
$\mathcal{F}$-torsion, 116
  free, 116
  submodule, 116
Function algebra, 165

Gauss extension, 71
Generalized Bezout, 148
Generalized left Bezout order, 148
Goldie prime ideal, 37
Good filtration, 198
Good reduction, 176
Graded extension, 75
Graded total valuation ring, 71
gr-maximal order, 106

Hopf valuation filtration function, 199
Hopf valuation function, 199

Ideal divisor, 160
Integral divisor, 159
σ-Invariant, 140
Isolated subgroup, 19
Isomorphic places, 12

Krull, 117
σ-Krull order, 140

Left Bezout order, 36
Left divorsial, 110
Left Gabriel topology, 115

Marubayashi and Van Oystaeyen, *Prime Divisors and Noncommutative Valuation Theory*, Lecture Notes in Mathematics 2059, DOI 10.1007/978-3-642-31152-9,

Left σ-derivation, 132
Left semi-hereditary, 36
Local components, 172
Local different, 172
Localized, 2
Loewy series, 119

σ-Maximal order, 140
Maximal order, 102
Moderate order, 203
M-system, 1

n-chain ring of S, 35

Ore extension, 132
O_v-lattice, 96

Pairs, 3
Partially ordered set, 19
Pointed, 24
Positive filters, 24
Preplace, 5
Prime, 1
(σ, δ)-Prime, 146
Prime divisors, 159
Primeplace, 5
Principal symbol, 32
Principle, 160
Proper, 159
π-Pseudoplace, 5, 8
Pseudovaluation, 153

Quotient ring, 102

Reduced rank, 119
Reduction of A with respect to Λ, 97
Rees ring, 91
Residue skew field, 77
Right Bezout order, 36
Right (left) order, 102

Semi heriditary, 36
Separated, 2
Skew polynomial ring, 132
Ψ-Spanned, 16
Specialization, 7
(σ, δ)-Stable, 145
σ-Stable, 140
Strong filtration, 88

Totally ordered, 19
Trivial, 116

Unique factorization ring, 143
Unramified, 16
Upper prime, 49

Valuation, 5
 filtration, 97
 form, 169
 ring, 11
 ring of D, 16
 vectors, 166
V-crossed product, 70

w-closed, 126
w-Noetherian, 126

LECTURE NOTES IN MATHEMATICS

Springer

Edited by J.-M. Morel, B. Teissier; P.K. Maini

Editorial Policy (for the publication of monographs)

1. Lecture Notes aim to report new developments in all areas of mathematics and their applications - quickly, informally and at a high level. Mathematical texts analysing new developments in modelling and numerical simulation are welcome.
   Monograph manuscripts should be reasonably self-contained and rounded off. Thus they may, and often will, present not only results of the author but also related work by other people. They may be based on specialised lecture courses. Furthermore, the manuscripts should provide sufficient motivation, examples and applications. This clearly distinguishes Lecture Notes from journal articles or technical reports which normally are very concise. Articles intended for a journal but too long to be accepted by most journals, usually do not have this "lecture notes" character. For similar reasons it is unusual for doctoral theses to be accepted for the Lecture Notes series, though habilitation theses may be appropriate.

2. Manuscripts should be submitted either online at www.editorialmanager.com/lnm to Springer's mathematics editorial in Heidelberg, or to one of the series editors. In general, manuscripts will be sent out to 2 external referees for evaluation. If a decision cannot yet be reached on the basis of the first 2 reports, further referees may be contacted: The author will be informed of this. A final decision to publish can be made only on the basis of the complete manuscript, however a refereeing process leading to a preliminary decision can be based on a pre-final or incomplete manuscript. The strict minimum amount of material that will be considered should include a detailed outline describing the planned contents of each chapter, a bibliography and several sample chapters.
   Authors should be aware that incomplete or insufficiently close to final manuscripts almost always result in longer refereeing times and nevertheless unclear referees' recommendations, making further refereeing of a final draft necessary.
   Authors should also be aware that parallel submission of their manuscript to another publisher while under consideration for LNM will in general lead to immediate rejection.

3. Manuscripts should in general be submitted in English. Final manuscripts should contain at least 100 pages of mathematical text and should always include
   - a table of contents;
   - an informative introduction, with adequate motivation and perhaps some historical remarks: it should be accessible to a reader not intimately familiar with the topic treated;
   - a subject index: as a rule this is genuinely helpful for the reader.

   For evaluation purposes, manuscripts may be submitted in print or electronic form (print form is still preferred by most referees), in the latter case preferably as pdf- or zipped psfiles. Lecture Notes volumes are, as a rule, printed digitally from the authors' files. To ensure best results, authors are asked to use the LaTeX2e style files available from Springer's web-server at:

   ftp://ftp.springer.de/pub/tex/latex/svmonot1/ (for monographs) and
   ftp://ftp.springer.de/pub/tex/latex/svmultt1/ (for summer schools/tutorials).

Additional technical instructions, if necessary, are available on request from lnm@springer.com.

4. Careful preparation of the manuscripts will help keep production time short besides ensuring satisfactory appearance of the finished book in print and online. After acceptance of the manuscript authors will be asked to prepare the final LaTeX source files and also the corresponding dvi-, pdf- or zipped ps-file. The LaTeX source files are essential for producing the full-text online version of the book (see http://www.springerlink.com/openurl.asp?genre=journal&issn=0075-8434 for the existing online volumes of LNM). The actual production of a Lecture Notes volume takes approximately 12 weeks.

5. Authors receive a total of 50 free copies of their volume, but no royalties. They are entitled to a discount of 33.3 % on the price of Springer books purchased for their personal use, if ordering directly from Springer.

6. Commitment to publish is made by letter of intent rather than by signing a formal contract. Springer-Verlag secures the copyright for each volume. Authors are free to reuse material contained in their LNM volumes in later publications: a brief written (or e-mail) request for formal permission is sufficient.

Addresses:

Professor J.-M. Morel, CMLA,
École Normale Supérieure de Cachan,
61 Avenue du Président Wilson, 94235 Cachan Cedex, France
E-mail: morel@cmla.ens-cachan.fr

Professor B. Teissier, Institut Mathématique de Jussieu,
UMR 7586 du CNRS, Équipe "Géométrie et Dynamique",
175 rue du Chevaleret
75013 Paris, France
E-mail: teissier@math.jussieu.fr

For the "Mathematical Biosciences Subseries" of LNM:

Professor P. K. Maini, Center for Mathematical Biology,
Mathematical Institute, 24-29 St Giles,
Oxford OX1 3LP, UK
E-mail: maini@maths.ox.ac.uk

Springer, Mathematics Editorial, Tiergartenstr. 17,
69121 Heidelberg, Germany,
Tel.: +49 (6221) 4876-8259

Fax: +49 (6221) 4876-8259
E-mail: lnm@springer.com